Public Policy and
Administration in Africa

Public Policy and Administration in Africa

Lessons from Nigeria

Peter H. Koehn

Westview Press
BOULDER, SAN FRANCISCO, & LONDON

Westview Special Studies on Africa

Copyright © 1990 by Westview Press, Inc.

Published in 1990 in the United States of America by Westview Press, Inc., 5500 Central Avenue, Boulder, Colorado 80301, and in the United Kingdom by Westview Press, Inc., 13 Brunswick Centre, London WC1N 1AF, England

Library of Congress Cataloging-in-Publication Data
Koehn, Peter H.
  Public policy and administration in Africa : lessons from Nigeria /
by Peter H. Koehn.
     p. cm. — (Westview special studies on Africa)
  Includes bibliographical references.
  ISBN 0-8133-7757-9
  1. Bureaucracy—Nigeria. 2. Political planning—Nigeria.
3. Nigeria—Economic policy. 4. Nigeria—Politics and government.
5. State governments—Nigeria. 6. Local governments—Nigeria.
I. Title.
JQ3090.K64 1990
354.669′01—dc20
                                                                          89-22572
                                                                             CIP

Printed and bound in the United States of America

The paper used in this publication meets the requirements of the American National Standard for Permanence of Paper for Printed Library Materials Z39.48-1984.

10   9   8   7   6   5   4   3   2   1

Contents

Tables and Figures

Acknowledgments

In the preparation of this book, I have been assisted by a number of individuals and institutions. The Social Science Research Council, the University of Montana, and the Institute of Administration at Ahmadu Bello University provided important support for my research on state and local policy issues and administrative practice. The actual field work could only be accomplished with the cooperation of those directly involved in the processes investigated. For their patience and time, I am most grateful.

Students at A.B.U. enrolled in the Administrative Management Training Course, the Advanced Diploma in Local Government studies, and various post-graduate programs taught this instructor a great deal. I learned even more about the subjects addressed in this book from my colleagues. Of particular benefit in this connection has been professional association and personal friendship with Olatunde Ojo, Joseph Shekwo, Lawrence Ega, Halidu Abubakar, and Festus Nze. Nigerian students attending the University of Montana graciously continued the educational process following my return from their vibrant country. A sabbatical award from the University of Montana proved vital in connection with some of the material presented in this volume. Special thanks are extended to John Paden, Cheryl Johnson-Odim, Hans Panofsky, and Richard A. Hay, Jr., for their hospitality and efforts to ensure that the period I spent as Visiting Scholar at Northwestern University during that sabbatical year would be rewarding.

Several chapters in this volume have been heavily influenced by collaborative work with other scholars, or by external criticism. For instance, Gavin Williams provided a set of helpful comments on an earlier version

of the agricultural policy discussion. During his tenure
as Visiting Fulbright Scholar at the University of Montana,
Olatunde Ojo coauthored the manuscript for <u>Africa Today</u>
that formed the basis for the chapter on management of
Nigeria's debt crisis. The first two chapters of this
book can be attributed in large measure to Krishna
Tummala's encouragement. Richard A. Hay, Jr., offered many
useful suggestions concerning the study of land allocation
and development planning. Richard Vengroff provided a
useful set of comments related to the implementation of
land-allocation policy in Africa. The late Abubakar Yaya
Aliyu contributed generously to the analysis of local
government in Nigeria. In addition, Henry F. Goodnow and
James R. Scarritt, both of the University of Colorado, have
greatly facilitated my work on public administration and
politics in Africa over the past twenty years. Richard E.
Stren's counsel played an important part in the decision to
produce this volume.

A number of individuals deserve recognition for their
assistance in the production stages of this book. At the
University of Montana, I relied heavily on the diligence
and patience of several typists. They include Susan
Matule, Sue Koehn, Susan Mowrer, Loretta Edwards, and my
wife, Aminata Khady Diop. It also has been a pleasure
working with Sally Furgeson, Rebecca Ritke, and Barbara
Ellington of Westview Press.

A decade has passed since I first arrived in Nigeria
fully cognizant of the vast, historically and socially
complex nature of the country. This book is intended to
share what I have learned about public policy and administration in Africa over the past ten years. One factor
stands out above all others in promoting that learning
experience. That is the support and confidence extended by
my Department Head at the Institute of Administration,
Professor A. Y. Aliyu. Taken in November 1987 from family,
colleagues, and country while in the prime of his life and
career, Yaya will never see this final product. It is my
fervent hope, nevertheless, that it merits his confidence.

P.H.K.
Missoula

Introduction

Chinua Achebe, the renowned writer, published a small book in 1983 that he entitled <u>The Trouble with Nigeria</u>. Achebe's tract treats administrative corruption, indiscipline, and the need for leadership. For similar reasons, this book could have been called <u>The Trouble with Public Policy and Administration in Africa</u>.

The conduct of public policy making and administration in Africa is fraught with controversy. Public service frequently attracts "the best and the brightest" of the country's labor force. Many profess commitment to development objectives and are eager to confront the most demanding challenges. Yet, the public servant is frequently characterized as self-serving, corrupt, power-hungry, and a burden on society. Finding his/her colleagues blamed, accused, and retrenched, the idealistic public administrator often reacts with loss of motivation and lowered morale.

Public Administration in Trouble

Public administration is indeed in trouble in Africa. Creative development policies and programs often are articulated, but far less frequently are seen through to completion. Some projects are abandoned in mid-stream; others benefit those with the least compelling needs. Citizen participation in planning and decision making receives lip service. The genuine delegation of authority is an exceptional event. Resources are scarce, evaluation is practically unknown, and maintenance is ignored. These conditions are fundamentally related to "the trouble with Nigeria" and other African states.

1

    Throughout the continent, one encounters calls for
public sector reform in the wake of widespread dissatisfaction with the way in which the people's affairs are
managed.[1] The basic will and capacity of state functionaries to serve as agents of development are increasingly
questioned. What accounts for the predicament in which
state employees currently find themselves? How can public
servants extricate themselves from popular disillusionment
and the political culture of cynicism? By applying a critical perspective to the Nigerian experience, this book aims
to assist students of African politics and comparative
administration who seek to comprehend the underlying
reasons for the prevailing malaise and to identify promising alternatives to existing approaches.

    In addressing these issues, Nigeria offers a particu-
larly revealing context. In structural terms, the country
possesses the largest public bureaucracy in Sub-Saharan
Africa and a system organized in ways that are quite familiar to students of public administration in other former
British colonies. Moreover, the ills and dilemmas of
public policy making and execution reported elsewhere on
the continent can be found in Nigeria -- often in more pronounced form. There are subjects, nevertheless, which can
be portrayed most fruitfully through explicit comparative
analysis. On such occasions, references are included to
the practice of public administration and the experience of
policy reform in selected anglophone and francophone
countries.

Public Sector Reform

    In the face of burdensome international debts and
domestic economic crises, most African governments have
adopted externally designed prescriptions for internal
structural adjustment (Hodges, 1988:22). The policy-reform
requirements recently imposed upon the state by the Western
financial establishment can be arranged in two broad categories: changes in (1) economic policy and (2) the public
bureaucracy. An entire chapter of this book (Chapter 4)
is devoted to analyzing the impact of externally promoted
changes in economic policy. The second category involves
demands for public sector reform, which commonly include
reductions in government expenditure, privatization of
parastatals, cutbacks in the size of the bureaucracy,
deregulation, decentralization, and more efficient public

sector management (see Vengroff, 1988a:1; Browne, 1988:5-6; Lancaster, 1988:31; 1987:226).

In alliance with technocrats in local financial institutions and potential investors, Western donor agencies have succeeded in using the appealing fix of further loans to secure nominal commitment from the state to an unpopular package of policy reforms (Lancaster, 1988:32-33; Loxley, 1986:100). Under General Ibrahim Babangida, Nigeria has ranked among the most diligent in its commitment to structural adjustment. What lessons can we draw from this effort regarding the likely impact of the requisite policy changes on the national economy? How effective are government actions in overcoming existing barriers to public sector reform and in rectifying critical problems of development administration in Africa?

In this regard, it is important to note that students of development administration typically identify an agenda for action that differs from the suggested reforms set forth by the International Monetary Fund and the World Bank. The priority concerns of scholars include issues of process and access in policy making and implementation, mobilizing local resources, empowering citizens for sustainable self-reliant development and participation in decision making, breaking bonds of dependency, preserving natural resources, and promoting equity in material allocations and service delivery. There are some areas of overlapping interest. Researchers and donors, as well as a growing number of government leaders, are concerned with excessive bureaucratic size (see, for instance, Abernethy, 1983:2-3; Browne, 1988:9; Vengroff, 1988b:7; Loxley, 1985:132), improving local productive capacity (Loxley, 1986:101), and decentralization (Rothchild and Olorunsola, 1983:8-9, 20). However, scholars tend to be more aware of and to place greater emphasis on political conditions and social structure as factors affecting administrative performance (see Honadle, 1982:175; L. White, 1987:190-199; Graham, 1988:1, 4-5; Callaghy, 1987:103, 105; Hyden, 1980; Loxley, 1985:120). Reflecting on the soft-state conditions which characterize many African countries, Donald Rothchild and Victor Olorunsola (1983:20, 7), for instance, caution against policy analysis which incorrectly assumes "an enormous capacity on the part of state agencies to put policies and programs into effect." Thus, after acceding to externally imposed conditions in a desperate search for foreign capital, most African governments have proven unable or unwilling to carry out the promised reforms (Browne,

1988:8-9). A "soft-state" understanding suggests the value
of focusing on the deeply entrenched obstacles which must
be overcome in order to bring about improved public management in Africa. Constraints on effective administrative
practice at local as well as national levels of government
are a recurring concern in this volume.

Critical Perspective

Public Policy and Administration in Africa brings a
straightforward critical perspective to bear on concrete
topics of current concern (see Fischer and Sirianni,
1984:5, 17). Other texts on Nigerian public administration
have been found "to skirt around the intense controversies
and to take government statements at face value" (Wilks,
1985:267). I have tried to avoid repeating these mistakes.

A critical perspective seeks to unmask and overcome
barriers to enlightened and meaningful action (see
Denhardt, 1984:171-173). The procedure is to confront and
challenge established methods of administrative thought and
behavior, identify contradictions between actors' goals and
prevailing institutional arrangements, assess alternative
approaches, and propose and advocate changes in policy,
process, and/or structure. The critical perspective
applied in these pages is guided by an underlying concern
for advancing the interests of the poorest and most disadvantaged social strata. It subjects the arguments about
development strategy advanced by government policy makers,
technical specialists, and donor agencies to careful scrutiny and exposes the weaknesses in many of them (see
Goulet, 1978:12). Class analysis and political economy
considerations are particularly useful in this regard (also
see Meillassoux, 1970:97, 108).

As a teacher of public administration at universities
in Africa and North America for nearly two decades, I have
learned from my students that it is rewarding intellec-
tually and useful practically to uncover the underlying
factors responsible for administrative weakness. Within
the public services, men and women grapple daily with a
multiplicity of deeply rooted factors that constrain
governmental performance. The African public servant must
deal with external forces and demands, pressures to engage
in unethical conduct, omnipresent social obligations (see
Hodder-Williams, 1984:172-173; Hyden, 1980:213; 1983) the
burdens of relying upon face-to-face communications and
personal relations (Rondinelli, 1981:617; Montgomery,

1986:213-215), the effects of uncertain export revenues,
volatile and vastly different regimes, shortages of well-
trained colleagues, a cynical and/or poorly educated
clientele, class and ethnic biases, and impediments of
organizational structure.

One must move beyond problem identification, however,
in order that both student and teacher are not left
unsatisfied, overwhelmed, and unprepared to deal with
future situations. When public administration is in dire
trouble, the persistent struggle to improve it only takes
on more urgent and serious form. The principal assumptions
which guide this book are (1) that public administrators
and policy makers in Africa are capable of engaging in
effective development administration, (2) that useful
alternatives to current procedures and practices can be
identified and implemented, (3) that such alternatives will
involve alterations in processes as well as in policies and
institutions, and (4) that the most fruitful alternatives
will be simple to introduce, small in scale, and partici-
pative in nature and design. In short, this is meant to be
a realistic but upbeat book. Given the nature of these
assumptions, moreover, special emphasis is placed upon
local-level administration in this study.

Process Focus

The chapters which follow address policy formulation
and execution, organizational structure, development
planning, and issues of personnel and financial administration. They work with all three levels of Nigeria's federal
system of government. Special attention is devoted to
process dynamics throughout the volume. The complete
policy-making process, from formulation through implementation (see Biersteker, 1987:291-292) and impact, is subjected to critical analysis.

Political and administrative processes shape the outcome of bureaucratic behavior. Process study involves the
analysis of rules and procedures, informal access points,
and the decisional criteria employed by gatekeepers. The
emphasis is on behavior and conflict rather than on formal
structure (also see Wilks, 1985:270). Students of public
administration and public policy need to know who is
involved and excluded at critical decision points, who is
in control of the process, and details about relative power
and influence. It also is important to identify the
interests which policy makers articulate and the values

6

they uphold (Goulet, 1978:12). In short, it is necessary to establish the political economy context for public management in Africa.

Bases for exclusion and barriers to wider participation are particularly revealing. The analyst endeavors to uncover prevailing patterns of systematic bias and the process features associated with such outcomes. The value of this approach becomes particularly apparent when the time comes to advance proposals for change which hold out the prospect that the public sector will become more responsive to the needs of the poor. As Robert Denhardt (1984:viii) puts it, "democratic outcomes require democratic processes."

Background

This book is based upon a decade of research, reflection, and writing. The field research period corresponds with the author's tenure from 1978-1980 as Principal Research Fellow in the Department of Research, Management, and Consultancy at the Institute of Administration, Ahmadu Bello University, Zaria. Public policy and administration are topics of abiding interest in Nigeria and the people I met proved quite willing to contribute their point of view on the research questions I pursued. With few exceptions, public servants responded fully even to the most sensitive of my inquiries. I benefitted greatly from the support and insights offered by my colleagues in the Department of Research, Management, and Consultancy (since renamed Local Government Studies), particularly the then Department Head and later Director of the Institute, Professor A. Y. Aliyu. Grants from the Social Science Research Council and the University of Montana also supported various aspects of the research reported here.

A diverse set of methodologies are employed in investigating the issues of policy and administration considered in this volume. The principal methods are political and historical analysis, elite interviewing, and work with data and information gathered from government documents.[2] Chapter 5 presents a systematic analysis of state land allocation records, while Chapter 6 is based upon local planning data. Different analytical approaches are utilized in order to enrich and enliven the book. In general, the style of presentation reflects the author's conviction that micro-level data and insights are more valuable in elucidating public policy orientations and administrative behavior in Africa than are aggregate macro-level statistics.

Organization

The book opens with two chapters that deal with the key structural and political issues which consistently engage the administocracy. First, the nature, growth, and performance of public administration in Nigeria is evaluated in historical perspective. Chapter 2 treats administrative involvement in policy making, a central component of the administocracy thesis. These initial chapters establish the vital nexus between public administration and public policy in Africa.

The next two sections are organized by level of government as well as by policy issue. National policy and administration are approached first. The issues considered here are agricultural production and management of international debt crises. These subjects rank among the most pressing concerns which confront national government officials in the Third World at the present time. They dramatically illustrate the nature of Nigeria's political economy and highlight the situation of dependency which forms a central aspect of the context within which African public policy makers operate.

The section on state and local policy and administration moves from larger to smaller units of government. Land-allocation processes at the state level are considered first. Throughout Africa, access to land is of vital consequence for rich and poor alike. The next chapter assesses state and local government involvement in planning development projects. Finally, Chapter 7 provides a grass roots discussion of organization, staffing, and training at the local level.

The conclusion analyzes the character and autonomy of the public services in Africa. In their policy-making capacity, have national, state, and local administrators acted as selfless public servants devoted to the interests of the citizenry and to promoting self reliance, or as a parasitic class of functionaries scheming to promote personal aggrandizement and/or the extraction of national resources by multinational corporate entities? Are things likely to change in Nigeria's Third Republic? Reform proposals, including the growing clamor for privatization, are reviewed within the context of the prevailing political economy. The Nigerian case provides the basis for the

concluding discussion of constraints on public sector
reform in Africa.

Applications

Practitioners, whether primarily involved in the formulation, implementation, or evaluation of public policy,
need to possess a thorough understanding of development
challenges at all levels of government. Moreover, the
ethical development administrator must be able and willing
"to think in terms of larger questions" (Denhardt,
1984:154; Goulet, 1978:12). From Nigeria's recent
experience with land allocation, development planning,
large-scale agricultural projects, and IMF conditionalities, there are important lessons for the public servant
employed by a local, state, or national agency as well as
for the student of comparative administration and African
politics. As Nigerians prepare for the Third Republic, a
critical review of recent administrative behavior and
policy-making practice also can provide a constructive
basis for reflection, deliberation, and proposals for
change.

In addition to acquiring specific insights with immediate application in tackling the problems at hand, my hope
is that students of development administration and political change who use this book will gain appreciation for the
importance of process study and the utility of critical
analysis. The practitioner's task is to use such knowledge
to lead public administration out of trouble.

NOTES

1. See, for instance, the conclusion reached by the
government-appointed Political Bureau (Nigeria, Political
Bureau, 1987:104-111) based upon extensive public testimony
(also Gould, 1988:1-2; Stren, 1988:217; Abernethy, 1983:2;
Nellis, 1980:416; A. Phillips, 1985:261).

2. Unless otherwise noted, the tables presented in
this book are based upon data collected by the author.

1

Public Administration in Historical Perspective

The contemporary practice of public administration in Africa has evolved from pre-colonial traditions, colonial influences, and post-independence experience. Although the details differ from country to country, the broad outline of the recent history of the career public service on the continent can be illustrated by reference to the Nigerian case. This discussion is most applicable, of course, to states which share the legacy of British colonial rule.

The nature and practice of public administration in Nigeria has changed considerably over the past 30 years. A small colonial service has been transformed into 22 indigenous civil services that jointly employ nearly a million persons. With the inclusion of the armed forces and local government staff, the total personnel strength of the Nigerian public bureaucracy currently approaches 2 million -- undoubtedly the largest by far in Sub-Saharan Africa. The organization and conduct of public administration since independence have been shaped in important ways by regime change, political culture and pressures, external influences and approaches, and reform efforts. This chapter describes and critically analyzes the principal developments in Nigerian public bureaucracy, from its colonial origin through the late 1980s.

PUBLIC ADMINISTRATION IN NIGERIA:
A HISTORICAL/STRUCTURAL OVERVIEW

The Nigerian governmental apparatus has been divided into several independent public services since 1954. The organization of separate regional (later state) civil services predated regional self government which, in turn,

9

preceded independence by from one (North) to three (East
and West) years. The regional governments succeeded in
recruiting many of the most highly qualified senior
Nigerian staff into their services prior to 1960 through
attractive pay offers and rapid advancement opportunities
(see, for example, Adebo, 1979:197; Udoji, 1979:201;
Ciroma, 1979:209). Phillip Asiodu (1970:128-129) concludes
that "in the place of one Civil Service which before 1954
had posted officers from one corner of Nigeria to another
regardless of ethnic origins, we now [1966] had five Civil
Services increasingly parochial and resentful of any
suggestion of Federal Civil Service pre-eminence, or of the
need for federal directives or leadership in the national
interest" (also see Oyovbaire, 1985:74). The independent,
competitive position of the regional/state civil services
has been preserved up to the present time.

In spite of their separate constitution, the formal
organizational arrangement of the Nigerian public services
has evolved in remarkably similar fashion. In broad
outline, this structural evolution has incorporated colonial secretariat organization, British administrative patterns, indigenous adaptations and practices, and (more
recently) selected ideas based on the conduct of administrative affairs in the United States.

The Colonial Secretariat Structure

Figure 1.1 depicts the major features of colonial
secretariat organization under British rule. Under this
organizational structure, all communications on questions
of policy to and from the governor had to be channeled
through the Chief Secretary (Akpan, 1982:146). The latter
played a dominant decision-making role and, supported by a
number of assistant secretaries and clerks, served as the
effective head of the civil service. The Executive
Council, presided over by the Colonial Governor and con-
sisting of all heads of departments (i.e., the Chief
Secretary, the Attorney General, the Financial Secretary,
the Commandant of the Nigerian Regiment, and the Directors
of Medical Services, Works, Transport, and Education),
decided high-level policy matters (Balogun, 1983:76).

Almost all higher civil servants were British
colonial officers (Adamolekun, 1978a:12; Balogun, 1983:77).
Residents and district officers served as local chief
executives assigned general responsibility, similar to
that exercised by the Chief Secretary in Lagos, for all

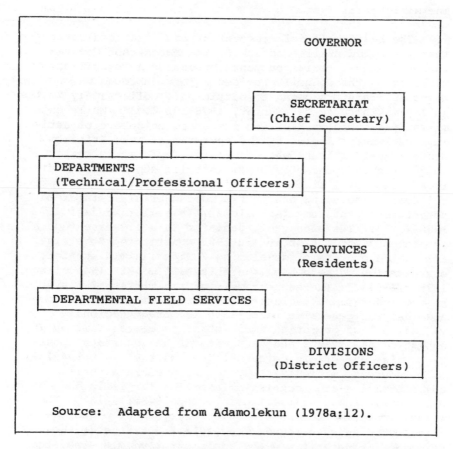

Source: Adapted from Adamolekun (1978a:12).

Figure 1.1 The Colonial Secretariat Organizational
Structure in Nigeria

government activities in their province or district. These
officers possessed considerable autonomy and discretion in
the field (Kingsley, 1963:312; Murray, 1978:105). Functional departments, formally directed in professional matters from London, organized their field administration
according to provincial and/or district boundaries. David
Murray (1978:91-92) explains that "to a large extent the
district officers and field officers of functional departments were each the agents of the Resident, for it was the
Resident who supervised and coordinated the activities of
these other officials, and who either handled himself, or
supervised the handling of, relations with the subordinate
institutions of [local] government."

The Ministerial System

The basic colonial secretariat organizational structure prevailed until adoption of the Macpherson Constitution of 1951. This document introduced a Council of
Ministers. The Council provided a forum through which
Nigerian ministers began to participate collectively in the
policy-making process (Murray, 1978:103-104). While each
ministry assumed work within its assigned sphere of activity previously handled by the secretariat, the British
professional officers who headed the departments continued
to operate independently of ministerial supervision until
1954. The small size of the minister's staff -- about five
officials, including one or two messengers -- reinforced
departmental autonomy (Adamolekun, 1978a:13; Adedeji,
1968a:2-3). The subsequent decision to amalgamate federal
departments and ministries in 1959 constituted "the final
act in the process of developing a Nigerian ministerial
organization patterned on the Whitehall model" (Adamolekun,
1978a:12-14).[1] The entrenchment of the ministerial system
at the federal and regional levels following independence
also had the general effect (with important regional
variations) of promoting administrative centralization and
enhancing the status and authority of headquarters' positions vis-à-vis field personnel (see Murray, 1978:104-139).

Under the ministerial system adopted in Nigeria,
administrative and technical ministries initially possessed
distinctive features (see Adamolekun, 1978a:15-16). The
principal difference between the two types of organizational structure surrounded the role and status of the
permanent secretary. In the administrative ministry, the
permanent secretary served as the chief administrator and
chief advisor to the minister. As chief advisor, the
"perm sec" would elaborate policies and help ascertain the
preferred means of carrying them out (Adamolekun, 1978a:
17-18; also see Kingsley, 1963:312-313). Responsibilities
as administrative head included interpreting policies, coordinating ministry activities, supervising functional execution and monitoring results, defending ministry budget
proposals, and upholding the ministry's interests in inter-
ministerial meetings and in relations with other agencies
and external groups. Until the 1988 Civil Service Reforms,
the perm sec also served as chief accounting officer, decided on the selection, placement, promotion, and discipline of lower-level officers (i.e., GL 01-06), and passed
recommendations on such matters with respect to higher officers (GL 07 and above) to the Civil Service Commission.[2]

In the technical ministry, the technical service directors shared executive responsibilities with the permanent secretary. However, the permanent secretary (along with his/her deputy and assistants) retained exclusive authority for personnel and financial matters and remained responsible for the overall coordination of ministry functions. When acting in the latter capacity, the permanent secretary would not give instructions regarding the way technical activities should be carried out or interfere with detailed professional work "'save in the most exceptional circumstances'" (cited in Adamolekun, 1978a:20).

Generalists vs. Specialists

Two important observations regarding the dynamics of administrative relations can be drawn from an analysis of Nigeria's initial ministerial organizational structure. First, the permanent secretary occupied a powerful policy-formulating and executing position in technical as well as administrative ministries. Second, the specialist class of civil servants[3] experienced deeply felt frustrations over policy and administrative subordination. With regard to policy formulation, professional officers criticized their administrative counterparts for lack of technical competence and understanding, relative youthfulness and inexperience (see Asiodu, 1979:79), and short tenure dealing with the specialized affairs of their particular ministry. The conviction that technical expertise and perspectives should prevail in policy making and functional supervision led professional officers to demand that the permanent secretary of all technical ministries be appointed from their ranks. Although professional officers are not formally barred from assuming the post, the prevailing practice in Nigeria has been to appoint administrative officers as permanent secretaries -- even in technical ministries (Balogun, 1983:81; Otobo, 1986:113; however, see Asiodu, 1979:80). In the first decade following independence, only six professional officers served as federal permanent secretaries -- in three ministries (Adamolekun, 1978a:19). This practice institutionalized conflicts between specialists and generalists in technical ministries over promotion to influential and prestigious posts (Adamolekun, 1978a:19-21; Ayida, 1979:223; Udoji, 1979:207; Tukur, 1970a:164-165; Adedeji, 1968a:9-10; Aluko, 1968:70, 77). While persistent dissatisfaction on the part of professional and technical officers (Balogun, 1983:160-161;

Nigeria, Political Bureau, 1987:105-109) rightly constitutes a matter of serious concern, requiring that specialists be appointed as the chief civil servant in certain
ministries and as the head of professional departments (see
Nigeria, Political Bureau, 1987:111) carries the risk of
entrenching narrow views and technocratic biases in policy
making.

Administrative-Executive Class Dichotomy

As adapted by the Nigerian civil services, the British
legacy and model of dividing the public bureaucracy into
separate classes produced additional conflicts within the
generalist ranks. Under the colonial tradition, general
service positions in Nigeria had been divided into two
clearly defined classes: the senior service and the junior
service. Racial barriers (until 1946), strictly enforced
entry qualifications based upon formal educational attain-
ments (see Asiodu, 1979:74-75), and the lack of any connection between performance in the lower class and promotion
to the higher one all served to reinforce the rigid distinction drawn between the two compartments (Fletcher,
1978:143-146; Asiodu, 1970:135).
In 1956, the four governments began to implement
recommendations contained in the Gorsuch Report for the
establishment of an executive class (with a corresponding
"higher technical" class) that would bridge the gap between
junior and senior officers. The Gorsuch Report aimed principally at eliminating the common practice whereby high-
level administrative and professional officers performed
routine executive work and at providing performance incentives for middle bureaucratic personnel by allowing promotion into the administrative class based upon a candidate's
record in the executive ranks rather than on competitive
examination. The duties to be assigned to executive class
officers involved "routine but responsible work" of a
general administration, personnel, or accounting nature.
In adopting the Gorsuch Report's recommendations, Nigerian
governments moved a step closer to the model and experience
of public administration in the United Kingdom (Fletcher,
1978:146-154).
It is not clear that either of the two principal
objectives advocated by the Gorsuch Report has yet been
realized in the Nigerian civil services, however. The

chief result of attempts to implement its recommendations
has been the creation of a new caste in the services. This
intermediate class of executive officers[4] has failed to
take over more responsible tasks from the administrative
class and to carve out unique functional roles. Most administrative grade officers continue to utilize executive
officers as chief clerks. As a result of their reluctance
to delegate responsibility (see Ayida, 1979:227), higher
officers in the Nigerian civil services still devote an
indefensible amount of time to routine work. In most
respects, then, the executive grades have continued to be
regarded and treated as a junior service rather than as a
middle management cadre. The junior status of the executive class has been further reflected in and perpetuated by
the vastly superior salaries and emoluments enjoyed by
administrative officers (Fletcher, 1978:171; Anise,
1980:29-33).[5] Failure to realize expectations raised by
adoption of the Gorsuch Report produced considerable
resentment within the executive officer ranks and exacer-
bated conflicts between the various classes that constitute
the Nigerian civil service (see Fletcher, 1978:158-75;
Ayida, 1979:219). With few executive officers appointed to
the administrative cadre as a reward for effective job performance, the civil services failed to fashion the connections envisioned by the Gorsuch Report between middle
management performance, training programs, and promotion
policies (Fletcher, 1978:172, 167-168, 175; Asiodu,
1970:135-137).

By the late 1970s, this situation had changed somewhat
with the introduction of national training courses specifically designed to facilitate the movement of executive
grade officers into the ranks of the administrative class.
For instance, roughly 100 executive officers drawn from
state and local governments and parastatal organizations in
10 northern states graduate annually from the year-long
Administrative Management Training Course (AMTC) conducted
by the Institute of Administration, Ahmadu Bello
University. Upon satisfactory completion of all course
requirements, most of these trainees secure transfers to
administrative grade posts within one year after returning
to their service. According to Allison Ayida (1979:219),
former secretary to the federal military government, "the
Udoji-Williams unified salary structure and the subsequent
adjustments [also] have made it possible for the executive
and technical staff to advance to Grade Level 13 as part of

the normal career expectations for those adjudged
unsuitable as senior management material for the Higher
Civil Service ...."

The titles and corresponding grade levels assigned to
administrative officers in the federal service since 1979,
arranged in descending hierarchical order, are: permanent
secretary (GL 17), principal secretary (GL 16), deputy
secretary (GL 15), undersecretary (GL 13), principal
assistant secretary (GL 12), senior assistant secretary
(GL 10), assistant secretary I (GL 09), and assistant
secretary II (GL 08). Table 1.1 shows the percentages
within each rank in the administrative class of sampled
personnel holding professional qualification, bachelor's
and advanced degrees, and no higher education degree

TABLE 1.1
Educational Backgrounds of Administrative Class,
Nigerian Federal Civil Service, 1978

Grade Level(N)	% No Degree	% Bachelor's	% Masters	% Doctorate	% Professional
GL 17 (44)	13.6%	61.4%	13.6%	2.3%	9.1%
GL 16 (28)	17.9	57.1	10.7	0.0	14.3
GL 15 (79)	16.5	63.3	6.3	5.1	8.9
GL 14 (38)	18.4	47.4	23.7	0.0	10.5
GL 13 (112)	8.9	70.5	8.9	3.6	8.0
GL 12 (125)	8.8	65.6	12.8	9.6	3.2
GL 10 (119)	14.3	73.9	5.9	5.0	.1
GL 09 (182)	8.2	73.6	10.4	1.1	6.6
GL 08 (275)	8.4	80.7	6.2	a	4.4
GL 08-17 (1002)	10.7	71.5	9.2	3.0	5.7

a = less than .1%
Source: Constructed by the author from data in Balogun (1983:121).

in 1978. We observe immediately that persons in the top
four grade levels (14-17) proved more likely at that time
not to possess a university degree (or professional qualifications) in comparison to administrative officers serving

in lower ranks. However, 12 per cent of all top-ranking
officers (GLs 14-17) had obtained a master's degree.

Bureaucratic Expansion

Enormous growth is characteristic of post-colonial
public administration in Africa (see Hodder-Williams,
1984:169-170; Stren, 1988b:217; D. O'Brien, 1979:213). In
the decades following independence, the Nigerian civil services experienced rapid expansion at all levels. From a
colonial service base of 1,100 at the time of the unification of north and south in 1914 and 39,100 established
posts in 1952, the total number of federal and regional
government personnel grew to 71,693 by 1960 (Murray,
1978:93; Adamolekun, 1978b:322).[6] By 1974, the Public
Service Review Commission estimated the total size of the
Nigerian public service (excluding local government, judi-
ciary, and military personnel) at approximately 630,000.
The number of parastatal employees nearly equaled combined
federal and state civil service employment by that time.
The armed forces contributed an additional 250,000 personnel in 1978 (Campbell, 1978:61, Anise, 1980:27). A nation-
wide manpower survey based on 1978-1979 local government
estimates identified roughly 386,600 established positions
at the LG level -- and this figure excludes general
laborers as well as district, village, and hamlet heads
(calculated from Orewa, 1978:Annex I:1, 4). By 1981, the
total number of public sector personnel approached two
million (Adamolekun, 1983:125). The number of federal
civil servants alone reportedly reached about 400,000 at
the end of 1983 (Otobo, 1986:102).

Much of the enormous expansion in Nigeria's public
service can be attributed to growth in state government and
parastatal employment due to political pressures, as well
as to increasing governmental activity at all levels.
Adedeji (1968a:6) asserts that "partisan political factors,
and particularly the need to accommodate their political
colleagues, led the heads of governments in the post-
Independence period to create more and more ministries."
The creation of twelve states out of the four former
regions in 1967 and the addition of seven new states in
1976 provided a major impetus behind expansion of the
public bureaucracy at this level. As a consequence,
moreover, the new state governments had to advance staff
into high-level positions overnight (Adebayo, 1981:147).
The establishment of additional LG units at the time of the

1976 local government reform and in the wake of the return
to civilian rule had the same effects on the third tier of
Nigeria's political system. Between 1966 and 1980, moreover, the number of public enterprises (utilities, banks,
development authorities, social service boards, commercial
and industrial operations) increased from about 70 to
nearly 800 (Adamolekun, 1978b:316-317; 1983:44-45, 125;
Williams and Turner, 1978:153; Anao, 1985:269).[8]

"Parkinson's Law" also explains some of the growth in
the size of the public sector. The number of hierarchical
levels at the top of the Nigerian civil services had
expanded by 1980 to accommodate the positions "undersecretary" and "principal assistant secretary" immediately
below the deputy permanent secretary (now retitled
"principal secretary") level. The total number of higher
officers in Nigeria's public services increased dramatically from an estimated 763 in 1964 (Harris, 1978:285) to
roughly 50,000 ten years later (Beckett and O'Connell,
1977:8).

The oil glut and the country's foreign exchange crisis
brought bureaucratic expansion to an end in the 1980s.
Indeed, the Buhari regime embarked on a major retrenchment
exercise (Otobo, 1986:102). With the exception of a few
departments and parastatals, the government imposed 15 per
cent across-the-board personnel reductions on both the
federal and state public services. In Bendel, the state
government dismissed some 5,000 civil servants in one day.
In Plateau State, authorities laid off all public employees
initially hired in 1982 and 1983 and made retirement com-
pusory for staff 50 years of age and older (Othman,
1984:458-459). By the end of August 1984, a sizeable
reduction-in-force had occurred.

Nigerianization and Indigenization

The accelerated replacement of colonial and expatriate
officers with Nigerian personnel constituted a major objective for the public services in the immediate pre- and
post-independence periods. Prior to the abolition of
racial barriers in 1946,[9] all administrative, professional,
managerial, and technical positions in the senior colonial
service had been filled by British nationals. By 1954,
Nigerians had been promoted or appointed to 19 per cent of
the senior service posts (Fletcher, 1978:143-144). Rapid
Nigerianization and breathtaking advancement characterized
the period 1957-1966 (see Chick, 1969:106). All govern-

ments directed their public service commissions to appoint
Nigerians to vacant public service posts.[10] During this
period, the public service commissions allowed a number of
candidates without university degrees entry into the administrative ranks (Nwanwene, 1978:201; Cole, 1960:334).
This would later lead to frustration over blocked channels
of promotion among their more qualified successors (Chick,
1969:106).

By 1960, Nigerians held 62 per cent of all senior
officers posts in the public services. The extent to which
the Nigerianization policy had been implemented by 1960
varied considerably among the respective services, however,

TABLE 1.2
Proportion Senior Officer Posts Occupied by
Nigerians in 1960: By Service

Service	% Senior Posts Held by Nigerians
Eastern Region	80%
Western Region	82%
Northern Region	28%
Federal Government	57%

Source: Constructed by the author from data in Nwanwene
(1978:201).

as Table 1.2 reveals.[11] Moreover, Nigerians filled only
one permanent secretary position in the western and
northern governments and only 3 out of 13 such posts in the
Eastern Region in 1959/1960. At the federal level in 1959,
Nigerians occupied 1 out of 14 permanent secretary posts, 2
of the 20 deputy permanent secretary slots, and 6 out of 34
senior assistant secretary positions (Cole, 1960:332, 335).
Two years after independence (i.e., in 1962), Nigerians
held all permanent secretary posts in the Western Region,
two-thirds at the federal level, more than half in the
Eastern Region, and four of these top civil service
appointments in the Northern Region (Kingsley, 1963:310-
311; Nwanwene, 1978:200-201).

In spite of the importance placed upon accelerated
Nigerianization of the public services, governments
continued to employ substantial numbers of expatriates at
high levels on a contract basis in the early 1960s --
particularly in the northern and federal services (see
Kingsley, 1963:311; Cole, 1960:333-334; Harris, 1978:285).
The recruitment of expatriate technicians and managers
increased rapidly during the 1970s. A 1979 study conducted
by Patrick Heinecke (1979:9-10) of higher civil servants
(GL 08 and above) located in the capital city of Kaduna
State revealed that expatriates comprised the following
proportions in five state ministries: Education (49%),
Works and Housing (44%), Finance (35%), Health and Social
Welfare (27%), and Agriculture and Rural Development
(20%). According to Heinecke's figures, the average
percentage of expatriate senior staff in these five
ministries (35%) is higher than the average for four of the
largest private firms operating in Kaduna (28%).[12] In
nearby Kano State, moreover, 88 of the 114 state government
doctors in 1980 were not Nigerian nationals (Stock,
1985:481).

The continued preference in the northern region/states
for expatriate officers employed on temporary contracts
must be understood in the context of the disproportionately
small number of indigenes of that region who possess the
minimum entry educational qualifications (see, for
instance, Chick, 1969:106).[13] As early as 1957, the Public
Service Commission of the Northern Region adopted an explicit policy of "Northernization" of the public service: "if
a qualified Northerner if available, he is given priority
in recruitment; if no Northerner is available, an expatriate may be recruited or a non-Northerner on contract terms"
(cited in Cole, 1960:334; also see Elaigwu, 1977:147). By
the time of independence, application of the Northernization policy indicated that "few Southerners would be
allowed to remain long in any conspicuous positions in the
service of the Northern Regional Government" (Cole,
1960:334). Upon the creation of the Mid-West Region in
1963, moreover, Chief Akintola's government dismissed all
Mid-Westerners (more than 3,400 employees) from the Western
Region's public service (Nwanwene, 1978:200-201). The
Eastern and Mid-western regional governments also pursued
extreme regionalization of their civil services, and the
personnel redeployment exercises which followed the
creation of new states occurred primarily on the basis of
the employee's state of origin (Ojo, 1980:51-52).

The goal of staffing higher service ranks exclusively
or preponderantly with indigenes of one's area continues to
be pursued by the states (Asiodu, 1970:129-132; Nwanwene,
1978:207).[14] The practice of giving priority in civil
service recruitment to individuals from one's own state or
region may be advantageous in terms of providing employment
and experience opportunities that might not otherwise be
available, in terms of formulating suitable policy recommendations and facilitating positive bureaucracy-public
relations (Price, 1975:216; Hyden, 1983:67), and in terms
of promoting closer relations between local politicians and
civil servants. On the other hand, recruitment, posting,
and promotion practices which discriminate against non
indigenes virtually eliminate opportunities for inter-state
labor mobility (Kingsley, 1963:316; Ojo, 1980:55, 61),
threaten to perpetuate bureaucratic exclusiveness and
divisiveness (Nwosu, 1977:55, 79, 121), and involve the
risk of inferior job performance (Asiodu, 1970:129,
132-133; Balogun, 1983:206). The emphasis on
self-sufficiency in all classes of personnel has, in a
number of states, perpetuated reliance upon expatriate
staff (particularly in certain professional ranks, such as
engineers, architects, planners, doctors), led to frequent
turnover among temporary employees, and resulted in a high
proportion of vacant posts and "acting" appointments
(Anosike, 1977:42; Ojo, 1980:51-52). A 1977 government
study of Nigeria's manpower requirements placed the level
of staff vacancies for most higher and intermediate-level
scientific and technical posts at between 40 and 55 per
cent and reported a 15 to 30 per cent vacancy rate in
administrative and other non-technical positions (cited in
Ojo, 1980:54; also see Ukaegbu, 1985:501; Ogbonna,
1984:238, 240).

At the federal level, the universities and certain
ministries and public corporations evolved into "the
preserve of one or another of the regional groups represented in the federal government" (Harris, 1978:302;
Luckham, 1971:218, 227-228). Historically, indigenes of
the northern states have been grossly underrepresented in
the federal civil service relative to the population size
and national political strength of the area. The extent of
this imbalance at higher administrative levels is revealed
by official figures showing that in 1960 when Nigeria
received its independence "only 29 out of 4,398 'officers
in C Scale and above' listed the Northern Region as their
'region of origin'" (cited in Cole, 1960:335).

By 1970, regional and political pressures produced greater attention and sensitivity to the goal of establishing a representative national public service (Adamolekun, 1978a:40; Kehinde, 1968:95-99; Adedeji, 1981:804).[15] Although firm figures are not available, indigenes from northern and western states certainly increased their representation in high-level administrative posts at the federal government level over the past two decades.[16] Few women occupy senior public management positions, however (Ikoiwak, 1979:179).[17] The underrepresentation of women in public management is an issue which has been neglected throughout Africa -- with detrimental effects for governmental performance.

Assessment

Following independence, the Nigerian civil services maintained certain distinctive structural features of British administrative tradition (Adedeji, 1968a:8, 6). In the face of rapid bureaucratic expansion, Nigerianization and politicization of the civil services, and increasing government involvement in development-oriented activities, however, controversy and criticism grew over the performance, status, and role of the public bureaucracy (Adedeji, 1968a:9-10; 1981:804; Asiodu, 1970:138-139; Garba, 1979:I).

Perhaps the most enduring feature of public administration throughout Nigeria, and elsewhere in Africa, is the central role which personal contacts play in getting the job done and in career success (also see Hyden, 1983:145-146). The conduct of administrative affairs is greatly influenced by personal preferences, loyalty considerations, and face-to-face interactions. Decisions are frequently reached on an ad hoc basis without reference to written documentation (Callaghy, 1987:97).

Understanding the central place of inter-personal relations offers the student of public administration and public policy making insights into behavior and outcomes which are not forthcoming from examination of formal structural arrangements. Inclinations and expectations are crucial in this regard (see Koehn, forthcoming). One legacy of colonialism has been a public service which simultaneously expects lucrative material rewards for high-status administrative positions, yet refuses to commit personal loyalty and dedication to an institution identified with imperial subjugation and exploitation

(see Oronsaye, 1984:37; Cohen, 1980:78-82). We find here
the roots of the trouble with public administration in
Africa.

In light of this historical background, efforts to
modify colonial administrative principles and practices in
accordance with national political circumstances and objectives constitute a particularly important feature of public
administration throughout the continent. The next section
analyzes such changes in the Nigerian context.

STRUCTURAL AND ADMINISTRATIVE INNOVATIONS AND ADAPTATIONS

Following the 1966 coup d'état which overthrew the
First Republic and allowed General Aguiyi Ironsi to assume
the post of Head of the newly imposed Federal Military
Government,[18] the regime context within which the Nigerian
public services must function has vacillated among different military administrations and civilian governments.[19]
The ruling military leadership and the Second Republic
Constitution thoroughly reorganized and revamped the upper
levels of the Nigerian political-administrative system.
The focus here is on the most important administrative
changes. This discussion also elucidates some of the crucial features of public administration under military rule
-- a common phenomenon in Africa.

The Military Regimes

Until Major-General Ibrahim Babangida assumed power,
Nigeria's military rulers vested supreme legal, policy-
making, and executive authority in two organs: the Supreme
Military Council (SMC) and the Federal Executive Council
(FEC). The SMC, composed of 24 ranking military officers
(including the Head of State and Commander-in-Chief of the
Armed Forces, and the Supreme Headquarters Chief of Staff)
constituted the paramount official decision-making structure. The military governors of the regions/states served
on the Supreme Military Council until excluded by General
Murtala Mohammed in 1975. After 1975, the SMC supervised
the work of the FEC and NCS (Oyovbaire, 1985:10, 133). The
military and civilian[20] commissioners who served as heads
of ministries made up the FEC, a body charged with over-
seeing policy execution and the coordination and administration of government programs (Adamolekun, 1978b:312).

A third organ, the National Council of States (NCS), consisted of the Head of State as chairman, all armed forces chiefs of staff, the Inspector-General of Police, and all state military governors. The Head of State personally appointed the military governors. They reported directly to him in classic chain-of-command fashion (Oyovbaire, 1985:138, 10).

With the abolition of all elected political offices and the institution of these new structural arrangements, the Nigerian civil services became directly accountable to the military leadership. Under military rule, the Head of State and the state military governors paid scant regard to the convention that all important policy matters should be sanctioned by the SMC and/or the Executive Council (Ayida, 1979:220). Augustus Adebayo (1979:9-10, 13) maintains that "from the inception of the military regime in Nigeria, it was basically accepted that the task of running the government was the sole responsibility of the Military and that where Executive Councils had been established, their role was merely advisory." He also shows that General Yakubu Gowon would impose decisions on the SMC that were nearly unanimously opposed by its other members through "cajolery," "veiled threat," or arbitrary action. In contrast, Generals Murtala and Obasanjo adopted more collegiate and collective decision making styles (Dent, 1978: 116). Campbell (1978:80) even asserts that an inner group of SMC members possessed veto power over a wide range of policy decisions in the post-1975 military cabinets.

The reputation of the federal civil service suffered a setback in early 1975 when, following a reorganization exercise which involved the creation of additional ministries and the appointment of a new head of service, General Gowon included a former commissioner and others recruited from the discredited state administrations among his 13 new permanent secretary appointees. Furthermore, "senior northern personnel seem particularly to have resented their continued exclusion from the inner circle of administrators grouped around the new service head" (Campbell, 1978:76, 72). In July of that year, General Murtala Mohammed assumed power following a coup d'état. He immediately removed all 12 military governors and most federal and state commissioners. Within a year, more than 10,000 civil servants, including the high-level administrative officers who had been closely linked to the Gowon regime, had been dismissed from federal and state ministries and parastatals (Aina, 1982:73).

Throughout the thirteen years of military rule in the 1960s and 1970s, the principal structural features of the ministerial system of administration remained essentially unchanged at the national and state levels (Adamolekun, 1978b:311). One important exception is the creation of the post of secretary to the military government/head of the civil service (SMG). The Murtala/Obasanjo regime strengthened the Cabinet Office and set up within it powerful "parallel departments which made extensive use of outside consultants and researchers and operated more as task forces than as executive agencies." The new offices, which frequently circumvented the existing ministerial bureaucratic structure, included the Political Department (responsible for issuing pronouncements on foreign and domestic policies), the Economic Department, the Public Service Department, and the Research Department (Balogun, 1983:126). Staffing these offices provided one means of responding to pressures for upward mobility among top bureaucrats; that is, by appointing more permanent secretaries than there are ministries (Otobo, 1986:112).

Military leaders often based appointment to top public service posts at the federal and state level on personal friendship and other subjective criteria. State Public Service Commissioners acquiesced in appointments that violated the regulations and norms of the civil service. Such appointments, coupled with the deficient qualifications of some military governors, weakened the sense of professionalism possessed by administrative officers and undermined discipline (Adebayo, 1981:170-171, 174-175).

Shortly after assuming power in August of 1985, Major-General Ibrahim Babangida abolished the SMC and replaced it with a 28-member Armed Forces Ruling Council (AFRC). General Babangida heads this new body, which had been reduced to 19 officers by 1987. Moreover, in a change in nomenclature, he took the unusual step of arrogating the title "president" (West Africa, 24 August 1987, p. 1615).

In place of a Chief of Staff Supreme Headquarters, one now finds a Chief of General Staff (CGS). The CGS assists the Head of State in headquarters administration and political affairs and, unlike the former Chief of Staff, exercises no responsibilities with respect to the armed forces. The Chief of General Staff is in charge of relations with the 21 state military governors, national research institutes, and the Public Complaints Commission. He also suggests appointments to the boards of federal government-

owned corporations and parastatals, and works on policies
involving local government (Thisweek, 20 October 1986).

Local Government Reforms

For most of the post-independence period, centralization rather than the genuine devolution of authority to the
local government level has been the prevailing trend
(Smith, 1985:188-190). Central government and parastatal
agencies have typically dominated local decision making in
francophone Africa (see You and Mazurelle, 1987:15, 18).
In the face of resistance on the part of powerful central
government bureaucrats, attempts at decentralization made
little progress in anglophone countries (see, for example,
Rondinelli, 1981:595-596, 612; Smith, 1985:189, 197). In
Tanzania, in fact, the central government had abolished
elected district and urban councils and assumed their
responsibilities by the mid-1970s (Samoff, 1986:2-4).

Currently, there is renewed interest in decentralization throughout Africa. This development is a consequence
of growing awareness regarding the centrality of local-
level politics in people's lives and increasing emphasis on
community mobilization for participation in self-help
efforts. For instance, Tanzania's national leadership
moved to resurrect elected local governments and cooperatives in the 1980s -- partly in response to popular
pressures (Samoff, 1986:5-8; also see Balogun, 1988:53).
The new interest in decentralization even is evident in
francophone Africa (You and Mazurelle, 1987:14). Richard
Stren (1988b:220) reports that "reforms during the 1980s
have widened the responsibilities of local urban councils,
loosened central controls, and created a measure of
democratically elected government in Dakar and in Abidjan."

In practice as well as policy, Nigeria has been in the
forefront of the move to return authority and responsibility to the grass roots. In the Nigerian case, emphasis
has been placed on revitalizing the existing local government system. The primary impetus for the devolution of
governmental authority has been military leadership (see
Graf, 1986:117; Koehn, 1988a:54).

The 1976 local government reform marked the beginning
of a striking reversal in regional/state domination over
governmental activity at the local level in Nigeria. The
reform established 301 independent LG units and assigned

each a nearly uniform set of exclusive and concurrent (with
state government) functions. Subsequently, Article 7 of
the Nigerian Constitution of 1979 guaranteed democratically
elected local government councils and granted LGs specific
principal and shared responsibilities.

In general, however, state government officials
resisted steps which would have transferred meaningful
authority to the local government level. State authorities
continued to control key staffing decisions, retained
extensive influence over policy making at the local level,
and remained heavily involved in LG budget preparation and
financial management. In the face of numerous constraints
on local revenue-raising, LGs grew dependent upon resource
allocations from above (see Aliyu and Koehn, 1982; Koehn,
1988b; Smith, 1985:197-198; You and Mazurelle, 1987:15).[21]

In 1988, Nigeria's Head of State initiated several
actions aimed at overcoming some of the major obstacles to
genuine devolution of authority to LGs which had surfaced
over the past decade. First, he announced in his 1 January
budget message that local governments would henceforth
receive their federal grant allocations directly rather
than via the state government institutions which had
diverted these funds in the past (West Africa, 7 March
1988, p. 401). General Babangida next mandated that all
ministries of local government be abolished and directed
state governments to refrain from undertaking any constitu-
tionally reserved LG functions (National Concord, 7 October
1988, p. 2). Although the October 1988 directive removed
from the political scene the state administrative body
which exercised the tightest control over LGs in the past,
it has not appreciably enhanced the autonomy of local councils because newly established local government departments
or directorates located in the military governor's office
quickly assumed state-wide monitoring and control functions
(Egwurube, 1989:31).

Another indicator of the increased importance placed
on community organization in Nigeria is the central role
assigned to local government councils in leading the way to
civilian rule during the transition to the Third Republic
(see Egwurube, 1989:38-40). The transition process commenced in December 1987 with the election of LG councils
along non-partisan lines (see, for instance, Ojo, 1988 and
Edoh, 1988). At the end of 1989, a new round of partisan
LG council elections will mark the start of a three-year
period of dyarchy (shared military and civilian
government).

Administrative Reform: The Udoji Measures

No discussion of administrative reform efforts in
Nigeria would be complete without reference to the impact
of the Public Service Review Commission's study and recommendations. After two years of inquiry conducted under
sweeping terms of reference, the Commission appointed by
the Gowon regime and chaired by Chief Jerome Udoji submitted its report (Nigeria, Federal Republic, 1974) in
September 1974. The major recommendations of the Udoji
Commission can be divided into two categories: (1) structural reforms designed to "increase the efficiency and
effectiveness of the Public Services in meeting the
challenge of a development-oriented society" and (2) the
regrading of civil service posts, changes in salary scales,
and the establishment of a unified, "harmonized" remuneration scheme for the public and private sectors.[22]

Integrated Project Management. The central component
in the Commission's recommended approach to strengthening
the country's administrative capacity to realize development objectives is its advocacy of integrated project
management in preference to the existing organizational
practice of dividing government work into separate functional ministries. The report maintains that the closed
ministerial system results in separation of the various
technical skills required to execute most development projects successfully and encourages practices that impair
coordination, the most appropriate assignment of responsibilities, and the fixing of accountability on major
undertakings that involve several ministries.[23]

The Udoji Commission's proposed approach to development administration calls for the creation of ad hoc task
forces or teams composed of relevant specialists drawn from
throughout the service for the purpose of designing and
managing the execution of specific development projects.
The principal advantages of this system, according to the
Report's authors, are greater "flexibility in solving
problems than is typically afforded by the traditional,
functional bureaucratic structures" and the coordinated,
concentrated application of expertise and resources that
may be in short supply (also see Adedeji, 1968a:9). Upon
completion of an assigned project, task force members would
revert to their former position in the bureaucracy until
called upon to participate in another undertaking. The
Commission also recommended that each government agency be
headed by a 'chief executive' patterned on the general
manager of a private corporation (see Balogun, 1983:160).

In its response to the Public Service Review Commission's report, the Gowon government chose to ignore these
recommendations for organizational and management reforms.
Integrated project management along the lines advocated by
the Udoji Commission in 1974 has not become a central feature of contemporary Nigerian public administration
(Adamolekun, 1982:21, 27). Recognition that the project
management proposals advanced by the Udoji Commission pose
a serious threat to the existing bureaucratic structure and
hierarchy (particularly to the authority exercised by generalist administrators) accounts in part for the federal
government's unwillingness to implement them (see Campbell,
1978:73; Williams and Turner, 1978:161).

There are sound practical objections to this approach
as well. In Nigeria, the agricultural development projects
sponsored by the World Bank have been run along integrated
project management lines similar to those recommended by
the Udoji Commission. Major administrative drawbacks have
characterized this approach to the management of rural
development projects -- including insulation from popular
local participation in planning and implementation, insensitivity to local needs and demands, lack of coordination
with established ministries, the overly complex and demanding nature of the newly introduced management tasks, and
failure to develop the institutional capacity of existing,
permanent administrative structures (see Wallace, 1980a:67;
Hyden, 1983:92-93; Teriba, 1978:43-44, 85; Adamolekun,
1978b:316-318; Kehinde, 1968:92-98; Adedeji, 1968b:151-152;
Okpala, 1977:157-166; Nwosu, 1977:68-69). When foreign
assistance is available, moreover, projects usually succeed
in attracting scarce qualified staff from elsewhere in the
public bureaucracy. The resulting staff exodus weakens
permanent administrative agencies (Nyaburerwa, 1988:69;
Korten, 1980:484). Externally supported projects also
experience great difficulty retaining the services of their
most capable personnel since trained and experienced managers are in short supply and MNC and donor pay scales and
benefits are more attractive (see Nellis, 1986:36).

Harmonization. A second, and related, major recommendation of the Udoji Commission concerned the public sector
salary structure. By 1970, public servants in Nigeria and
elsewhere in Africa, particularly senior management personnel, generally received considerably lower salaries and
less attractive benefits than their counterparts working
for large private enterprises. The lucrative material
rewards offered by multinational firms and international
development agencies such as the World Bank led public

administrators to demand equivalent benefits for themselves (see Abernethy, 1983:15). The managers of parastatals achieved some success in this regard. As the salary structure for public enterprises became far more attractive than that of the federal civil service (see, for instance, Makgetla, 1986:413), they managed to drain competent staff away from the ministries. Those who did not leave the civil service resented the higher remuneration obtained by their counterparts working in public corporations (Otobo, 1986:113). In response to these developments and growing competition over highly qualified personnel, the Adebo and Udoji Commissions carried similar charges to explore areas in which the harmonization of public and private sector wages, salaries, and other conditions of employment would be desirable and feasible (Williams, 1976a:32; Asiodu, 1970:140-144; Williams and Turner, 1978:160).

Under considerable pressure from public servants at all levels, the Gowon administration reacted more favorably to the Public Service Review Commission's grading and salary proposals than it did to those aspects of its report aimed at improving management and performance within the public sector (Akinyele, 1979:237; Ayida, 1979:223). Specifically, the FMG accepted the Udoji Commission's recommendations regarding the introduction of a unified grading structure and harmonization of public and private sector salaries. The FMG's initial actions in both areas in early 1975 stirred up additional, competing demands and triggered a series of strikes and protest demonstrations that forced the faltering Gowon regime to modify certain aspects of the original regrading scheme and to grant further salary increases.

The Udoji Commission recommended that the multitude of differentially graded and remunerated posts existing in the various civil services be systematically reclassified under a single, unified position and salary scheme composed of 17 grade levels. The Commission based its specific position classification proposals on the results of systematic job analyses conducted within each civil service that revealed which posts possessed essentially equivalent duties, responsibilities, and required qualifications. The FMG not only adopted the unified grading scheme proposed by the Udoji Commission for the civil services,[24] but extended it to encompass the entire public sector (Adamolekun, 1978b: 332-334; Williams and Turner, 1978:161).

In January 1975, the FMG acted favorably on the Udoji Report's recommendations that the wages and salaries of certain classes of civil servants be increased dramatically

in order to bring about comparability with private sector
pay scales. The government doubled many salaries, including the rates paid to daily laborers at the bottom of the
scale and to top-level administrators. The Gowon administration also decided to backdate the increases to April 1,
1974 and to exempt half of the arrears from taxation.
Moreover, the FMG implemented an even more generous civil
service pension scheme than the Udoji Commission had proposed and substantially raised the rate at which it paid
supplementary allowances (Campbell, 1978:73; Ayida, 1979:
227-228; Anise, 1980:26; Williams, 1976a:32; Williams and
Turner, 1978:166). These outcomes of the Udoji exercise
offer an excellent illustration of "weak political leadership and excessive administrative power" (Adamolekun,
1982:32; also see Chapter 2).

The combined effect of the regrading exercise and the
government's differential "catch-up" salary awards triggered a spate of protests on the part of workers who perceived that they would suffer an erosion in their status
and/or income relative to the categories of civil service
personnel most favored by the new policies. Public service
employees in technical and professional positions complained that the government's regrading of posts worked to
their disadvantage, and joined intermediate-scale civil
servants in protesting the smaller salary increases they
had received relative to the administrative cadre. Workers
in public corporations and the private sector, who had been
excluded from the initial "Udoji" awards, agitated for comparable salary increases.[25] Confronted by a series of private and public sector strikes and other labor actions, the
Gowon regime eventually capitulated. The Government allowed private employees to negotiate wage settlements comparable to the Udoji awards (with full arrears), granted
specialist public servants a greater measure of parity with
administrators, and provided that meritorious professionals
could earn as much as the chief executive of their agency
(Campbell, 1978:73-74; Williams and Turner, 1978:161, 166:
Williams, 1976a:32, 51; Dent, 1978:111; Collins, Turner,
and Williams, 1976:187; Adamolekun, 1982:19).

The long-term impact of the Udoji Commission has been
considerable, although the results have differed in important respects from those envisioned in its report and recommendations. In the federal service, operation of the
unified grading scheme, ministerial reorganizations, and
expansion of the number of directorships at the head of
newly created specialized divisions enhanced promotion
prospects for professional and technical personnel.[26]

Maintenance of the uniform pension scheme facilitated
inter-service mobility (Adamolekun, 1982:25-26). Neither
the imposition of a unified grade structure nor the attempt
to harmonize salaries produced the intended effect of attracting qualified private sector personnel into the public
services, however (Nigeria, Federal Republic, 1974:34;
Asiodu, 1970:144; Ayida, 1979:222). Indeed, inter-sectoral
mobility has continued to flow overwhelmingly in the opposite direction during the post-Udoji years, with high-
ranking professionals and administrators exhibiting the
greatest propensity to leave government work (Ayida, 1979:
222, 226-227; Stock, 1985:477; Balogun and Oshionebo, 1985:
303; Ofoegbu, 1985:140; Adamolekun, 1982:27; also see
Nellis, 1986:31).

The government's failure, in spite of its generous
Udoji awards, to alter this situation must be viewed in the
context of a political economy characterized by rapid expansion of multinational corporate operations in the midst
of an oil boom, indigenization measures that dramatically
improved the position of Nigerian businessmen, and perceived skilled manpower shortages (see Collins, Williams,
and Turner, 1976:186-187, 191-192; Collins, 1977:134-146;
Anosike, 1977:28). In the face of such conditions, the
realization of comparable public and private conditions of
employment constituted an elusive policy objective. Indeed,
privately employed professionals and managers still manage
to pay lower taxes, receive more lucrative fringe benefits,
and secure higher incomes than their public service counterparts (Forrest and Odama, 1978/79:127; Joseph, 1978:235;
Asiodu, 1970:143; Heinecke, 1979:1-3; Nigeria, Political
Bureau, 1987:110; Ihimodu, 1986:229; Otobo, 1986:124).[27]

Between 1975 and 1979, the Murtala/Obasanjo regime
essentially froze the unified salary scheme established
following the Udoji review.[28] The new civilian regime
enacted a minimum wage of 100 naira per month and increased
the salary structure accordingly in 1980, but only in the
lower grades (GLs 01-06). In 1981, it raised the minimum
annual salary at GLs 01-03 to 1500 naira (Adamolekun,
1983:122; Otobo, 1986:118). Table 1.3 presents the basic
monetary dimensions of Nigeria's unified salary scheme for
the public services following these revisions. As of
December 1983, 323,453 persons earned GL 01-06 salaries in
the federal civil service. There were 52,307 federal
employees at GL 07-12, and 5,866 managers occupied GL 13-17
positions (Otobo, 1986:102).

The price of nation-wide uniformity in salary
structure is reduced flexibility in public personnel

TABLE 1.3
Unified Salary Scheme for the Nigerian Public Services

Grade Level	First Step (₦ p.a.)	Last Step (₦ p.a.)	Annual Step Increment (₦ p.a.)
01	1,500	1,500	–
02	1,500	1,500	–
03	1,500	1,524	–
04	1,500	1,752	42
05	1,740	2,172	72
06	2,196	2,772	96
07	2,832	3,552	120
08	3,564	4,464	150
09	4,668	5,640	162
10	5,760	6,732	162
11	6,744	7,284	180
12	7,404	8,052	216
13	8,064	9,024	320
14	9,168	10,128	320
15	10,296	11,328	516
16	11,568	12,720	526
17	12,996	14,268	636

*There are 7 steps for GLs 01-10, 4 steps for GLs 11-14, and 3 for GLs 15-17.

administration (Balogun and Oshionebu, 1985:303). While the civilian government successfully resisted labor pressures to increase the minimum wage to 300 naira per month, it eventually acceded to demands for special treatment on the part of university employees and to other pressures for amendments to the unified salary structure. With the aid of a nation-wide strike, university staff became the first to succeed in pulling out of the unified service (see Balogun, 1983:197-198; Bangura, 1986a:55). The recommendations of the Presidential Commission on Salary and Conditions of Service of University Staff (Cookey Commission) led to the creation of the enhanced University System Scale (U.S.S.). The U.S.S. substantially increased basic salaries for university employees without allowing professors to earn more than permanent secretaries (Otobo, 1986:119).[29]

Similar protest actions and demands by parastatal employees led to the creation of the Presidential Commission
on Parastatals (Onosode Commission) in May of 1981. The
government accepted the Onosode Commission's recommendations that (1) the unified grading system be separated and
distinguished from the unified salary system and (2) that
parastatals with 'autonomous employer status' be authorized
to determine pay for jobs -- subject to a ceiling set by
the government (Adamolekun, 1983:58-59; Otobo, 1986:119-
120, 124).[30] The race for comparatively higher salaries
had recommenced.

The Second Republic

The 1979 Constitution introduced several lasting
structural changes in the Nigerian administrative state.
In the first place, Articles 139 and 177 provided for the
appointment of special advisors to the president and each
governor. Persons appointed to such posts held temporary
political rather than civil service appointments and served
"at the pleasure of" the president/governor. The authority
and influence of many special advisors rivaled or exceeded
that exercised by federal ministers/state commissioners.
For instance, the President's Special Advisor on Budgetary
Matters, who also acted as Director of Budget, assumed the
leadership role in the budget preparation process
(Adamolekun, 1983:133-134; also see Uwazurike, 1987:250).
The emergence of a powerful executive office of the
president/governor constitutes an important development
in the upper reaches of Nigerian public administration (see
Ofoegbu, 1985:124; Lewis, 1985:133). Moreover, the role
of special advisors, the nature of the presidential system
of government, and constitutional prohibitions (Articles
135, 173) against concurrently holding the office of
minister/commissioner and member of a legislative body, all
constituted further moves away from the British-inspired
doctrine of collective ministerial responsibility -- which
had never been strictly adhered to in Nigeria (Adamolekun,
1978b:327).
The Second Republic Constitution also split the office
of SMG (at both federal and state levels) into the two separate posts of secretary to the government and head of the
civil service. The person appointed to fill the latter
office must be a member of the respective federal or state
civil service (Articles 157 and 188). Beyond this, the
Constitution remained silent regarding the two positions.

The absence of any precise legal definition and division of responsibilities for the two offices allowed ample opportunities for conflict and variability in role performance. In general, the secretary to the government assumed the more powerful political and policy advisory aspects of the former SMG's role (i.e., acted as "chief" advisor) and the head of the civil service became a secondary figure who concentrated on personnel management under the new system (Akinyele, 1979:239-240; Bach, 1980:2; Okoli, 1978/1979:19, 10; Akpan, 1982:137, 187-188; Ofoegbu, 1985:124).

Major variations emerged during the Second Republic in federal and state government appointment policies and practices as state governors embarked on different approaches in response to local circumstances and preferences. The eagerness with which they took advantage of the flexibility provided by the 1979 Constitution gave rise to considerable diversity in administrative practice within the federation and, in most states, to the creation of a "much wider band of quasi political appointments ..." (Murray, forthcoming; Oyovbaire, 1980:269-270; Akpan, 1982:138).

The politicization of the top echelons of the administrative machinery under Nigeria's multi-party political system militated against greater cooperation among, no less unification of, the federal and state civil services (see Asiodu, 1970:133). This is best illustrated by the negative reaction which greeted the presidential liason officers, whom Shehu Shagari designated as special advisors to the president, in the non NPN states (Nwabueze, 1985:74, 252-257). The increasingly political nature of appointments also increased the likelihood that newly elected presidents and governors (or later-day military rulers) would replace incumbent permanent secretaries and heads of departments (as well as the secretary to the government) with fresh appointees of their own choosing (see Kolo, 1985:147).[31] In support of this approach, Joseph Garba (1979:I) maintains that "if the fortunes of the principal advisers of Government are tied to the fortunes of the government itself, civil servants may be more committed to the successful execution of the government policies." With the advent of Nigeria's presidential/gubernatorial system of government, in any event, "permanent secretaries were no longer permanent" (Ofoegbu, 1980:5). Vice President Alex Ekwueme (1980:8) aptly characterized the permanent secretary under the new system as "among the least permanent or most unpermanent of public officers."

In practice, perm secs had not been permanently

assigned to a particular federal or state ministry since independence. The relatively rapid rotation of generalist permanent secretaries is a distinctive characteristic of African public administration (Montgomery, 1987:350). In Nigeria, until the 1988 Civil Service Reform, a small group of individuals rotated as permanent secretary from ministry to ministry, or in and out of the cabinet office (Adamolekun, 1978a:21). Augustus Adebayo, for instance, served as perm sec in six different posts within the ten-year period between 1963 and 1973 (Adebayo, 1981:131-133). One result of this system is the frequent appointment of high-level generalist administrators who possess no expertise in and may not even be conversant in the primary subject matter jurisdiction of their ministries (see Udoji, 1979:207-208).[32] Consequently, fresh appointees are forced to rely heavily on advice tendered by the professional experts in their new department (Adebayo, 1981:132-133).

Questions immediately arose in 1979 regarding the fate of those permanent secretaries not reappointed (or removed from office) by a newly elected chief executive. A permanent secretary in the office of the federal head of service, E.O. Olowu, announced in February 1980 that government officials who took up political appointments (as special advisors and directors as well as perm secs) would not automatically be deemed to have resigned or retired from the civil service and could elect to remain a member of the service (New Nigerian, 25 February, 1980, p. 24).[33] Thus, officers who previously belonged to the civil service retained tenure and other accumulated rights and benefits, and could be reassigned upon completion of such political assignments to administrative positions that fell outside the presidential/gubernatorial appointment provisions of Articles 157 and 188 of the Constitution (see Ofoegbu, 1980:5; Bach, 1980:3; Kolo, 1985:148).[34]

In a unique and bold step, the framers of the 1979 Constitution moved to ensure that federal and state government agencies reflect "the federal character of Nigeria." Specifically, Article 14 (sections 3 and 4) provided that:

> the composition of the Government of the Federation or any of its agencies and the conduct of its affairs shall be carried out in such manner as to reflect the federal character of Nigeria and the need to promote national unity, and ... there shall be no predominance of persons from a few states or from a few ethnic or sectional groups in that government or in any of its agencies. The composition of the Government of a

State, a local government council, or any of the agencies of such government or council, and the conduct of
the affairs of government or council or such agencies
shall be carried out in such manner as to recognize
the diversity of the peoples within its area of
authority and the need to promote a sense of belonging
and loyalty among all the peoples of the Federation.

Similar provisions governed the appointment by chief executives of ministers/commissioners, permanent secretaries,
and agency heads (Articles 135(3), 157(5), 173(2), 188(4)).
With respect to career public servants, M. J. Balogun
(1983:206-207) emphasizes that the 'federal character'
clause "does not imply removing incumbents of positions to
make room for people from areas hitherto underrepresented
in an organization ...." In fact, he adds, "Section 39(1)
of the 1979 Constitution specifically forbids discrimination on grounds of ethnic origin, sex, religion or political affiliation." Nevertheless, Sam Oyovbaire (1985:
244-245) raises the provocative point that the principle of
federal character "contains negative consequences for the
identity, culture and loyalty of a growing category of
Nigerians who are born, bred and socialized outside of
their parents' states or communities of birth, and who are
therefore, in constitutional and political terms, going to
be without states or communities with which they are truly
and culturally identifiable."

The "federal character" clause has exerted a discern-
able impact on Nigerian public administration. For
instance, eight states did not place indigenes among the
federal perm secs holding office at the start of the
Shagari administration in 1979. Within a year, each state
(exept Borno) could boast at least one federal permanent
secretary. Three states were represented by more than one
indigene: Bendel (8), Ondo (4), and Sokoto (2) (Musa,
1985:117).

In August 1980, however, the Governor of Sokoto State
charged that staffing of the Nigerian National Petroleum
Corporation did not reflect the federal character of the
country. According to the Governor, NNPC had only 13 indigenes of Sokoto (President Shagari's home state) among its
roughly 1,000 employees (New Nigerian, 8 August 1980,
p. 16). The state-by-state breakdown of employees working
for another federal parastatal (Nigeria Airways) in 1980
shows the agency in violation of the "no predominance of
persons from a few states" provision of Article 14. A
total of 413 employees (5.3% of the total) per state would

constitute a perfectly equal distribution. The data
reported in Table 1.4 reveal that 4 states are considerably
above this figure, with 14 states showing less than propor-
tionate representation on the Airways staff. Indigenes

TABLE 1.4
State of Origin of Nigeria Airways Employees
(September 1980)

State	No.	% Total	State	No.	% Total
Anambra	375	4.8%	Kwara	268	3.5%
Bauchi	42	.5	Lagos	305	3.9
Bendel	1,263	16.1	Niger	34	.4
Benue	302	3.8	Ogun	1,351	17.2
Borno	100	1.3	Ondo	581	7.4
Cross River	307	3.9	Oyo	780	11.3
Gongola	208	2.6	Plateau	69	.9
Imo	1,076	13.7	Rivers	157	2.0
Kaduna	152	1.9	Sokoto	39	.5
Kano	354	4.5	Expatriates	93	1.2

Total = 7,856

Source: Constructed by the author from data in New
Nigerian, 18 December 1980, p. 1.

from 4 northern states (Niger, Bauchi, Sokoto, and Plateau)
comprised less than one per cent of the total personnel
force employed by the public enterprise (also see Marenin,
1987:26-27; New Nigerian, 17 December 1984, pp. 1, 3). On
the other hand, persons from the northern states predominate among the chairmen of important public commissions and
in the diplomatic corps (Nwabueze, 1985:306-307).[35]
     The military regime which replaced Shehu Shagari's
government continued to adhere to the federal character
principle at the national level. Thus, Major-General
Muhammudu Buhari appointed an indigene of Bendel State
as head of the civil service and a member from every other
state to the newly formed Federal Executive Council
(Rothchild, 1987:124). Formally or unofficially, governments in Kenya, Ivory Coast, Cameroun, Guinea, Ghana, and

Malawi also have sought to achieve geographical balance in
top-level positions (Rothchild, 1987:124).

At the state level, pressures for preferential local
hiring have increased as trained manpower becomes available. To cite one example, Sokoto State attempted for a
time to restrict hiring to indigenes of that state. In
1980, moreover, the Kaduna State House of Assembly passed a
resolution requesting that at least 50 per cent of the
staff deployed to work in federal government field agencies
be indigenes of the state in which such offices are located
(New Nigerian, 28 October 1980, p. 1). Thus, representation problems persist in spite of the federal government's
reported decade-long practice of "giving preference to the
indigenes of the states in which federal offices and agencies are located" (Adamolekun, 1983:127). Conflicts also
occur over the intra-state distribution of government jobs
(see Nwabueze, 1985:307). For instance, charges that indigenes of their area had been excluded from positions of bureaucratic power bolstered the arguments advanced in 1987
on behalf of creating the new states of Katsina and Akwa
Ibom (Nigeria, Political Bureau, 1987:172-173). The military governor of Bauchi State from 1985 to 1988, Chris
Garuba, made some powerful enemies by attempting to redress
the lock on key state civil service positions maintained
until that time by indigenes of Katagum LGA (Abu, 1988:21).
Efforts to achieve intra-state representation in appointments are rendered exceedingly complex by the great variety
of available ways of defining one's background.

1988 Civil Service Reforms

The next watershed in Nigeria's post-coup administrative reform movement occurred in 1988. In his budget address that year, General Babangida announced a number of
"fundamental changes" in the civil service structure.
Known as the Civil Service Reforms of 1988, the newly introduced measures generally extended and solidified the
reform directions embarked upon in the 1979 Constitution.

First, the 1988 Civil Service Reforms revised and
standardized ministerial organization and conferred the
title "chief executive and accounting officer" on the minister rather than the permanent secretary (Nigeria, Federal
Republic, 1988:7).[36] Permanent secretaries are now
referred to as "director-general" and serve as deputy to
the minister. Section I(7) of the Implementation Guidelines for the 1988 Reforms contains important language that

"the Minister shall delegate a substantial part of his administrative and financial functions and authority to the
Director-General who should be fully involved in the key
decision-making processes." The Guidelines (pp. 7-8)
further confirm that directors-general are political
appointees who serve at the "pleasure of the President" and
must resign with the government that appoints them.

Figure 1.2 presents the standardized ministerial
structure adopted in 1988. Each ministry and extra-
ministerial agency is required to have departments of personnel management, finance and supplies, and planning,
research, and statistics. There can be no more than five
operating departments. Departments are subdivided into
divisions, divisions into branches, and branches into sections (Nigeria, Federal Republic, 1988:10). In even more
fundamental changes, the "posting of officers from one
Ministry to another will no longer be allowed"[37] and most
personnel matters are to be decided by intra-ministerial
boards and committees in accordance with uniform guidelines
issued by the Federal Civil Service Commission. The Implementation Guidelines (p. 7) also created a National Planning Commission and transferred the Budget Department from
the Ministry of Finance and Economic Development to the
President's Office. These changes move Nigerian public
administration even closer to U.S. practice and farther
away from British administrative patterns.

In addition, the 1988 Civil Service Reforms re-tackle
the sensitive issues of promotion opportunities for professional and intermediate-cadre officers, and federal character. On the first score, Section III(18) of the Guidelines
provides (p. 18) that "every civil servant" is eligible to
rise to the top-most career post in his/her ministry (i.e.,
GL 17, director). Special promotion examinations have been
introduced for advancement from GL 06 to 07 and for entry
into the senior-management cadre at GL 14.

The 1988 Reforms revised the federal-character principle by distinguishing between selection and promotion.
Specifically, under Section IV (1 and 3.1) of the Guidelines (pp. 15-17), the Federal Civil Service Commission re-
tains responsibility for hiring staff at GLs 07-10 in
accordance with the federal-character principle. There-
after, however, the criteria for promotion focus on merit
principles and explicitly exclude considerations of geographical balance. There is no mention of federal character in the article describing ministry appointments to
positions at GL 11 and above.[38] At the lower end of the
hierarchy (GLs 01-06), junior staff are no longer to be

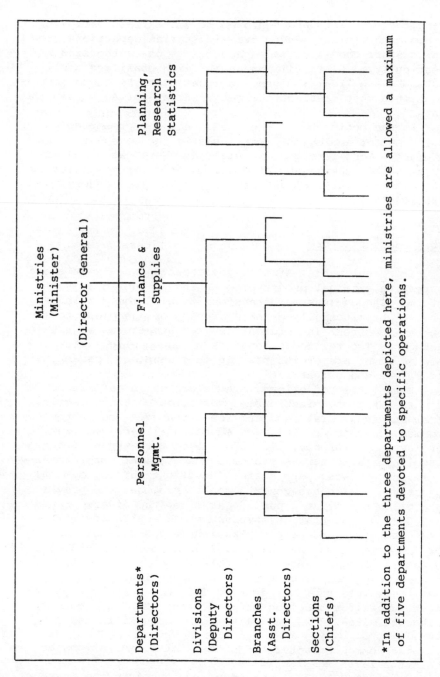

Figure 1.2   Post-Reform Ministerial Structure in Nigeria (1988)

*In addition to the three departments depicted here, ministries are allowed a maximum of five departments devoted to specific operations.

recruited centrally and deployed to state field offices.
Within each ministry and extra-ministerial department, the
Junior Staff Committee in each state is now authorized
(Section IV, 3.6) to fill vacancies with qualified appli-
cants who are "real indigenes of that State." These ap-
pointments must "reflect the geographical spread within the
State."

At the beginning of the 1990s, equal representation in
public sector employment (particularly for women) remains
an elusive long-term goal in Nigeria. Here, as elsewhere
in the world, the legacy of discrimination is most glaring
in the higher ranks of the career service (see Peters,
1980:106-107).

Public Sector Reform

In recent years, African governments have experienced
increasing external pressure to engage in public sector
reform. The demands for transformation of the public sector, which principally emanate from donor agencies, form a
central component in foreign-inspired structural adjustment
packages. Two related dimensions of the public sector reform movement are of interest in this chapter: parastatal
reductions and privatization.

Parastatal Reductions. There are two broad categories
of parastatal organizations: production-centered (agricultural and industrial) and service-oriented. The latter
provide economic, social, and educational services (see,
for instance, Ihimodu, 1986:224-225). Both types of parastatal typically pursue economic (development, profitmaking
or revenue-raising) and social (distribution, employment)
objectives. Although certain goals are likely to prevail
in a given agency at a particular time, the diverse mandate
possessed by parastatals involves considerable administrative complexity and presents an inherent source of tension
for management (see McCullough, 1988:3; Nellis, 1986:35).
In Zambia, for instance, parastatals emphasized profitability as the principal criterion for operation in order to
meet their international debts. This had serious negative
consequences for the long-term development of the country
(Makgetla, 1988:18-20, 23; also see Hodder-Williams,
1984:154).

The common structure of parastatal organizations at
the state level in Nigeria is well described by Ifeyori
Ihimodu (1986:225-228). The governor appoints the board of
directors and most chief executives (general managers).

Civil servants, usually from the supervising ministry, frequently comprise nearly half of the board's membership (see
also Phillips, 1989:439). The commissioner of the parent
ministry and/or the permanent secretary in charge of parastatal affairs within the governor's office issues policy
directives to the board of directors and monitors implementation and performance. The dual relationship with ministry and governor's office is an obvious source of confusion,
conflict, and delay (Ihimodu, 1986:225-226; Kolo, 1985:149,
155; Lewis, 1985:133).

African political leaders have exhibited a penchant
for creating public commercial enterprises and turning over
ministry functions to parastatal bodies. Parastatals have
assumed vast responsibilities in such diverse functional
areas as housing, water supply, insurance, scholarships,
electricity, sewage treatment, transportation, satellite
communication, agricultural production, hotels, breweries,
and mining (Nigeria, Political Bureau, 1987:62; Lewis,
1985:133; Kolo, 1985:149; Adedeji, 1988a:60; Nellis, 1986:
12; Ihimodu, 1986:225; Cohen and Koehn, 1980:298; Makgetla,
1988). John Nellis (1986:vii, 5) conservatively estimates
that there are about 3,000 public enterprises (excluding
regulatory agencies and statutory boards) in Africa.

With technical assistance from the World Bank and
other donors, African regimes have dedicated considerable
attention and resources to strengthening public enterprises
(e.g., R. O'Brien, 1979:123). They often recruit foreign
nationals to staff key management positions in public corporations (Hodder-Williams, 1984:153). By and large, all
their efforts have been to no avail. Parastatals continue
to be plagued by enormous losses rather than to record
revenue-enhancing profits (Nellis, 1986:ix, 17-18).[39]
Most also are characterized by partisan, corrupt, expensive, and ineffective management (ibid., ix, 18, 35-36;
Nigeria, Political Bureau, 1987:62; Ihimodu, 1986:232,
234; Rondinelli, 1981:623; Anao, 1985:269, 283; Adamolekun,
1983:50-51; Otobo, 1986:123; L. White, 1987:139; Bates,
1987:244-245; Hodder-Williams, 1984:154, Forrest, 1986:
16-17; Hyden, 1983:100-101).[40] In the retrospective
assessment of Julius K. Nyerere, by replacing local
government and cooperatives with parastatal organizations,
Tanzania "'ended up with a huge [bureaucratic] machine
which we cannot operate efficiently'" (Gauhar, 1984:830).[41]

In spite of the "bad press" which parastatals have
recently received, available evidence suggests that it is
more difficult to eliminate them or reduce their role in
the economy than it is to criticize them. In Kenya, for

instance, the emphasis is on "redirecting" government subsidies "toward foreign exchange-generating parastatals and
away from basic goods state enterprises" (Lehman, 1988:15).
The government of independent Zimbabwe actually increased
subsidy support for parastatals and expanded the number of
state-owned enterprises in the 1980s. Its holdings include
banks, manufacturing industries, and agricultural operations (ibid., pp. 16, 23-4). In Nigeria, the Kwara State
government managed, through dissolution and merger, to
reduce the number of parastatals from 26 to 20 after 1983
(Ihimodu, 1986:223; also see Nellis, 1986:45).

Privatization. With regard to state-owned enterprises
(parastatals of the production-centered variety), one currently encounters calls for "privatization" everywhere on
the continent (Mkandawire, 1988:19-20; L. White, 1987:139;
McCullough, 1988:1; Brooke, 1989:4). The constituency
which supports privatization in Africa (retired bureaucrats
and military officers, indigenous businesspeople, politicians and public administrators with capital) is growing in
numbers (Mkandawire, 1988:21, 23-24). Nevertheless, the
record of actual state divestiture is less impressive than
official pronouncements imply in Senegal, Kenya, and Ghana
(Wilson, 1988:26; also see Nellis, 1986:45, 47).[42] The
most extensive privatization exercise has occurred in Ivory
Coast. It had affected 28 industrial firms by 1987
(Wilson, 1988:26).

In the face of mounting pressures from its international creditors, Nigeria also engaged in a major privatization exercise. Where the Buhari administration advocated
gradual and selective privatization and rejected nearly all
proposals for commercialization of public utilities, General Babangida pledged in his 1986 budget address that the
federal government would "'divest its holdings in agricultural production, hotels, food, beverages, breweries,
distilleries, distribution, electrical and electronic ap-
pliances and all non-strategic industries'" (cited in
Nigeria, Political Bureau, 1987:63). Nigeria's July 1988
privatization decree implemented this pledge. Under the
decree's provisions, the government is to sell 67 state-
owned companies and to divest itself of controlling interest in 21 other enterprises (including the infamous Nigeria
Airways). Another 25 parastatals, including major utilities, are to be fully or partially "commercialized" -- that
is, forced to survive without government subsidies or with
reduced support (New York Times, 15 August 1988, p. D4;
Christian Science Monitor, 13 July 1988). There are even
plans to privatize the Nigerian National Petroleum Company

(NNPC) and hints that modifications of Nigeria's indigenization decrees may be forthcoming (Africa Analysis, 15
April 1988, p. 5).

Along with privatization, the public sector reform
measures required by structural-adjustment programs have
included reductions in public sector employment. Many
African governments took steps in this direction in the mid
1980s, although personnel expenditures continue to consume
a high proportion of most budgets (Nsingo, 1988:82). For
instance, Sudan dismissed 9,000 national railway employees
in 1986 and Senegal engaged in extensive cutbacks of public
employees working on rural and industrial projects
(Fromont, 1988:94) Other parastatals in Niger and Sierra
Leone experienced substantial workforce reductions in
recent years (Nellis, 1986:31). In most cases, public sector employees "have been laid off without any plans being
drawn up for their reemployment" (Fromont, 1988:94).

In the rush to privatize, there is a tendency to over-
look the accompanying social and political costs. In the
first place, the dismantling of government structures could
result in the complete abandonment of vital state services
(Frisch, 1988:70). In addition, divestment from highly
profitable commercial enterprises removes revenue from
government coffers that has been used to fund essential
rural and urban services for the poor. At the same time,
privatization opens new opportunities for foreign investors
and multinational firms to extract resources and capital
and further erodes prospects for economic self reliance
(see Nigeria, Political Bureau, 1987:63; Browne, 1988:9).
Thus, foreign interests, mainly French firms, purchased
approximately half of the parastatals privatized in Ivory
Coast (Wilson, 1988:26). The international entrepreneurs
willing to risk investing in public enterprises in coup-
prone African countries typically "demand high rates of
protection and mechanisms to allow them to recoup their
total investment in extremely short periods" (Nellis,
1986:47).

Moreover, the termination of government subsidies
usually results in higher prices for consumers and increased unemployment (Lehman, 1988:24). In light of the
weakness of the African state, privatization is likely to
turn public enterprises into unregulated and non-responsive
private monopolies. In the absence of effective regulation, essential services could be disrupted and the
public's welfare ignored (Nellis, 1986:44-45). In sum, the
benefits and costs of privatization and commercialization
are not shared equally by all classes.[43]

CONCLUSIONS AND LESSONS

Changes in political economy promoted relentless
expansion in the size of the public bureaucracy throughout
Africa and expanded the privileged economic opportunities
available to insiders (see Meillassoux, 1970:106). In
many countries, the armed bureaucrats have strengthened
their position within the ruling class. Increases in
bureaucratic power and rewards have not been accompanied,
however, by noticeable improvements in the performance
of public servants or gains in public sector productivity
under civilian or military regimes. A reliable indicator
that this is the case in Nigeria is the continued erratic
provision of such essential services as water, electrical
power, telecommunications, and the maintenance of law and
order (Otobo, 1986:123; Stren, 1988b:222).

The Nigerian experience with the 1975 purge,
reductions-in-force in 1984, and privatization since 1986
suggests that cutbacks in public service personnel exert
negligible positive long-term impact on bureaucratic
performance. In immediate response to the mass dismissal
exercise, "the personal performance of public officials
improved enormously, people arrived on time, worked harder
and spent less time on private business" (Dent, 1978:123;
also Campbell, 1978:81). There is no evidence, however,
that the purge, the subsequent establishment of new
investigatory and adjudicatory institutions (including a
Public Complaints Commission, a Permanent Corrupt Practices
Investigation Bureau, an ombudsman, public accounts
committees, legislative investigation committees, and
special tribunals to deal with allegations of
maladministration[44]), and the promulgation of formal codes
of conduct[45] have served as long-term deterrents to
corruption, abuse of authority, lack of accountability,[46]
or laxity in the discharge of responsibilities[47] by public
servants (Dent, 1978:117,123-124; Zahradeen, 1980:5;
Campbell, 1978:82; Williams and Turner, 1978:167;
Adamolekun, 1983:192, 210-211; 1982:27; 1985b:321-322;
Ayeni, 1987:309, 315). Indeed, Adamolekun (1978b:325-326)
concludes that "the balance sheet of all these new measures
suggests that no qualitative change has occurred in ... the
behaviour and performance of the public servants." This
outcome is consistent with Robert Price's analysis
(1975:38-41, 182-183, 208, 216-219) regarding
institutionalization of the status components and not the
role aspects associated with organizational position.

The 1975 purge did unsettle the previously entrenched security of tenure enjoyed by Nigerian civil servants (see New Nigerian, editorial, 4 February 1980, p. 1). In conducting the exercise, the military ignored existing Civil Service Rules and resorted to what the SMG at the time has characterized as "revolutionary legality" (Ayida, 1979: 224). The Murtala administration eventually conceded that its purge had brought about "'panic and uncertainty in the ranks of serving officers'" and terminated the program in November 1975 because of its "'unsettling effect on the services'" (cited in Campbell, 1978:81-82; also see Dent, 1978:120). Subsequent to the 1979 elections, a fresh series of allegations surfaced in the press concerning the dismissal, forced retirement, and unjust reassignment of public officers by some newly elected state executives as part of efforts to consolidate the position of their political party within the civil service (New Nigerian, 4 February 1980, p. 1). The Buhari regime imposed the most sweeping RIFs in 1984. According to experienced civil servants, external analysts, and even one former Head of State (General Obasanjo), such attacks on security of tenure have resulted in civil services plagued by low morale, reluctance to assume responsibility and render advice, an impugned public image, and increased difficulty attracting and retaining qualified manpower (Adedeji, 1981:805; 1988a:63; Ayida, 1979:224-226, 229; Ciroma, 1980:5, 7; Adebayo, 1981:87, 150; Akinyele, 1979:238, 241; New Nigerian, editorials, November 29, 1978 and February 4, 1980, p. 1; Garba, 1979:I; Tahir, 1977:255; Adamolekun, 1979a:194; Campbell, 1978:82).

Students and practitioners of public policy and administration in Africa need to be concerned about deteriorating performance and insensitivity to the needs of the poor. The failings of the public service account in large measure for the state of perpetual crisis which consumes governments across the continent. Over the long term, however, the problem of poor performance in the public sector will not be resolved by purges, forced retirements, and privatization. Deeper structural and attitudinal constraints need to be addressed.

In Nigeria, the military has never enforced a disciplinary scheme systematically related to administrative and management performance nor replaced dismissed civil servants with a new institutional core of development administrators able to act independently of parochial and foreign pressures (see Okoli, 1980:12; Otobo, 1986:123). Two clear lessons emerge from this historical review of

Nigerian public administration. First, more serious attention must be given to linking pay, promotion, and evaluation to job performance, initiative, and competence (see
Otobo, 1986:123-125; Vengroff, 1982:16, 18, 21; 1988b:4;
L. White, 1987:110; Esman and Uphoff, 1984:280).[48] Second,
governments must ensure that organizational conditions
exist which will heighten loyalty and commitment to the
public interest on the part of civil servants entrusted
with the crucial tasks of development administration (see
Ikoiwak, 1979:247-248; Koehn, forthcoming). The next
chapter on bureaucratic involvement in policy making
provides further reason to be concerned about the orientations and behavior of career public officials.

NOTES

    1. However, former departments (renamed divisions)
continued to act in a relatively autonomous fashion within
certain ministries (Cole, 1960:33).
    2. Under regulations issued by President Shehu
Shagari in 1980, the Federal Civil Service Commission delegated full appointment power up to GL 07 to permanent
secretaries and heads of extra-ministerial departments.
Other federal policies authorized officials in the latter
categories to offer temporary appointments at GL 08 and 09
and to employ certain specialist professional staff
(medical officers, architects, engineers, surveyors, land
officers, and accountants) up to GL 11 (New Nigerian, 28
October 1980, p. 1).
    3. The specialist class consists of public employees
possessing "professional, scientific and technical qualifications, including ancillary technical staff" (Adedeji,
1968a:6). They include scientists, engineers, doctors, and
agricultural officers. Generalists normally have been educated in the humanities and social sciences (Nigeria,
Political Bureau, 1987:107).
    4. Ranks within the executive class are arranged in
the following descending hierarchical order: Principal
Executive Officer (PEO), Senior Executive Officer (SEO),
Higher Executive Officer (HEO), Executive Officer (EO),
Assistant Executive Officer (AEO).
    5. The middle (executive) levels of the bureaucratic
structure benefitted least from the Udoji awards discussed
later in this chapter (Anise, 1980:26; Dent, 1978:111).
    6. Cole (1960:323) sets total public employment in
1958 (including local government staff and all daily

laborers) at the much higher figure of 302,200. Ayida (1979:223) places the approved establishment for the Federal Civil Service at 26,000 in 1952/53, and at nearly 200,000 (excluding parastatals) in 1976/77.

7. The Udoji Commission placed the overall strength of the public sector at 658,030, with the state civil services contributing 161,000 of the total (cited in Adamolekun, 1978b:322).

8. On the creation and early expansion of public corporations in Nigeria, see Cole (1960:328); Adedeji (1968b: 151); Hyden (1983:96-97). Hodder-Williams (1984:153-154) describes the enthusiasm for parastatal organization which characterized post-independence African states.

9. Williams (1976a:27) traces Africanization of the colonial service to the Foot Commission Report of 1948.

10. According to Kingsley (1963:310), district officer positions usually were the first to be Nigerianized.

11. Cole (1960:335) explains that in light of the Northern People's Congress' strength within the Federal Parliament at the time and the fact that "posts vacated by expatriates would almost certainly be filled by Southerners," it is "understandable why Nigerianization ... proceeded more rapidly in the Eastern and Western Regions than in either the Northern Region or Federal Government." Also see Adebo (1979:197).

12. On expatriate employment among certain multinational corporations operating in Nigeria, see Ake (1985: 187). Also see Ojo (1985:145, 187).

Expatriates from France still filled 3 per cent of all public sector positions in Senegal in 1975. They were particularly influential in the Office of the President and in the Ministries of Finance, Public Works, Public Health, Rural and Industrial Development, and Interior (R. O'Brien, 1979:118-120).

13. On the early neglect of appropriate in-service training programs, see Fletcher (1978:163-165); Asiodu (1970:136); and Anosike (1977:30, 33, 36-37, 41-43, 46). Programs offered at several Nigerian universities, such as the AMTC (Administrative Management Training Course), ADPA (Advanced Diploma in Public Administration), and MPA, have somewhat rectified this situation in recent years (see Adamolekun, 1974:11-19).

14. Ojo (1980:52-62) shows that the compulsory National Youth Service Corps (NYSC) scheme, which deploys the vast majority of graduates from the universities and other institutions of higher learning to states other than their own, has fostered a somewhat greater willingness on

the part of state governments to employ non-indigene graduates, and of the latter to accept such offers, since its
establishment by the FMG in 1973. However, Marenin
(1987:25) found that since the economic decline of the mid
1980s, any vacant state government positions have been
reserved for indigenes.

15. Markovitz (1977:319) also maintains that after
the civil war, "the federal government reinstated almost
all of the Ibo civil servants who joined the secession
...."

16. According to one source, the Yoruba "dominated
and controlled key positions in the Federal Civil Service,
Foreign Service, and Federal Corporations" by the late
1970s (Ikoiwak, 1979:176-178).

17. In the 1979 elections, only the PRP committed
itself to fair representation for women. In Kano State,
however, the PRP government subsequently appointed one
woman as perm sec (education) and one woman to each of the
10 parastatal boards (Callaway, 1987:380, 387-388).

18. Ironsi's rule proved short-lived, largely due to
opposition among all northern groups to his hasty moves to
institute a unitary system of government and unify the
separate regionally based civil services under a single
Public Service Commission (Ostheimer, 1973:62-63; Feit,
1968:190-191; Chick, 1969:106n; Luckham, 1971:252-278;
Asiodu, 1970:131; Dudley, 1968:x; Williams and Turner,
1978:144; Bennett and Kirk-Greene, 1978:15; Welch and
Smith, 1974:127-128).

19. These are: the Gowon regime (1966-1975), the
Murtala/Obasanjo regime (1975-1979), the Second Republic
(1979-1983), the Buhari administration (1984-1985), the
Babangida period (1985-present), and the Third Republic
(1992).

20. General Ironsi never appointed civilian commissioners and kept surviving former politicans out of
formal government positions entirely. He vested direct
political responsibilities in six senior military officers,
including himself as Head of the Military Government and
Supreme Commander of the Armed Forces (Luckham, 1971:254).
General Gowon first appointed civilians to head ministries
and extra-ministerial departments following the division of
Nigeria into 12 states and the secession of the Eastern
Region in 1967 (Bienen and Fitton, 1978:29, 51). By 1974,
civilian commissioners comprised a majority on the FEC,
although the new council appointed by Gowon in January 1975
reversed the balance in favor of military officers. The
Murtala/Obasanjo administration appointed academics or

technical experts rather than former politicians as its 11
civilian (out of 25 total) commissioners (Campbell,
1978:71, 80; Balogun, 1983:126).

    21. Patricia Stamp (1986:22, 29) shows that a similar
situation prevails in Kenya. Also see Hodder-Williams on
francophone as well as anglophone Africa (1984:174-178).
During the 1970s, the Tanzanian central government abolished urban local government and directly administered
cities. Julius Nyerere later expressed considerable regret
over this decision because of the costs associated with
promoting "'top-heavy bureaucracy'" (Gauhar, 1984:830).

    22. Among its less drastic recommendations, the
Commission emphasized the ongoing need for training public
servants at all levels and proposed replacing the "confi-
dential reporting system" with an "open reporting system"
under which each senior officer's annual written performance evaluation must be countersigned by the subordinate, who may record any disagreements with the report
(Adamolekun, 1978b:324). Akpan (1982:190) holds a negative
view of the open reporting system. In any event, the new
system has not resulted in changes in personnel actions
(Adamolekun, 1982:29).

    23. On the inefficiency and bureaucratic delays
caused by rigid adherence to the ministerial chain of command and by inter-ministerial conflicts, see Asiodu
(1970:138-139).

    24. With the exception of the permanent secretary
level, however, the Government maintained the "superior
gradings and salaries of posts in the federal civil service
vis-a-vis posts in the civil services of the states..."
(Adamolekun, 1982:19).

    25. Even lower-level government employees and higher
civil servants evidenced dissatisfaction over the results
of the regrading exercise. According to the SMG at the
time, "ironically, one of the most dissatisfied groups from
the Udoji rankings were the Federal Permanent Secretaries
themselves" (Ayida, 1979:227; also see Campbell, 1978:73).

    The FMG had assured military personnel that they could
expect salary increases comparable to those obtained by the
civil services (Campbell, 1978:74-75). Anise (1980:27)
suggests that the rates of salary increase received by
military personnel exceeded those secured by civil servants, although the government refused to release figures
on implementation of the Udoji awards within the armed
forces.

    26. In spite of the Udoji reforms, conflicts con-
tinued between administrators and professionals --

particularly at the state government level (see Adamolekun, 1982:31).

27. The Shagari administration made an ineffectual attempt to regulate professional incomes in the private sector between 1980 and 1982 (Otobo, 1986:118-119, 119n). During the Second Republic period, moreover, at least seven states failed to pay public school teachers in timely fashion (Forrest, 1986:16).

28. Compare the table found in Adamolekun (1978b:323) with Table 1.3. The government allowed lower-level wage earners a modest 1 to 7 per cent increase in the 1977-1978 fiscal year (Joseph, 1978:235). The FMG abolished the 50 naira per month car-basic allowance in both the public and private sectors effective April 1, 1979 and revised all salary scales upward by 300 naira per year in its place (see Anise, 1980:36n; Balogun and Oshionebo, 1985:303-304).

29. The staff of polytechnics and advanced technical and teachers colleges did not fare as well at the hands of the Adamolekun Commission (Otobo, 1986:120).

30. Both the 1971 Adebo Commission and the 1974 Udoji Commission had unsuccessfully recommended earlier that public industrial and commercial corporations be granted increased managerial flexibility and autonomy in the deter-mination of staffing and salary schemes (Williams and Turner, 1978:160-161). Professor Adamolekun shares this viewpoint. He argues (1982:22) that "there is no com-pelling reason for marrying a unified salary system to a unified grading system and there is no justification for imposing a uniform system on all public organizations regardless of their peculiarities."

31. Chief Udoji, incidentally, advocated (1979:206) that permanency and pensionability end at GL 14 in the interests of political responsiveness and enhanced civil service efficiency (also see Balogun, 1983:164).

32. It may also account, in part, for the relatively young age of many permanent secretaries throughout Nigeria's post-independence history -- including the super perm secs under General Gowon (see Balogun, 1983:116-117).

33. However, Articles 135(5), 143(1a), 173(4), and 181(1a) of the Constitution clearly prohibited the appoint-ment of civil servants as federal ministers, members of established federal executive bodies, state commissioners, and members of established state executive bodies.

34. This practice is commonly followed in France (Bach, 1980:4). Parallels also exist with provisions in the U.S. Civil Service Reform Act of 1978. According to

Alan K. Campbell (1979:95), "career executives who accept a
presidential appointment requiring Senate confirmation will
continue to be covered by the pay, award, retirement and
leave provisions of the Senior Executive Service, and are
entitled to placement back into the service when they leave
their presidential appointment."

35. As a long-term measure aimed at increasing the
share of civil service posts held by indigenes of the historically disadvantaged northern states, the Shagari administration introduced the practice of "setting quotas for
admission to federal universities and allocating post-
graduate scholarships on a proportional, if not an extra-
proportional, basis" (Rothchild, 1987:124).

36. Adamu Fika, the last federal head of service,
protested this change on the grounds that it invites
compliance with political pressures and that whenever state
governors made commissioners the accounting officer in the
past "'accountability suffered much negligence ...'"
(Akinrinade, 1988:16). The Political Bureau (1987:110,
112) put forth a less compelling argument in favor of
altering the previous arrangement. The members of this
body maintained that "as accounting officers, the administrative heads often place unnecessary bureaucratic
obstacles to quick execution of ... [noble] projects."

37. Instead, "each officer, whether a specialist or
generalist, will now make his career entirely in the
Ministry or Department of his choice, and thereby, acquire
the necessary expertise and experience, through relevant
specialized training and uninterrupted involvement with the
work of the Ministry or Department" (Section III, 2). The
1988 Guidelines encourage horizontal and vertical mobility
within a ministry, however. Moreover, "administrative
officers in general administration shall now specialize in
one area of Management, e.g., Personnel, Planning, Budgeting, Finance, Research, Statistics, etc., which will
enhance their deployability into other relevant departments
of the Ministries" (ibid., p. 11).

38. The influential Professor 'Ladipo Adamolekun had
argued (1985a:333) in favor of advancement strictly on a
merit basis. In this event, however, appointment to key
administrative cadre posts could continue to occur on the
basis of equal representation by state, with the employees
of each state "selected by the merit principle."

39. Robert Bates (1987:245-246) describes how the
fiscal arrangements governing Uganda's Coffee Marketing
Board have encouraged the inflation of operating costs.
Since the central Treasury covers any losses and seizes

any "profits," the Board finds it advantageous "to generate higher salaries, inflated payrolls, lavish offices, excessive travel allowances, and other prerequisites."

Ironically, public enterprises in Marxist Ethiopia, most notably Ethiopian Airlines and the Ethiopian Telecommunications Authority, rank among the few profitable and efficient "success stories" on the continent (Nellis, 1986:24-25; Africa Research Bulletin, Economic Series, 30 September 1988).

40. Dean McHenry (1984:273) maintains that persons in government who seek to advance the cause of privatization have worked to ensure that public corporations do not succeed in Nigeria (also see Ihimodu, 1986:232).

In a more far-reaching critique, Neva Makgetla (1988:19-28) shows that the Zambian government favored those parastatals with powerful foreign partners that diverted investment from development goals in pursuit of short-term profits. State control over the copper mines proved particulary costly. Nationalization of the mines required the assumption of high-interest international credit. In addition, reliance on complex, capital-intensive manufacturing technologies forced the Zambian government's parastatal holding company, Zimco, to enter into management and technical assistance contracts with multinational corporations and to defer to its foreign partners. Management's preference for imported inputs and expatriate employees added to the outflow of capital resources. Concomitantly, the government failed to nationalize powerful banks and engineering firms -- "companies that, though extensive linkages with other sectors, exert substantial influence on the economy as a whole."

41. Moreover, Tanzanian parastatals often are "dominated by foreign partners" and they facilitate "private accumulation by individuals in the governing class" (Samoff, 1981:292).

42. Another type of privatization has occurred more extensively, however. In many large African cities, private entrepreneurs have informally and incrementally begun to provide infrastructure and services. In terms of urban transportation, for instance, "the trend in the 1980s is for a large public bus company to be supplemented by an informal system of minibuses and/or private cars which provide a more accessible service (particularly to peripheral low-income areas) at a slightly higher price" (Stren, 1988b:232, 218).

43. In a careful analysis of informal and incremental privatization in urban Africa, Stren (1988b:220, 243) finds

that "the distribution of benefits to the urban poor as a result of privatization tends to be most benign under conditions where start-up costs for small-scale entrepreneurs are relatively low and there are few possibilities for monopoly control of the services." These conditions tend to prevail in terms of urban transport and housing and to be absent in the case of water supply and refuse disposal. Stren also points out that some individual private service providers are relatively poor and argues that the state, as presently constituted, "is largely inimical ... to the needs of the poor."

44. Including the Code of Conduct Bureau and the Code of Conduct Tribunal called for by the 1979 Constitution (Articles 15-20 of the Fifth Schedule). Also see Articles 82 and 83 on the investigatory powers of the National Assembly.

Neither the Code of Conduct Bureau nor the Code of Conduct Tribunal actually functioned during the Second Republic (Nigeria, Political Bureau, 1987:216; Diamond, 1984:49-50, Nwabueze, 1985:298-301; Oyovbaire, 1987:23). Adamolekun (1983:129) reports that the Permanent Corrupt Practices Investigation Bureau achieved little prior to its abolition in 1979. The legislature never established or activated public accounts committees (West Africa, 19 October 1987, p. 2068). Moreover, effective external auditing cannot be accomplished when government expenditure accounts are from one (national) to four (state) years in arrears (A. Phillips, 1985:265-266).

45. See the text of the "Oath for Public Officers" reprinted in Adamolekun (1978b:325-326) and the "Code of Conduct for Public Officers" (Fifth Schedule to the 1979 Constitution).

46. On this score, the chairman of the Public Accounts Committee of the House of Representatives charged in 1981 that the federal government's financial accounts from 1975 through 1979 had never been audited by the military government, and that the Post Office's accounts had not been audited for over nine years (West Africa, No. 3326, 27 April, 1981, p. 946; also see Cohen, 1980:83-84).

47. Ukaegbu (1985:505-508) finds the "culture of unexplained absence" more widespread among the middle and lower ranks than among the higher ranks of top-level scientific and professional personnel. This situation is attributable in part to the relative lack of delegation which characterizes the Nigerian bureaucracy.

48. The 1988 Civil Service Reforms incorporate important steps in this direction. In particular, they call for

the introduction of a new employee evaluation scheme which emphasizes performance criteria. All employees, up to the level of director/head of department, are to be evaluated principally according to the concrete achievements of themselves or the units reporting to them in relation to prescribed performance standards and specific targets. The ministry's annual report must likewise focus on performance output and results in meeting assigned targets. It remains to be seen whether these changes in evaluation methods will result in significantly improved performance in practice as well as "on paper."

RECOMMENDED READING

A classic and comprehensive study of public administration around the world, recently updated, is Ferrel Heady's Public Administration: A Comparative Perspective, 3rd edition (New York: Marcel Dekker, Inc. 1984). For an explicit Third World concentration, turn to Krishna K. Tummala's Administrative Systems Abroad (Washington, D.C.: University Press of America, Inc., 1982). U.S. textbooks tend to devote little attention to public administration overseas -- particularly in Africa. A good choice for the reader interested in U.S. public administration is David F. Schuman and Dick W. Olufs, Public Administration in the United States (Lexington: D.C. Heath, 1988). Another, with a chapter on comparative administration, is Gerald E. Caiden, Public Administration, 2nd edition (Pacific Palisades: Palisades Publishers, 1982). A persuasive positive portrayal of the civil service can be found in Charles T. Goodsell, The Case for Bureaucracy; A Public Administration Polemic, 2nd edition (Chatham: Chatham House Publishers, Inc., 1985). For a critical perspective, consult Robert B. Denhardt, Theories of Public Organization (Belmont: Brooks/Cole Publishing Company, 1984). A useful, albeit exclusively western-focused, anthology is Frank Fischer and Carmen Sirianni (eds.), Critical Studies in Organization and Bureaucracy (Philadelphia: Temple University Press, 1984). One of the best books for the student of development administration is Coralie Bryant and Louise G. White, Managing Development in the Third World (Boulder: Westview Press, 1982). A more technical and detailed study is White's Creating Opportunities for Change: Approaches to Managing Development Programs (Boulder: Lynne Rienner Publishers, 1987).

Two classic studies of public administration in Nigeria are David J. Murray (ed.), <u>Studies in Nigerian Administration</u>, 2nd edition (London: Hutchinson & Co., Ltd., 1978) and Adebayo Adediji (ed.), <u>Nigerian Administration and the Political Setting</u> (London: Hutchinson Educational, 1968). "Must reading" for understanding military rule is Robin Luckham's <u>The Nigerian Military; A Sociological Analysis of Authority & Revolt, 1960-67</u> (Cambridge: Cambridge University Press, 1971). The best of the more recent, although already dated works on the topics addressed in this chapter are 'Ladipo Adamolekun, <u>Public Administration; A Nigerian and Comparative Perspective</u> (London: Longman, 1983) and M.J. Balogun, <u>Public Administration in Nigeria; A Developmental Approach</u> (London: Macmillan Nigeria, 1983).

2

Bureaucratic Involvement
in Public Policy Making

In Africa, public servants have consistently played a major role in the processes of policy initiation and execution. The exact scope of bureaucratic involvement in policy making ebbs and flows in response to historical, legal, and political changes in the environment of public administration at the federal, state, and local government levels.

This chapter traces the impact of colonialism on administrative behavior and analyzes the policy-formulating and executing roles and relationships which prevailed during Nigeria's First and Second Republics and under different military regimes. The focus here is on the policy-making process and the bases and extent of bureaucratic participation in it. The concluding chapter of the book presents a critical assessment of the consequences of administrative involvement in public policy making.

While it is not unusual for strategically located mid-level public administrators to influence policy outcomes, the focus of attention here is on the role played by high-level administrative officers. In Nigeria, administrators at GL 12 and above constitute the policy-generating cadre and comprise the bureaucratic elite of the country (Otobo, 1986:103). They include perm secs and deputy perm secs, directors and deputy directors of professional departments, secretaries to LGs, and the general managers and assistant general managers of parastatal organizations (see Asiodu, 1979:75).

HISTORICAL AND STRUCTURAL ORIGINS

British administrative practices and conventions bear considerable responsibility for the particularly active involvement of higher civil servants in public policy making which one encounters in anglophone Africa. In the first place, the broadly defined role assumed by administrative officers in policy formulation and in the provision of political advice can be seen as a legacy of the predominantly administrative British colonial system of government (Lofchie, 1967:48; Ciroma, 1979:215; Adedeji, 1988a:61). Thus, Nigeria inherited a tradition of "administocracy,' or rule by career administrators (see Balogun, 1983:76; Wilks, 1985:274).

In the years following independence, a ministerial conception of government structure, under which appointed career officials are found in virtually all posts in the upper reaches of the hierarchy (except as ministers) and, therefore, control most of the key positions in the decision-making process, has been sustained as the prevailing model of administrative organization (Maloney, 1968: 120, 122; Ayida, cited in Bienen and Fitton, 1978:50). This structural arrangement facilitates extensive bureaucratic involvement in policy formulation, advocacy, and execution. Finally, career administrative officials readily adopted a norm of administrative behavior which not only permits, but expects that ranking civil servants will express a strong normative position on issues and will take the initiative in developing public policy alternatives and implementation strategies and in advising their ministers "on the full implications of policy options open to the Government" (Ayida, 1979:219; 1968:64; Adamolekun, 1978a: 12-14, 18; Nwosu, 1977:64; Wilks, 1985:274; Harris, 1978: 288-290; Asiodu, 1970:126; Bach, 1980:1; Adedeji, 1968a:10; 1968b:146; Adebo, 1979:198; Udoji, 1979:205; Ciroma, 1979:214-215; Asabia, 1968:114-115).

FIRST REPUBLIC PATTERNS

Under the Whitehall model of administration, as adapted in Nigeria, the permanent secretary served as the chief advisor to the minister and as the chief administrative officer. As paramount advisor, the permanent secretary should engage in "'the elaboration of ... policies and plans and assist in the determination of the best means of carrying them out'" (cited in Adamolekun, 1978a:17-18;

also see Kingsley, 1963:312-313). This includes "firmly
indicating" preferred options (Asiodu, 1979:75-76). As
administrative head, the permanent secretary is responsible
for interpreting policies, coordinating ministry activities, supervising functional execution and monitoring
results, acting as chief accounting officer, defending
ministry budget proposals, and upholding the Ministry's
interests in inter-ministerial meetings and in relations
with other agencies and external groups.

In practice, higher civil servants (particularly
permanent secretaries) played more than an advisory role in
the public policy-formulation process. Administrative
officers rarely received political guidance or leadership
from their ministers (Adebayo, 1981:75). Adebayo (1979:
5-6) maintains that "the 'average' politician, especially
in the [First Republic] civilian regime, conceived his role
as approving or disapproving whatever proposals or recommendations were placed before him by his Permanent
Secretary." According to P.C. Asiodu (1979:76), when the
perm sec "felt very strongly that a wrong decision was
about to be taken ..., [he] had the right, and occasionally
exercised it, to see the secretary to the prime minister
and, if necessary, the prime minister himself to brief
him." Moreover, policy papers initiated and prepared by
permanent secretaries frequently formed the basis for
Executive Council deliberation and sanction, and career
administrators drafted and tightly controlled ministry and
agency budgets (Aliyu, 1979:8; Ayida, 1979:220, 228;
Harris, 1978:292-293; Adebayo, 1979:8-9; Phillips, 1981:2;
1989:427).

In sum, higher civil servants have acted as central,
and often dominant participants in the policy-formation
process since the early stages of Nigeria's political
history (Adebayo, 1979:4, 9). Among the factors accounting
for this situation are political instability and delegation
of broad discretionary authority over policy and program
implementation. Lack of experience and educational
qualifications on the part of ministers,[1] the absence of
legislative scrutiny, and the better informed, more expert
and realistic perspectives reputedly brought to bear on
development issues by members of the administrative and
professional classes also figure prominently in the
explanations and justifications for bureaucratic
encroachment in the policy-making domain advanced both by
practitioners and by external analysts (see Lofchie,
1967:47-48; Adebayo, 1979:4-9, 14; Maloney, 1968:122-124;
Ayida, 1979:220; Udoji, 1979:204, 206; Aluko, 1968:77;

Ciroma, 1979:215; Akinyele, 1979:233-234; Adedeji,
1968b:146-147; Nwosu, 1977:64-65; Balogun, 1983:80; Bienen
and Fitton, 1978:46, 49; Phillips, 1981:2). In addition,
concentration of functions at the federal government level
promoted enlargement of the central bureaucracy. Increased
bureaucratic size, in turn, "raised the degree of
dependence on that [bureaucratic] machine by the political
executive, and expanded the scope of discretion exercisable
by men who were not politically accountable" (Achimu,
1977:170).

In principle, the execution of public policies has
been the exclusive preserve of Nigerian public servants.
Some political figures have shown scant regard for this
convention, however. Indeed, First Republic ministers
regularly interfered in the details of administrative
activity for political and/or personal reasons (Murray,
1968a:18-21; Adedeji, 1968b:141-142, 147; Adamolekun,
1978a:36; Udoji, 1979:204). According to the critics,
ministerial intervention in the policy-implementation
realm subverted professional criteria in decision making,
impaired administrative efficiency, promoted frustration
and resentment within the public bureaucracy, and further
accentuated role conflicts between political and
administrative-class officers (Okunoren, 1968:105-106,
Harris, 1978:294; Akinyele, 1979:234; Adamolekun, 1978a:36;
Adedeji, 1968b:141; Aliyu, 1979:8).

At the local government level, portfolio councillors
engaged in direct administrative supervision under arrangements in effect in most northern states prior to 1976. The
uniform local government reform edicts issued by all 19
states in 1976 restricted the supervisory councillor to the
exercise of "general political but not executive direction
over such department or group of departments ... as may be
assigned to him by the council." Nevertheless, in roughly
30 per cent of the 138 LGs in 10 northern states surveyed
by the author in 1978,[2] supervisory councillors continued
to "assign specific tasks and/or issue instructions
directly to department staff" and to "decide on the
deployment, discipline, and/or promotion of department
staff."

THE MILITARY REGIMES

From 1966 through 1979, Nigeria's military rulers
vested formal policy-making authority in the Supreme

Military Council (SMC). Under General Gowon's regime, state military governors, most of whom were relatively junior and inexperienced officers, formed a majority of the members on the Supreme Military Council. After General Murtala Mohammed assumed power in 1976, he relegated the governors to representation on the advisory National Council of States (Dent, 1978:112, 115; Williams and Turner, 1978:167; Campbell, 1978:69). In August 1987, the Babangida regime replaced the SMC with a 28-member Armed Forces Ruling Council (AFRC). Unlike the prior SMC, there are no civilians on the AFRC.[3]

Military rulers created the single post of secretary to the government and head of the civil service (commonly abbreviated as SMG) at the apex of the federal and state bureaucratic hierarchies. Appointees quickly built the office of secretary to the military government into the most powerful and prestigious administrative position. Self-designated as a "super perm sec," the SMG possessed unrivaled access to the Head of State and the military governors. He served as secretary to the Federal Executive Council. The secretary to the federal military government even sat in on most Supreme Military Council meetings (Okoli, 1978/79:10, 12).[4] In addition to serving as chief advisor to the military leadership, federal and state SMGs exercised important supervisory powers over the top ranks of the civil service. According to the former military governor of Oyo State, Brigadier General David M. Jemibewon, the SMG functioned as "chairman of the meeting of permanent secretaries, the administrative staff advisory committee, the senior management staff committee, and the committee on service matters" (cited in Okoli, 1978/79:11).

In a further effort to centralize control over state government operations, military governors appointed a number of permanent secretaries to newly created posts within their own office. According to Adebayo (1979:10), "some State Governments during 1966-1975 had as many as seven permanent secretaries concentrated in the office of the Military Governor, ... responsible to the Governor for almost the entire range of governmental activities -- economic, administrative, political, commercial and industrial." By 1977, nine permanent secretaries served as heads of Cabinet Office departments and units at the federal government level (Adamolekun, 1978b:314). This practice, which many elected chief executives continued during the Second Republic, had the effect of reconstituting a powerful secretariat office and reducing

the authority and range of functional responsibilities
possessed by ministries and departments (Adebayo, 1979:10;
Adamolekun, 1978a:12).

Ironsi

Military rule draws higher public servants deep into
policy-formulating roles. According to Robin Luckham
(1971:203-204), "the Ironsi regime relied heavily on its
permanent secretaries, and virtually all important
decisions were taken by a narrow group of half a dozen
military leaders, together with a handful of civil servant
advisors."[5] In the absence of commissioners, General
Ironsi vested ministerial powers in the Federal Executive
Council (Ayida, 1979:228).[6] Nevertheless, the Head of the
Military Government and the FEC delegated broad authority
to promulgate subsidiary statutes as well as extensive rule
making powers to the permanent secretaries, who acted as
the de facto political heads of their ministries (Luckham,
1971:225n; Adamolekun, 1979c:211; Balogun, 1983:89).
Federal civil servants also participated in Executive
Council meetings on an ad hoc basis under Ironsi and
top-level administrative officers at the regional level
"sat in the Executive Councils as of right ..." (Luckham,
1971:254-255).

Gowon

By all accounts, the direct involvement of public
servants in the policy-making arena reached its zenith
under General Yakubu Gowon (Asiodu, 1970:126; Bienen and
Fitton, 1978:42-46; Aliyu, 1979:7). Most military officers
lacked education on a level comparable to higher civil
servants and had no experience with governmental processes
and public decision making. These factors, coupled with a
reluctance to appoint former politicians to authoritative
or advisory posts in the new regime and the absence of any
organized public constituency, left the Head of the Federal
Military Government and the military governors dependent
upon the political advice and policy alternatives
articulated by civil servants (Gobir, 1970:161; Maloney,
1968:120-122; Williams and Turner, 1978:153; Bienen and
Fitton, 1978:46-47; Nwosu, 1977:64; Adebayo, 1979:10;
Lewis, 1985:131-132). Thus, General Gowon is reported to

have relied primarily upon suggestions made by federal permanent secretaries and the SMG rather than on recommendations tendered by his ministers or by the military officers serving on the Supreme Military Council. As an indication of the extent of the power possessed by high-level public administrators, Adebayo (1979:14) reports that state military governors would travel to the Lagos residences of federal permanent secretaries in order to lobby for their support on matters "that might in due course come before the Supreme Military Council, the Federal Executive Council or even directly before the Head of State" (also see Bienen and Fitton, 1978:43, 47; Achimu, 1977:169-170).

Through personal access, "super perm secs"[7] and other ranking civil servants secured the backing of the Head of State (and the military governors) for their proposals in advance of SMC and FEC meetings. Joint authorship with the chief executive practically guaranteed that the policy memoranda advanced by public servants would not be substantially changed by the Council. Whenever necessary, permanent secretaries, who held standing invitations from the Head of State to participate fully in Supreme Military Council and Federal Executive Council deliberations,[8] utilized that privilege to counter any objections raised by council members to memoranda they had drafted. On occasion, they urged the adoption without alteration of proposals that their own ministers already had spoken against (Adebayo, 1979:11-14; Campbell, 1978:71; Bienen and Fitton, 1978:45-46; Dent, 1978:118).

Under the Gowon administration, then, policy advising, initiating, formulating, advocating, and defending became the paramount preoccupation of the permanent secretary. Moreover, "super perm secs" and other top-level civil servants overtly usurped the powers of their ministers/commissioners and visibly dominated both the shaping and determining of government policies and the allocation of public resources (Bienen and Fitton, 1978:43; Garba, 1979:I; Maloney, 1968:124; Adebayo, 1979:11; Aliyu, 1979:7; Collins, 1980a:323; Oyovbaire, 1980:273; Collins, Turner, and Williams, 1976:185-186, 192; Ikoiwak, 1981:100; Yahaya, 1980:129). High-ranking public sector personnel also moved into positions as chairpersons and members of the board of directors of government corporations (Akinsanya, 1976:63-65). Major-General Joseph N. Garba (1979:1) confirmed on the eve of the return to civilian rule that "during the nine years of the Gowon regime, senior civil

servants literally held sway over decision-making, and some
of them could in fact over-rule their commissioners and get
away with it."

Under Yakubu Gowon, in particular, many career civil
servants took advantage of the opportunity provided by the
replacement of politicians with military men, who shared .
common administrative values and depended upon their
support and expertise, to expand their already powerful
role in fashioning public policies (Bienen and Fitton,
1978:29, 48; Feit, 1968:186, 190; Nwosu, 1977:65).
According to A. Y. Aliyu (1979:7), "decision-making in all
spheres of governmental activity, including the formulation
and implementation of policies, [and] allocation of
resources tended to reflect the preferences and values of
bureaucracy." During this period of fusion of the
leadership from both civilian and military bureaucracies,
"executive control over the conduct of administration
proved almost non-existent" (Adamolekun, 1983:180).

However, Gowon's policy of more directly involving
civil servants in running the affairs of state[9] eventually
led to dissention within the officer ranks of the armed
forces based on perceptions that the military bore the onus
of responsibility for policies primarily shaped by civil
servants.[10] Role expansion concomitantly drew top
administrative officers into an increasing number of
conflicts with army officers and more tightly linked their
position and the reputation of the public services to the
fate of the ruling military faction with whom they had
forged an alliance (Campbell, 1978:69, 71-72, 75; Collins,
Turner, and Williams, 1976:185-186; Bienen and Fitton,
1978:52; Akinyele, 1979:237; Adamolekun, 1982:30).

Murtala/Obasanjo

The Murtala/Ojasanjo regime attempted to check
bureaucratic "excesses" and to narrow the scope of civil
servants' authority in the policy-making realm. Following
the July 1975 coup, General Murtala Mohammed conducted an
unprecedented, sweeping purge of all ranks in the public
services.[11] In the Nigerian case, widespread recognition
that higher civil servants had become the dominant actors
in the policy-making arena only heightened their
vulnerability to dismissal or involuntary retirement on
grounds of corruption, inefficiency, insensitivity, or
disloyalty (Dent, 1978:155-156, 119-124; Balogun, 1983:91;
Bennett and Kirk-Greene, 1978:24; Bienen and Fitton,

1978:52-53; Campbell, 1978:81; Collins, Turner, and
Williams, 1976:188; Adamolekun, 1978b:325; Oyovbaire,
1980:272-274).

General Murtala also immediately banned permanent
secretaries from Supreme Military Council and Federal
Executive Council meetings unless specifically invited to
attend and participate (Turner, 1978:190; Williams and
Turner, 1978:163, 167; Aliyu, 1979:7; Balogun, 1983:128;
Campbell, 1978:80). The same change occurred at the state
level (Adebayo, 1981:87-88). High-level public servants
continued, nevertheless, to be preoccupied with policy
formulation under the new regime (Ciroma, 1979:214;
Phillips, 1981:5-6). Their involvement in the policy-
making process simply assumed less conspicuous forms.
During this period, appointed ministers and commissioners
continued to lack a popular base of political support. In
the absence of political parties, career officials easily
resisted the demands placed on the bureaucracy by weak and
fragmented interest groups. Regular, nation-wide meetings
of SMGs and permanent secretaries also served to enhance
their policy-initiating role under the Murtala/Obasanjo
regime (see Onyeledo, 1980:11; Nwosu, 1977:75-76).[12]

Although the Murtala/Obasanjo administration exercised
firmer political direction over the budgetary process at
the federal and state levels than its predecessors had
exerted, civil servants continued to be vitally involved.
High-level public managers controlled the initial stages of
budget formation, dominated inter-ministerial deliberations
on expenditure and revenue proposals, and defended their
policy recommendations at the executive council or cabinet
level (A. Phillips, 1985:256-259).

The policy-formulation process did differ from that
which prevailed under General Gowon in that the new rulers
often based major decisions on the reports and recommendations of appointed commissions or panels. This practice
has continued under the Babangida regime (Nigeria,
Political Bureau, 1987:115). The usual procedure involves
higher civil servants in the Cabinet Office, together with
the Secretary to the Government, in recommending commission nominees to the Head of State (Adamolekun, 1982:4).
Professionals, technical experts, university professors,
and highly regarded public administrators have dominated
these policy-study bodies (see, for instance, Adamolekun,
1982:3). Women, working class representatives, and persons
displaying radical tendencies have been largely excluded
(Nigeria, Political Bureau, 1987:115). Panel recommendations affected a wide range of a fundamental policies --

including state creation, revenue allocation, establishment
of the new federal capital in Abuja, and public sector
salaries and organization -- during the period under
consideration (Yahaya, 1979:266). And, civil servants
exercise wide discretion when implementing accepted
recommendations (Takaya, 1980:68; Adamolekun, 1982:33).

During the Murtala/Obasanjo administration, then, the
increasing importance placed on the kinds of professional
expertise and information controlled by public administrators strengthened their strategic position in the
policy-formation process (Okoli, 1978/79:10-11; Achimu,
1977:169; Nwosu, 1977:65; Bienen and Fitton, 1978:49). The
"institutional revolution" proclaimed by Allison Ayida (SMG
from 1975 to 1977) emphasized technocratic control over
decisions affecting the allocation of public resources
(Williams, 1980:92-95).[13] The frequent rotation and
relatively high turnover among permanent secretaries and
other high ranking administrative officers that persisted
during this period did limit the ability of bureaucrats to
claim possession of specialist knowledge and experience in
situations requiring technical decisions (Adamolekun,
1978a:21; Cohen, 1979:294-295,300-301; Udoji, 1979:207-208;
Nigeria, Political Bureau, 1987:109; Akpan, 1982:190).
Nevertheless, General Olusegun Obasanjo's 1977 affirmation
that the "role of the civil servant is to initiate policies
and to offer professional and technical advice to the
government..." indicates that no fundamental changes had
occurred in prevailing norms and expectations and suggests
that his administration merely sought a return to pre-1966
administrative practices (see Adamolekun, 1978b:314).

LOCAL GOVERNMENTS

Further evidence that administrators continued to
determine public policies throughout the military rule
period is provided by an investigation of council-staff
relations at the local government level (see Aliyu and
Koehn, 1982:38-52). In principle, the 1976 local
government reform greatly circumscribed the role of the
chief administrative officer (secretary) by assigning
nearly exclusive authority over local policy formulation to
the council. Voters elected three-fourths of the members
of each newly created LG council either directly or
indirectly on a non partisan basis in 1976. Most
councillors remained satisfied to ratify proposals
initiated and submitted by the secretary and/or department

heads. The informed responses of local government
officials from 134 out of the 152 LGs in the 10 northern
states to the statement "in [my] local government, the
local government secretary and/or department heads
formulate most important policies and the council rubber
stamps them" suggest that the 1976 reform measures met with
mixed results in terms of actually changing the traditional
nature of relations among the two sets of officials. A
majority of the respondents disagreed with the statement in
only 65 of the reporting LGs.

Moreover, many local government secretaries (SLGs) in
the 1976-79 period viewed their roles in the familiar terms
of the Resident or Divisional Officer (see Elaigwu, 1979:
24) and, therefore, endeavored to dominate the policy-
making process.[14] In terms of ability to influence local
policy making, the SLG towered over all other local
government staff by virtue of his formal and informal
authority. In addition to serving as chief executive
officer of the LG, state edicts provided (Kaduna State,
Local Government Edict 1976, Article 87, Sections 3-5)
that:

> The Secretary shall advise the council and its
> committees on all matters upon which he considers his
> advice is necessary, including the Standing Orders
> of the council and Local Government legislation, and
> shall be entitled to attend all meetings of the coun-
> cil and its committees.

> The Secretary shall advise ... the chairman of
> the council on all matters appertaining to ... [the
> latter's office].

> The Secretary shall perform all such other
> functions as may from time to time be assigned to him
> by the Government.

Such provisions granted the SLG ample opportunity to
initiate policy recommendations and to play a decisive role
in local decision making.

When coupled with the SLG's typically superior
educational qualifications, long record of government
experience, and central location in the local government
information and communication network, the available formal
powers allowed the chief executive officer to dominate
local policy making in most instances. Some secretaries
performed this role overtly, through open involvement in

general council and/or Finance and General Purposes
Committee (FGPC) debate and decision making, while others
preferred to operate "behind the scenes." In the Bauchi
Local Government, for instance, an informal caucus met in
the morning of nearly every working day during 1978 and
1979. Only the Council Chairman, the 3 supervisory
councillors, and the SLG participated in caucus meetings.
The secretary would brief other caucus members on events
and problems arising during the previous and present day
and any participants who had been on tour would report on
their findings. On matters requiring action, the caucus
would pursue one of three alternative courses: (1) make a
decision on its own; (2) take the matter to the FGPC, which
would either act on it or, if the matter required the full
council's approval, refer its recommendation to that body;
or (3) in exceptional cases, take the matter directly to
the general council. According to the Chairman of the
Bauchi Local Government Council (interview, 28th December,
1978), "important issues" always would be taken to the
Finance and General Purposes Committee after being
dis-cussed in caucus. Later, they would be raised before
the entire council, which invariably adopted the
committee's recommendation concerning how the matter should
be treated. The use of the caucus in the Bauchi Local
Government Area is described in detail because it
illustrates an institutionalized processual development
that allowed the local government secretary to play a
decisive "behind the scenes" role in local policy
formation. Even those caucus decisions not implemented
immediately and directly stood an excellent chance of
becoming FGPC recommendations and council policy. And, no
regulation barred the secretary from full participation in
and control over caucus meetings and decisions.

However, changes in the local government system
adopted by the Babangida administration in 1987 on the
recommendation of the Political Bureau (1987:93, 138)
portend a major shift in the SLG's role. Directly electly
LG chairpersons now possess executive powers and
responsibilities. When this change is coupled with the
direct allocation of federal revenues to LGs which
commenced in 1988, as well as with the election of
councillors along political party lines which is slated to
occur at the end of 1989, prospects brighten for the
development of popular, independent, and effective
grassroots units of government.

THE 1979 CONSTITUTION AND SECOND REPUBLIC DEVELOPMENTS

The return to civilian rule in 1979 marked renewed efforts to establish political control over the bureaucracy. The Second Republic Constitution substantially undermined the formal standing of higher civil servants in both policy formulation and policy execution. Constitutional provisions (Articles 139, 177) authorizing the appointment of special advisors to the president and the governors, as well as a secretary to the government expected to serve as the overall political advisor to the chief executive (see Akinyele, 1979:239-240; Bach, 1980:2; Okoli, 1978/79:19, 10), added powerful non-career competitors with whom permanent secretaries and the head of service had to vie when attempting to influence decision making. Special advisors held temporary political rather than career civil service appointments and served "at the pleasure of" the president/governor. The authority and influence of special advisors rivaled or exceeded that exercised by many federal ministers/state commissioners.

Furthermore, most permanent secretaries no longer acted as the executive head of their ministries. Federal ministers and state commissioners assumed this responsibility as the representative of the chief executive under Articles 136 and 174 of the 1979 Constitution (Akinyele, 1979:241; Bach, 1980:2; Ofoegbu, cited in Phillips, 1981:14; Akpan, 1982:186). Dafe Otobo (1986:117n) reports that this situation allowed ministers to exceed their budgets while ignoring most approved projects. In addition, they "pushed out the top bureaucrats from tender boards and participated themselves in the award of contracts ...." These developments occurred in spite of President Shehu Shagari's clear determination that the permanent secretary "'has direct responsibility for ensuring that all expenditures of funds in the Ministry's votes are not only proper but in accordance with the purposes for which the funds were voted'" (cited in Akpan, 1982:186).[15]

The 1979 Constitution also altered the status of the top echelon of the public bureaucracy by allowing elected chief executives to appoint individuals from outside the career civil service as permanent secretaries, heads of extra-ministerial departments, and secretary to the government (Articles 157, 188). With the return to civilian rule, the newly elected chief executives made important personnel changes in these ranks. In certain states, governors appointed a number of individuals from outside the civil service (most notably from the universities) as permanent secretary or secretary to the government, although the majority of appointees continued to be civil servants (Bach, 1980:2-3; Ciroma, 1980:5; Ekwueme, 1980:8-9; Murray, forthcoming; also see New Nigerian, editorial, 4 February 1980, p. 1). These appointments did not require legislative consent, a provision which implicitly granted the chief executive power to transfer and remove from office those individuals serving as permanent secretaries, heads of departments, secretary to the government, and head of the civil service (Ofoegbu, 1980:5; also see Shehu Shagari's arguments supporting this interpretation before the Constituent Assembly as reported in Phillips, 1981:9-10). According to Aliyu (1979:4), the power given to the president/ governor to appoint and remove all top supervisory officials in the public bureaucracy greatly circumscribed the authority of civil service commissioners.[16]

During the Second Republic, the primary criteria for appointment as permanent secretary became explicitly political. According to Ray Ofoegbu (1980:5), the critical bases for appointment were "competence, loyalty and total commitment." These changes constituted a decided shift away from British conventions of bureaucratic neutrality and anonymity (Heady, 1966:48; Adamolekun, 1978b:327) toward the deeper politicization of top administrative ranks that characterizes the U.S. and French presidential systems of government (Maloney, 1968:119; Ostheimer, 1973:122).[17] As a condition for holding appointment in the Second Republic, permanent secretaries, heads of extra-ministerial departments, and secretaries to the government were "expected to pursue with absolute commitment the manifestoes, programmes, and policies of the Chief Executive" (Akinyele, 1979:239).[18] President Shagari's appointment of non public servants, including unsuccessful NPN candidates for elected office, as presidential liason officers (PLOs) proved to be particularly objectionable in the 11 states not controlled by NPN governors (see Elaigwu, 1980:9-10, 24; Adamolekun, 1983:104; Oyovbaire, 1985:248-249).

By 1980, numerous conflicts and considerable confusion already had arisen over the policy-formulating role of permanent secretaries at the federal and state levels (see Phillips, 1981:14). On October 31, 1980, President Shagari

felt compelled to invite all federal permanent secretaries
to the State House for a second briefing. On this occasion, the President again emphasized that the presidential
system of government placed new limits on the authority of
civil servants (New Nigerian, 1 November 1980, p. 1).

Permanent secretaries played even more prominent
policy-making and implementing roles in Kaduna State. In
that state, the NPN-dominated legislature rejected all of
PRP Governor Balarabe Musa's cabinet nominees on four
occasions before voting to remove him from office under the
impeachment provisions of the Constitution in late June
1981.[19] The House of Assembly approved most nominees of
the new chief executive (former Deputy Governor Abba Rimi)
on 16 November 1981 and Kaduna State's first commissioners
under the Second Republic were officially sworn in one week
later (New Nigerian, 1 January, 1982, p. VII). During the
25-month political deadlock when the NPN majority in the
Kaduna House of Assembly steadfastly refused to approve any
of the PRP chief executive's nominees for commissioner,
permanent secretaries performed all of the commissioners'
functions. Higher civil servants exercised varying degrees
of influence over public policy formation in the other
eighteen states (Onyeledo, 1980: 5, 9, 15; Lar, 1980:7;
Ofoegbu, 1980; 1985:125; Lewis, 1985:134).

In general, however, the 1979 election results offered
state public servants an excellent opportunity to negotiate
legislative branch acquiescence with the practice of active
bureaucratic involvement in policy initiation. Voters
elected many of their former colleagues, who could be
counted on to be sympathetic and cooperative, to state
legislatures. One revealing study of the occupational
backgrounds of persons elected to the house of assembly in
Bauchi, Benue, Borno, Cross River, Imo, Kano, Kaduna, and
Kwara states found that former civil servants (including
teachers) constituted a majority of the legislators in one
state (Kwara), a plurality in another (Benue), and the
second largest group (after businessmen) in all of the
remaining states with the exception of Kano (Aliyu,
1980b:14).

Elected chief executives expressed greatest dissatisfaction with the bureaucratic apparata they inherited in
terms of the policy-execution function. Many permanent
secretaries displayed their unwillingness to accept a
subordinate position when it came to administrative
operations by overtly and covertly resisting ministerial
supervision of the policy-implementation process. In his

1980 meeting with the permanent secretaries, President
Shagari directed his strongest criticism at federal civil
servants who "obstruct the smooth and speedy implementation
of the programmes of this administration ..." (New
Nigerian, 1 November 1980, p. 9; also see New Nigerian,
editorial, 4 November 1980, p. 1).[20] Public enterprises
and other parastatals generally exercised the greatest
independence from political direction and control in the
execution of policy.[21]

RECENT DEVELOPMENTS

The regime of Muhammudu Buhari, which replaced the
Second Republic, emphasized the managerial capacity of the
state. The subsequent Babangida administration "is dominated by the military-bureaucratic alliance ..." (Marenin,
1988:226). In Pita Agbese's (1988b:10) assessment, "its
willingness to resort to repression and the economic
advantages that have accrued to the senior military officers as state managers since the first military coup in
1966 have combined to make the military the most powerful
faction of Nigeria's ruling class." These characteristics
provide the background for the following discussion of
recent administrative developments.
Nigeria's military rulers of the 1980s introduced
several structural changes affecting the policy-making
process. In the first place, the 1988 Civil Service
Reforms announced by Ibrahim Babangida limited permanent
secretaries (now directors-general) to service that is
coterminous with the administration that appoints them.
The explicit intention behind this move is to make the
office more political and, thereby, ensure that occupants
will be committed and responsive to the new government's
goals and policies (West Africa, 7 March 1988, pp. 397-
398). Second, the minister and not the director-general
explicitly assumes responsibility as chief accounting
officer. On the other hand, the minister is required to
delegate a "substantial part of his administrative and
financial functions and authority to the director general
who should be fully involved in the key decision-making
processes" and the latter "shall deputize for the Minister
whenever he is away from office" (Nigeria, Federal
Republic, 1988:8-9). Moreover, the increasing
specialization and professionalization encouraged by reform
provisions (pp. 7, 11) calling for civil servants to spend
their entire career within a single ministry or

extra-ministerial department and for administrative officers to specialize in one management area can only enhance the ability of career public administrators to influence policy decisions from within (see Peters, 1989:186).

The Babangida regime also brought intellectuals into the policy-making process in three capacities. Visibly and vocally, university staff played a leading role in newly introduced "national debates" over economic and foreign policy and the transition to civilian rule. Second, General Babangida appointed many academics to key government posts (Marenin, 1988:223, 226). The most obvious example of this tendency is the appointment in 1986 of 11 university staff members to the 15 final seats on the influential Political Bureau. In 1987, the Political Bureau submitted an extensive scholarly report and a series of sweeping policy recommendations. The Government accepted many of the Political Bureau's ideas -- with the exception of proposals stemming from its members' radically different conception of Nigeria's relationship to the global economy. Certainly, the intellectuals on the Bureau played an important role in shaping the broad outlines of the Third Republic political system (see Koehn, 1988a:51-56).

Finally, the Babangida government maintained the practice of cultivating close ties with national "think tanks" and university bodies that focus on economic policy, international affairs, and administrative reform. The most prominent of these policy-study institutions are the Nigerian Institute of Statistical and Economic Research (NISER), the Nigerian Institute of International Affairs (NIIA), the National Institute for Policy Strategic Studies (NIPSS), and the Institute of Administration at Ahmadu Bello University. The intellectuals associated with these institutes are able to influence public policy through participation on the Presidential Advisory Committee (PAC), the submission of applied research studies, the formulation of program guidelines, the provision of advice on policy formulation, implementation, and evaluation to the high-level officials of the various governments with which they are linked, the sponsorship of conferences on selected policy topics, and by occasional direct media appeals (see, for instance, Uwazurike, 1987:150-183; Newswatch, 30 November 1987, p. 17; Ostheimer and Buckley, 1982:300).

None of the changes in the policy-making process introduced during the Buhari and Babangida regimes can be viewed as fundamental. In Nigeria, and elsewhere in

Africa, the public bureaucracy continues to occupy a
central role. "Econocrats" (Akeredolu-Ake, 1985:36) of the
domestic and foreign variety have strengthened their grip
on the policy-making establishment.[22] In terms of policy
outcomes, "neglect of the lower classes ... remains a
dominant theme" and there has been no alteration in the
basic dependent "structural relation of Nigeria to the
global economy" (Marenin, 1988:227).

CONCLUSIONS AND LESSONS

It is now widely recognized by scholars and practitioners alike that career public administrators are deeply
involved in setting public policy. African bureaucrats
play major roles in policy initiation and throughout the
policy-formulation stage (see Gould, 1980:xiii; Schumacher,
1975:xx, 73-74). For this reason, David Abernethy (1983:
14-15) and others (Saasa, 1985:311-312, 317) argue that
bureaucratic behavior must be analyzed "in interest group
terms, as expressions of probably the most influential
input structure in the political system." Career officials
also influence public policy outcomes as a result of the
broad discretionary authority they exercise when
interpreting and carrying out political decisions (see, for
instance, L. White, 1987:216; Peters, 1989:11). These
unchecked roles provide a strong indication of the
relentless growth of administocracy.

In addition to their influential place in policy
making, bureaucrats at all levels of office throughout the
continent are active and adept at engaging in "non-decision
making." They do this by "checking the entry of demands"
from competing actors in the political environment (Saasa,
1985:311). Public administrators in Africa have been
particularly effective in blocking or diverting pressures
and incipient demands on the part of the rural and urban
poor. Few institutional mechanisms exist for meaningful
citizen participation in shaping the public-policy agenda.
For most peasants, the ability to influence local and
low-level bureaucrats "remains minimal, let alone their
capabilities of influencing those making major policy
decisions" (Migdal, 1974:224; also see Samoff, 1981:292).

As a direct consequence of the prevailing narrow base
of participation, "there is a relatively high potential
for miscalculations in policy-making" (Saasa, 1985:317).
This situation is aggravated by the overcentralization of

governmental functions (Rothchild and Olorunsola, 1983:9) and by reliance on expatriate advisors and external analysts who lack understanding of and sensitivity to local conditions and objectives, and push foreign agendas (Hyden, 1986:22-23; Helleiner, 1986:8).[23] Two African countries which have experienced particularly extensive foreign involvement in post-independence policy making are Senegal and Kenya. French technical advisors continued to staff key policy-shaping and resource-allocating posts in the bureaucracy of Senegal into the 1970s (Schumacher, 1975: 73-74). In Kenya, and increasingly in Senegal as well of late, representatives of foreign donor agencies and international creditors have visibly and extensively shaped dimensions of public policy that are related to "development" broadly defined. Indeed, the International Monetary Fund is in a position to control economic policy in Kenya through the influence it exerts over the policy-making process within the Central Bank and the Ministry of Finance (Lehman, 1988:12).

The ascendancy of external involvement in policy making on the part of multinational financial institutions across the continent is due to growing indebtedness and vulnerability to outside influence on the part of African governments (see Gould, 1980:59:122; Lancaster, 1987:223). Africa's recent history of extensive foreign intervention has not produced highly acclaimed performance records. Indeed, donor and creditor agencies cannot escape blame for the poor state of economic policy on the continent and the top-down, narrowly based, and non-participatory nature of the process (Hyden, 1986:22).

In realistic terms, the crucial issue is not bureaucratic participation in the policy-making process, but the extent of the influence exerted by public administrators. Recognition that careerists and/or foreign advisors are excessively powerful lends support to efforts directed at reducing their involvement in public policy formation. However, the lack of effective political control over ministry civil servants and parastatal managers which one encounters throughout Africa (see, e.g., Nellis, 1980:417) renders such endeavors problematic.

The experience to date in Nigeria is not encouraging in this regard. The legacy of the colonial administraive state weighs heavily at all levels of government. Nigerian chief executives frequently are unable to establish political control over a large and powerful bureaucracy that is accustomed to internal direction and views itself as the

principal guardian and advocate of professional and public
interests (see Achimu, 1977:169-170; Ayida, 1979:226;
Oyinloye, 1970:148-150; Tahir, 1977:256; Nwosu, 1977:64-65;
Ofoegbu, 1980; Anise, 1980:35; Oyovbaire, 1980:268). This
perspective is pointedly articulated in Adedeji's contention (1968b:146) that politicians "must learn to appreciate
that civil servants very often do know better than either
ministers or the electorate and that it is their public
duty to bring the test of practicality to politicians'
enthusiasm"; in Oyinloye's assertion (1970:150) that "if
the civil servant feels strongly that his principles and
the public interest will be compromised he should mobilize
other forces to see that the policy is revised"; and most
brazenly in Lawson's convictions (cited in Nwosu, 1977:65)
that the public servant should "utilize all legitimate
means [including "judicious use of legitimate blackmail"]
to persuade the minister (or Commissioner) to accept the
policy proposals he submits ..." and that "it would be
sheer irresponsibility on the part of the civil servant ...
to proceed to implement a decision he believes to be
wrong."[24] Few public servants adhere to the point of view
expressed by Ntieyong Akpan (1982:186) in the following
passage:

> Whatever might be their inexperience (and many of
> them, irrespective of high educational qualifications
> or even experience in politics, are inexperienced
> in public administration), they are the personal
> representatives of the heads of government and members
> of the Cabinet and so better placed to appreciate the
> directions of policy emphasis, modifications or
> changes which the Civil Service executives need to
> know in order to fulfil [sic] their duties. For that
> reason the decision of the Minister/Commissioner,
> where there is divergence between him and the
> permanent Civil Service (say, in interpretation of
> policy) must prevail.

There has always been considerable overlap in political and administrative role performance at the upper levels
of government in Nigeria. In fact, the notion of permanent
secretaries acting as deputy ministers/or deputy commissioners is basically consonant with prevailing conventions
and prescriptions (see Aliyu, 1979:11; Oyeyipo, 1979:6;
Tukur, 1970a:177).[25] A clear-cut minister/deputy definition of roles both more accurately reflects the dual nature
of the functions which holders of the two posts have tended

to exercise in practice and effectively underscores the
critical place of political control in the policy-making
process. Reinforcing the latter principle constitutes a
particularly important component in efforts to prevent
career administrators from establishing a dominant place in
the policy-making arena.

Since public administrators have been deeply and
consistently involved in making public policy in Africa,
they bear a major share of the responsibility for the end
product. What types of policy decisions are career public
servants associated with as a class or interest group?
What ends tend to be pursued by the bureaucracy through
policy formulation and implementation? These are the
central questions addressed in the chapters that follow.
With reference to some of the most pressing policy concerns
on the continent, we shall both examine how public administrators shape authoritative determinations and critically
assess substantive outcomes. The question "who benefits?"
will provide the overriding focus of investigation into
issues of international debt management, agricultural
policy, land and housing allocation, and local development
planning.

NOTES

1. Adebayo (1981:77) adds time constraints due to
preoccupation with constituency matters to this list.
2. The methodology employed in the northern states
local government survey cited here and elsewhere in this
volume is described in Aliyu, Koehn, and Hay (1982:12-15).
3. The AFRC's initial membership is reported in the
New Nigerian, 30 August 1985, p. 1.
4. Following the 1975 coup, the new military leadership initially retained Allison Ayida as secretary to the
federal military government, but denied him "access as of
right" to SMC meetings (Dent, 1978:118).
5. General Ironsi never appointed civilian commissioners and kept surviving former politicians out of
government positions entirely. He vested direct political
responsibilities in six senior military officers, including
himself as Head of the Military Government and Supreme
Commander of the Armed Forces (Luckham, 1971:254). General
Gowon first appointed civilians to head ministries and
extra-ministerial departments following the division of
Nigeria into 12 states and the secession of the Eastern

Region in 1967 (Bienen and Fitton, 1978:29, 51; Phillips, 1989:429).

6. Under General Ironsi, the FEC consisted of military men, along with the Attorney General, the Inspector General of Police and his deputy. By 1974, civilian commissioners comprised a majority on the FEC, although the new council appointed by General Gowon in January 1975 reversed the balance in favor of military officers. The Murtala/Obasanjo administration appointed academics or technical experts rather than former politicians as its civilian commissioners (Campbell, 1978:71, 80).

7. For a list of super perm secs, see Otobo (1986: 112n).

8. Gobir (1979:161) and Bienen and Fitton (1978: 46-47) report that the SMG and certain permanent secretaries attended executive council sessions at the state level as official or ex-officio members under the Gowon regime. In contrast, Adebayo (1979:12) maintains that higher civil servants only participated at state executive council meetings at the discretion and express invitation of their commissioners. Adebayo generally portrays state government permanent secretaries in a more subserviant manner than other authors do (e.g., 1981:83). For instance, Balogun (1983:89-90) contends that "military rule in the Gowon days did not go beyond the posting of military governors to state capitals. The ministries and departments, the parastatal organizations, local and provincial administrations - all these were under the control of the civilian bureaucracy."

9. An arrangement that parenthetically proved similar in important behavioral respects to the colonial pattern of administration (see Feit, 1968:188-189; Dent, 1978:109).

10. On the other hand, as Bienen and Fitton (1978:29) point out, "military regimes put heavy burdens on civilian administrators. The military devolves on civil servants many tasks and decisions that it itself does not want to take."

11. At the top of the federal system, General Murtala retired 5 permanent secretaries and the chairman of the Public Service Commission (Dent, 1978:119).

12. On the other hand, one can also find examples of military governors who became personally involved in routine administrative affairs -- such as the allocation of government quarters (see Adebayo, 1981:173).

13. See the critical discussion of integrated project management presented in Chapter 1.

14. Similar findings are reported by Rondinelli (1981: 613) for Sudan.

15. Adamolekun (1983:200) suggests that the idea that the minister/commissioner be the accounting officer is justified by the direct accountability of the President/ governor under the presidential system.

16. In January 1981, the Governor of Niger State demoted two permanent secretaries without consulting the state Civil Service Commission. This action led to a week-long work boycott by numerous civil servants who viewed it as part of a move that favored one ethnic group over others in high-level government postings (New Nigerian, 6, 9, 10 January 1981, p. 1). On inter-ethnic conflict in Niger State, see Kolo (1985:144).

17. In practice, of course, the Nigerian public ser-vices (and civil service commissions) have never been totally apolitical and neutral. See Nwanwene (1978:207); Panter-Brick (1978:330); Harris (1978:301-303); Ciroma (1980:5); Onajide (1979:29); Wilks (1985:274).

The Civil Service Rules (formerly General Orders) prohibit Nigerian public administrators from holding political party office, publicly supporting or opposing a party or its programs, and campaigning for candidates for elective office (see Okoli, 1978/79:8).

18. Chief Akinyele (1979:239-240), SMG in Oyo State from 1976 to 1978, expresses skepticism that a "Whitehall-type civil service" can operate under a "White House-type government." Oyeyipo (1979:6) places emphasis on what permanent secretaries should not do. They should not "overstep the bounds of [the] general executive programme or push a policy that will be injurious to the electoral fortunes of the Chief Executive ... and his Party." Ciroma (1980:5) warns that "what is of concern and should be avoided is to be outright partisan in prosecuting develop-ment by subordinating the claims of one section to another for no just cause."

19. See Diamond's analysis (1981:9-11) of the background and nature of this conflict.

20. General Obasanjo had expressed similar frustra-tions with bureaucratic resistance to change and the reluc-tance of public servants to devote themselves fully to new programs (see Adamolekun, 1978b:315).

21. On the negative consequences resulting from this situation, see Teriba (1978:43-44, 85); Adamolekun (1978b:316-318); Kehinde (1968:92-98); Adedeji (1968b: 151-152); Okpala (1977:157-166).

22. In a provocative address on the "State of the Third World" delivered to those attending the March 1985 International Development Conference in Washington, D.C., Altaf Gauhar, editor of South magazine, charged that economists are responsible for a large part of the current policy mess. According to Gauhar, economists have "subverted democratic processes" by "assuming responsibility for identifying people's needs and what needs to be done to improve the lot of the people." Autocratic rulers, who hang on by reference to the "magic" of economic theories rather than through political support, have found a convenient ally in the economist. Finally, Gauhar maintains that the inappropriate and unfortunate economic theories applied in the Third World would not have become preeminent without the World Bank's financial support.

23. The World Bank and the IMF, for instance, "rely largely on visits of one or two weeks by missions from Washington to gather information and prepare their posi- tions on the economic conditions and needed changes in individual countries" (Lancaster, 1987:232).

24. As one would expect from this discussion, citizen involvement in policy making is virtually non existent in Nigeria. One formidable barrier to citizen participation is lack of access to governmental information, including the unavailability of published documents and the lack of vernacular versions (Adamolekun, 1983:213; Marenin, 1988: 222).

25. The Civil Service Reforms of 1988 took a large step toward formal recognition of this situation by pro- viding that "each Minister shall exercise his powers in full consultation with his Director-General who should be seen as the deputy to the Minister" (Nigeria, Federal Republic, 1988:8, emphasis mine).

RECOMMENDED READING

Several classic texts treat bureaucratic policy making in the U.S. context. The most useful and incisive are Francis E. Rourke, Bureaucracy, Politics, and Public Policy, 3rd edition (Boston: Little, Brown, and Company, 1984); Francis E. Rourke (ed.), Bureaucratic Power in National Policy Making, 4th ed. (Boston: Little, Brown, and Company, 1986); and Kenneth J. Meier, Politics and the Bureaucracy; Policymaking in the Fourth Branch of Government, 2nd ed. (Monterey: Brooks/Cole Publishing

Company, 1987). For a thorough discussion of the influential and political roles of policy analysts in government,
readers are directed to Arnold J. Meltsner, Policy Analysts
in the Bureaucracy (Berkeley: University of California
Press, 1976).

Although studies which specifically focus on
bureaucratic involvement in public policy making in the
Third World are rare, most scholars address the subject as
part of a broader discussion of public administration or
public policy. A comparative (largely Western) work is
B. Guy Peters, The Politics of Bureaucracy, 3rd ed. (New
York: Longman, 1989; especially Chapters 6 and 8).
For historical background on the situation in Nigeria, con-
sult Augustus Adebayo, "Policy-making in Nigerian Public
Administration, 1960-1975," Journal of Administration
Overseas 18 (January 1979):4-14.

3

Agricultural Policy: Bureaucratic Interests, Environmental Impact, and Socio-Economic Outcomes

Agricultural issues pose some of the most serious and persistent challenges confronting national policy makers in Africa. One-fifth of the continent's population are estimated to be chronically undernourished. Imports are increasingly relied upon to meet food needs. In nearly every year in the decade of the 1980s, production increases have failed to keep up with population growth rates. The stark statistics underscore both the importance of agricultural policy decisions and the failure of prevailing approaches.

The debate over agricultural policy involves several fundamental considerations. Policy advocates who emphasize production support projects which differ significantly from the approaches recommended by those who focus on problems of distribution. The issue of which crops a country's farmers should cultivate often constitutes a subject of intense controversy. While some stress the importance of food for domestic consumption, others advocate the cultivation of inedible commodities for export markets. Land use and reform, infrastructure and credit provision, the availability of water and other inputs, the rural labor cycle, and the needs of pastoralists and the landless constitute other volatile and interrelated issues of agricultural policy. In the background is the political economy context which shapes the formulation of agricultural policy. Three crucial considerations in this regard are the role of foreign donors and multinational corporations, the interests of the bureaucratic segment of the ruling class, and the extent and form of peasant empowerment.

This chapter is particularly concerned with declining food-crop production and the social and environmental impact of externally inspired agricultural development

schemes. Throughout Africa, bureaucratic officials at the
national level have cooperated with the representatives of
foreign firms and agencies in establishing national agricultural policy. Large-scale irrigation schemes and
integrated rural development projects constitute two common
outcomes of such collaboration. With primary reference to
the Nigerian experience, we analyze the nature and impact
of both types of undertakings in the pages that follow.
The principal objective of this analysis is to pinpoint
the beneficiaries of prevailing agricultural development
policies. Then, we shall consider the alternative
approaches that are available to national policy makers in
Africa who are committed to environmental conservation, the
equitable distribution of resources, and sustainable
self-reliant development.

THE NATURE AND IMPACT OF AGRICULTURAL POLICY IN NIGERIA

One of Nigeria's most pressing needs is to increase
the production of crops for domestic consumption. The
country's food-import bill has risen rapidly in the face
of declining domestic food production and increasing
population (see Gusau, 1981:74-75, 79; Williams, 1981b:
76-82; Freund, 1979:96-97; Forrest, 1981:246; Central
Bank reports cited in Toyo, 1986:232-233). Projections
based on Club of Rome data forecast an $11 billion annual
deficit in Nigeria's food trade by the year 2001 if
malnutrition is to be eliminated by the end of the century
(cited in Shaw and Fasehun, 1980:561).

National Agricultural Policy

The Federal Military Government began to take domestic
food production more seriously when confronted in the mid-
1970s by a growing agricultural trade deficit, escalating
food prices in politically influential urban centers, and
the prospect of further dependence on certain Western
countries for required food supplies. Nigeria's Third
National Development Plan (1975-80) referred to agricultural development as the country's highest priority.
Nevertheless, the crop-production sector received less than
10 per cent of annual state and federal government budget
allocations under the Murtala/Obasanjo regime (Forrest,
1977:77-79; Wallace, 1978/79:55; Oculi, 1979:63-64; Anise,
1980:19).

The technocratic approaches to agricultural development initiated and promoted by the FMG during this period
essentially ignored both the small-scale peasant farmers
who produce the bulk of Nigeria's food crops and the
environmental dangers inherent in a strategy guided by the
overarching principle that "big is beautiful" (also see R.
White, 1987). Large-scale irrigation schemes and rural
development projects, commercial farms operated by parastatal organizations, and expanded fertilizer distribution
and credit schemes constituted the primary ingredients in
the federal government's agricultural policy for the Third
Plan period (Forrest, 1981:239-250; Wallace, 1981b;
Williams, 1976b:136-137; 1980:158-161; Oculi, 1979:70-71;
Beckman, 1984). In addition, the Land Use Decree of 1978
opened the door to "widespread state acquisition of land
for large-scale agricultural schemes, whether directly
funded or as joint ventures with foreign capital" (Watts,
1983: 495; also see Kaduna State, 1981:6-39; Wallace,
1980b:2-7, 11; Andrae and Beckman, 1986:228; Kirk-Greene
and Rimmer, 1981:78-79; Turner, 1986:38n; Toyo, 1986:237).

The FMG launched five specific types of activities in
the mid-1970s aimed at stimulating food and/or cash crop
production: (1) river basin development schemes (irrigated
farming); (2) integrated rural development projects
(designed and assisted by the World Bank); (3) the National
Accelerated Food Production Programme (provision of a coordinated package of inputs); (4) the Operation Feed the
Nation (OFN) campaign (distribution of highly subsidized
chemical fertilizer supplies for application on staple food
crops);[1] and (5) agricultural credit programs (operated
primarily through the Nigerian Agricultural and Cooperative
Bank set up in 1973). These programs reflect several common, linked features. They are based on the assumption
that widespread adoption of Western technology offers the
key to rapid increases in tropical agricultural production.
They rely extensively on imported machinery, expertise,
chemical fertilizers, pesticides, and "improved" seeds.
They endeavor to involve producers more deeply in commercial farming. Finally, they tend to be capital-intensive.[2]

In terms of socio-economic results, these programs
failed to generate significant increases in food or cash-
crop production, and left Nigeria even more heavily
dependent upon imports and external forces than prior to
their implementation. Furthermore, they undermined the
ability of the vast majority of the country's rural
populace to remain agriculturally productive and self-
sufficient, both in terms of their impact on land and

political-administrative relations and through neglect of
peasant cultivators and pastoralists and the diversion of
desperately needed resources into the hands of a small
group of relatively wealthy farmers, private businessmen,
military officers, and state officials (see Wallace,
1978/79:61, 66, 71; 1981a,b; 1979b:7; 1980a:75; Nwosu,
1977a:37-38; 1977b:141-143; Matlon, 1979:10n; Nuru, 1980:
233; Sano, 1983:31-32, 51; Ter Kuile, 1983:53-54; Forrest,
1977:79-80; 1981:239-250; Oculi,1979:69-70, 73; King,
1981:279; Awa, 1980:9-14; Freund, 1979:97; Palmer-Jones,
1980:2, 10-11; Dudley, 1982:119-120, 270; Clough and
Williams, 1984; Beckman, 1984; Watts, 1983:493, 496-498;
Nigeria, Political Bureau, 1987:36; West Africa, 21
September 1987, p. 1834).

Of the five types of agricultural programs instituted
by the FMG, the irrigation schemes and the World Bank-
designed and assisted projects have received the greatest
emphasis and absorbed the highest levels of government
funding (Forrest, 1981:240-241; Palmer-Jones, 1984; Sano,
1983:39). In view of the scope of and significance officially attached to these projects, particularly in the
savanna region in the north of Nigeria, it is important to
explore their environmental and socio-economic impacts in
some detail. The next sections of this chapter, therefore,
take a critical look at a major river basin irrigation
scheme in Sokoto State and a model integrated agricultural
development project in Kaduna State.

Irrigation Projects

Under the Murtala/Obasanjo military administration,
massive irrigation projects attracted the largest
proportion of federal government spending on agriculture.
The FMG established seven river basin development
authorities in the northern states. Each authority is
charged with undertaking an expensive irrigation scheme
aimed at increasing domestic food and cash-crop production
by making water available for agricultural purposes.
Through irrigation and double cropping, the official
expectation is that small farmers will grow and market more
and different crops, and improve their economic standing in
the process (Wallace, 1978/79:60; 1981a:59-60; Garba,
1979:II; Oculi, 1979:71). Evidence gathered through
independent research efforts indicates, however, that the
overall impact of the projects has diverged considerably
from this expectation. Moreover, the authorities have

failed to devote attention and resources to mitigating the adverse environmental consequences which have resulted from these undertakings. These findings are illustrated here by reference to one of the schemes, the Bakalori project at Talata Mafera in Sokoto State.

A bank consortium with ties to the Fiat Corporation provided the main impetus for the Bakalori irrigation scheme (Watts, 1983:496). An Italian firm designed most of the project from Rome in the absence of any reliable studies of local farming systems and socio-economic relations. Without prior consultation with the people living in the affected area,[3] the Sokoto Rima River Basin Development Authority authorized the commencement of construction work in the early 1970s. By 1977, the Bakalori dam had been built and a large area had been flooded.

The first burden of the project fell upon the people of Maradun. Some 15,000 inhabitants of this area had lost their homes, their town, and their farmlands to the dam site. In addition to failing to secure promised new farm plots, the people displaced to make way for the dam and the project headquarters did not receive the compensation they were legally entitled to for the loss of their economic (i.e., fruit or nut bearing) trees and for crops that could not be cultivated over three wet seasons from 1977 to 1980 (Wallace, 1980c:4-5, 8-9, 36, 44; 1981a:61-65; Andrae and Beckman, 1985:132).[4] Gunilla Andrae and Bjorn Beckman (1985:131-132) found New Maradun a partial ghost town when they visited the resettlement site in 1983. Much of the town's population had "scattered in all directions in search of alternative sources of income" (also see Wallace, 1980b:2-3).

Elsewhere on the Bakalori project, contractors leveled about 3,500 hectares of land for a surface irrigation scheme, using laser-beam technology and bulldozers. The fertile topsoil disappeared, and the construction work removed virtually all of the protective and regenerative tree cover (see Hyden, 1986:24 on this problem). The plots affected lie within the region of Nigeria designated as most highly prone to desertification (Nigeria, NCAZA, 1978:III-3). Yet, the project staff planted no new trees. Those who held farm lands in this area did not receive compensation for the loss of three wet season crops, expropriation by the Authority of 20-25 per cent of their land as a contribution for road and canal construction, destruction of their economic trees, and the seizure of grazing land that had supported their livestock. Moreover,

the project's double-cropping plans (rice and wheat)
required the application of massive quantities of chemical
fertilizers and mechanized ploughing, seeding, spraying,
sprinkling, and harvesting (Beckman, 1984). Each of these
mechanized techniques had destructive consequences for the
soil (Andrae and Beckman, 1985:106-113, 132). The combined
impact of unmitigated alterations in environmental conditions and neglect of the economic consequences associated
with project intervention in the surface irrigation area
forced many small landholders into debt; some have been
compelled to rent or sell their land cheaply to large-scale
farmers from the locality or to wealthy outsiders engaged
in absentee farming (Wallace, 1981a:61-64; 1980c:20, 37).

The third type of land adversely affected by the construction of the Bakalori dam is the rich _fadama_ along the
banks of the Sokoto river. The river no longer overflows
its banks downstream of the new dam. As a result, dry season farming has been abandoned on an estimated 20,000 hectares of previously highly productive _fadama_. Project
staff paid no compensation or programmatic attention to the
villagers affected, many of whom have been denied a vital
source of food and income. Damming of the Sokoto river
also deprived some downstream villages of their water supply and fishermen of their livelihood (Wallace, 1980c:17,
36-37). In the long run, "downstream from extensive irrigation projects the water may become too saline for further
use, unless expensive desalinization measures are undertaken" (U.S., C.E.Q., 1980:35, 101; also see Le Houerou
and Lundholm, 1976:220-221; Ambroggi, 1980:104-106). In
addition, Fulani cattle herders traditionally crossed the
project area with huge herds each dry season. Although
project officials decided that Fulani herders must no
longer traverse the irrigation scheme, "no alternative
arrangements or plans have been made to accommodate their
very real needs" (Wallace, 1980c:17, 45; cf. Dash, 1981
for a Senegalese case).

An overall performance assessment of the Bakalori
project reveals the extent to which the costs associated
with large-scale irrigation schemes in the northern states
of Nigeria outweigh whatever benefits they have brought
about. At great environmental and economic expense,
including the sacrifice of 20,000 hectares of productive
fadama land in the dry season and an outlay of roughly $.3
billion in public funds (Wallace, 1981b; New Nigerian, 28
February 1980, p.1), the Sokoto Rima River Basin Development Authority managed to place a total of 1,000 hectares
of land under irrigated cultivation by the 1979/80 season.

However, local farmers had become so disaffected with various aspects of the Bakalori scheme that they refused to participate in irrigated farming, and erected road blocks aimed at disrupting work on the project in late 1979. In February 1980, more than 5,000 angry farmers effectively blocked all access to the dam. As many as 200 protesters may have lost their lives when police stormed the roadblocks on April 26 (see Andrae and Beckman, 1985:115-116). The confrontation resulted in a complete stoppage in the flow of irrigation water, destruction of nearly all the crops which project staff had planted on the 1,000 hectare area, and the payment of an additional 30 million naira to Impresit Bakalori, the foreign firm which had constructed the dam, to cover project losses and delays.

Work on the project recommenced in May 1980 after some concessions by the Authority to farmers who had been displaced or had lost crops (Wallace, 1980c:1; 1981a:65; New Nigerian, 28 February, 5, 10 March, 17, 23, 29, 30 April, 5, 14 May, 19 June, 1 July, 3 September, 1980; Sunday Triumph, 4 July 1982; Usman, 1982). However, only a fraction of the irrigated area had been brought under cultivation by mid 1983. Andrae and Beckman (1985:97-98, 123) estimate that total net annual expenditure on the Bakalori irrigation scheme, exclusive of overhead costs, amounted to 2,000 naira per hectare.

Integrated Rural Development Projects

The integrated rural development project constitutes the second type of major agricultural undertaking launched by the FMG in several states. Staff appointed by the World Bank planned, managed in a largely autonomous fashion (Sano, 1983:38), and evaluated each of the seven projects initiated in Nigeria by 1979. They adopted a common design that has been popular with the Bank since 1973. Donors favor the pre-packaged project approach because it ensures control over major policy decisions and facilitates expatriate involvement in the implementation and evaluation stages (see Robertson, 1984:73, 121-125; Paul, 1982:7). In Nigeria, an autonomous administrative structure operated each agricultural development project (ADP). The staff included "about one dozen expatriate technicians and managers, including the project director" (I.B.R.D., 1981:53). Indeed, the World Bank contends that liberal recourse to expatriate managerial and technical skills by the Nigerian federal and state governments constituted one of the main

factors contributing to the "success" of the ADPs
(I.B.R.D., 1981:53).

The principal ingredients in the integrated package
approach are agricultural inputs (new seed varieties, fertilizers, pesticides, and farm machinery), extension and
credit facilities, marketing assistance, and infrastructural development -- including the construction of dams and
rural feeder roads (see Nigeria, FADP, 1977; Wallace,
1980a:62, 65-66, 77n; Stryker, 1979:330-331; I.B.R.D.,
1974:31). World Bank loans provide 33-51 per cent of the
funds used to finance project activities. These loans must
be repaid in foreign exchange over a twenty-year period at
annual interest rates between seven and eleven per cent.
In each contract negotiated with the World Bank, Nigeria is
required to expend 40 per cent of the loan abroad -- on
salaries and emoluments for expatriate staff, tractors,
other vehicles and equipment, fertilizers, pesticides, etc.
(Wallace, 1980a:62-63; also see Olinger, 1979:105-106;
Oculi, 1979:65-66; Lappé, Collins, and Kinley, 1980:90-92;
Stryker, 1979:326-334; Williams, 1981a:35-38).

The FMG initially commissioned two integrated agricultural development projects in 1975. One of them, the
Funtua Agricultural Development Project (FADP), is located
on some of the most fertile land in Kaduna (now Katsina)
State. FADP encompasses an area of 7,590 square kilometers. Expenditure on the project amounted to 70 million
dollars over the period 1975-80, with 51 per cent of the
funds derived from the World Bank loan, 43 per cent from
state government funds, and the rest supplied by the
federal government. The stated objectives of the Funtua
project are to increase agricultural productivity and small
farmers' income. The FADP design is devoted to bringing
about change in the traditional farming system through top-
down, technocratic intervention. It concentrates on the
coordinated provision and application of technology packages in order to achieve targeted increases in crop production. The principal components in the package are
high-yield seeds which require irrigated cultivation and
intensified applications of inorganic fertilizers, herbicides, pesticides, and expert advice. Tractor hire units
have been established, and the efforts of large-scale
farmers to secure commercial loans for heavy equipment
purchases have been assisted by the project's farm management section. The main crops for which packages are provided at Funtua are 'yellow' maize, groundnuts, cotton, and
an "improved" sorghum (D'Silva and Raza, 1980:283-284;

Wallace, 1980a:64-65; Forrest, 1977:80; 1981:245; Sano,
1983:73-74; Beckman, 1984; Clough and Williams, 1984).

World Bank officials and FADP staff operated on the
assumption that an approach to agricultural production
primarily based upon the adoption of "improved" inputs and
modern farming techniques would be a necessary and sufficient condition for increasing domestic food supplies
and improving living conditions for the rural masses. The
project ignored other crucial local issues, including the
distribution of land, wealth, and power, rural health
conditions, the needs and practices of pastoralist groups,
the role of women, the place of non-farm economic activities, marketing arrangements, relative prices of different
crops, prevailing political arrangements and bureaucratic
orientations, rural industrial development, and environmental stress (Wallace, 1980a:64, 66, 76; Oculi, 1979:71;
Beckman, 1982a:32; Matlon, 1979:104; Clough and Williams,
1983; 1984).

It is not surprising, therefore, that FADP produced
results that departed dramatically from the stated
objectives. Brian D'Silva is one of the few independent
researchers to conduct a carefully designed study of FADP's
impact on agricultural production during the first three
years following its inception. He reports that, in spite
of the massive infusion of resources (including an extension worker-to-farmer ratio that "has not been achieved
even in most developed countries," low interest loans for
tractor purchases, and government price supports), project
yields have consistently been disappointing for the
improved sorghum, groundnut, and cotton crops. Moreover,
key inputs provided by this expensive project (particularly
fertilizer, credit, and extension visits) have been disproportionately concentrated on large-scale and "progressive"
farmers (including strategically placed urban businessmen,
public servants, and district heads), while the bulk of the
farmers in the area, who are classified as small-scale and
traditional, have been neglected (D'Silva and Raza, 1980:
284-295; Wallace, 1980a:69, 75; 1981b:248-249; 1981a:68-69;
Clough and Williams, 1984; Williams, 1986:18; Watts,
1983:502-504).

FADP's Chief Project Evaluation Officer explicitly
stated in 1978 that project staff "'prefer the trickle down
approach from farmer to farmer, accepting that some will
thereby benefit later than others. As a consequence of
this preference, we concentrate on our notorious 'progressive farmers'" (cited in D'Silva and Raza, 1980:289; cf.

Wallace, 1981b:248; I.B.R.D., 1981:53). Those monitoring
trends in the project area reported increasing sales and
rental of land to wealthy individuals, growing absentee
farming by businessmen and civil servants, and rising
income disparities (Oculi, 1979:67-68, 71; D'Silva and
Raza, 1980:291, 295; Wallace, 1981a:68-69; Beckman, 1984;
Shenton and Watts, 1979:61; Kaduna State, 1981:I, 28-29).
Furthermore, the concentration of scarce agricultural
inputs on the project area necessitated that the needs of
farmers living elsewhere in Kaduna State be neglected
(Wallace, 1980a:70, 73-74).

The Funtua project also has placed increasing pressures on the land, although its long-term environmental
consequences are less visibly striking than those inflicted
by the large-scale irrigation schemes. FADP is situated in
an area where farmers traditionally practice complementary
mixed cropping of cereals and legumes, rely heavily on
manure to maintain soil fertility, and apply inorganic
fertilizers only to priority crops. Richards (1985:68-70)
shows that intercropping possesses many advantages as a
farming strategy -- including promoting crop growth,
reducing pest, weed, and disease attack, providing more
easily protected fields, and flattening labor demands.
However, all of the technology packages introduced on the
Bank-assisted scheme promote monocropping.

The high-yield sole crop varieties emphasized by FADP
require applications of chemical fertilizers in quantities
that threaten to produce serious soil-acidification
problems in the area. FADP's most dramatic impact on the
farming system in the project area has been the application
of inorganic fertilizers. The quantity of subsidized
fertilizers (predominantly superphosphate and calcium
ammonium nitrate varieties) sold by the project's farm
service centers increased from about 9,000 metric tons in
1976/77 to more than 18,000 metric tons in 1977/78 (Gana,
1980:5; D'Silva and Raza, 1980:15; Nigeria, FADP, 1977).
Beckman (1984) concludes that "the ADPs are to a very large
extent organisations for the distribution of fertilisers."
The plant varieties introduced also are less resistant to
local weeds than are traditional crops. This situation
encouraged commercial farmers to use potentially harmful
chemical herbicides for the first time. The substitution
of highly mechanized cultivation for hand labor further
disturbs the delicate ecological balance of the area
(D'Silva and Raza, 1980:291; Abalu and D'Silva, 1979; Gana,
1980:5; Nigeria, FADP, 1977; Norman et al., 1979:29, 32,
56-59, 76, 112; Matlon, 1979:17, 22, 86).

Summary Assessment

The findings reported here with respect to the Bakalori irrigation scheme and the Funtua ADP are not unique. Rather, they are illustrative of the results associated with both types of agricultural projects throughout Nigeria. The adverse outcomes of the Bakalori project are related to the technocratic, control-centered approach to agricultural development in the north of Nigeria promoted by the river basin development authorities (see also Bernal, 1988:92-104, on Sudan). Nigeria's other irrigation schemes have shown similar tendencies to abuse natural resources, particularly land (also see R. White, 1987, for a Senegalese study that reaches similar conclusions). Only a minute proportion of the small-farmer population has been served by these projects, while heavily subsidized yields have consistently been lower than projected and crop production has declined precipitously on downstream floodlands (see Andrae and Beckman, 1985:97-98, 107-108, 131, 133). The main beneficiaries have been large and absentee farmers, foreign contractors, project staff, and the high-level public servants, military officers, and businessmen who serve on the governing boards of the river basin development authorities. Undeterred by the economic and environmental implications, the FMG set in motion additional programs aimed at establishing 274,000 hectares of irrigated land at an estimated cost of $4 billion, or $14,400 per hectare (Wallace, 1978/79:67-72; 1981a:242-247; Palmer-Jones, 1980; Sano, 1983:51).

The primary impact of ADPs on the local farming system has been the increasing application of state-subsidized chemical fertilizers. Other expensive imported inputs have been introduced, and monocropping strategies have been emphasized. While some large, "progressive," and absentee farmers have utilized project inputs to their advantage and others desire to secure them, disappointing yields have characterized the project areas (Sano, 1983:53-57, 66-67; Beckman, 1984).

Nigeria's military rulers and external creditors embarked on essentially the same mid-1960s green revolution strategies that had already resulted in adverse social and economic as well as environmental consequences elsewhere in the Third World (see Stryker, 1979:329-331; Dahlberg, 1979:49, 67, 90; Stahl, 1974:129, 135-138; Kang, 1982: 192-195, 200-205; Kerkvliet, 1979:137-138; Richards, 1985:122). When the familiar warning signs of acute and unattended stress on the land began to surface on the

massive irrigation schemes and integrated agricultural
development projects launched in the northern states,
officials showed no concern and made no effort to modify
prevailing policies and approaches or to mitigate their
impact.

Agricultural Policy Developments During the Second Republic

The civilian political party leaders who replaced
Nigeria's military rulers in October 1979 pressed for the
introduction of "modern" technological inputs on an even
wider scale in the agricultural sector.[5] Shortly after
President Shehu Shagari assumed office, he declared an
intention to make Nigeria self-sufficient in food and cash
crops by 1985 (New Nigerian, 22 May 1980, p.1). In
January 1980, the Shagari administration commissioned a
team of Nigerian and World Bank experts "to examine in
depth the food situation in Nigeria and to identify broad
programmes and strategies that will lead to an early
achievement of self-sufficiency" (Gusau, 1981:74). Nine
months later, President Shagari announced that Nigeria
would realize this objective through a multi-billion naira
"Green Revolution Programme."

The central components in the Green Revolution Programme set forth by President Shagari with assistance from
the World Bank advisers were incorporated into Nigeria's 82
billion naira Fourth National Development Plan (1981-85).[6]
Not surprisingly, the program concentrated on expansion of
agricultural development projects and large-scale irrigation schemes. The Federal Minister of Agriculture reported
that the Shagari administration had been so "impressed by
the success of the integrated agricultural development
projects ... started on a pilot basis in Funtua, Gombe and
Gusau ..." that it decided "to establish similar projects
in all [19] states of the Federation by 1983" (Gusau, 1981:
74-75). The plan allotted 2.3 billion naira to this undertaking, with roughly 800 million naira in project costs to
be financed through IBRD loans. By the end of 1981, the
World Bank had lent Nigeria $500 million in connection with
the eleven FADP-style projects which had been initiated at
that time (Sano, 1983:42). In addition, "an Accelerated
Development Area Programme (ADA) is to be implemented for
residual areas not yet encompassed by the ADPs. The
programme would apply the core elements of the ADP approach
such as improved extension, input distribution and rural
feeder roads in a simplified package which could later be

upgraded to full ADP status" (West Africa, 16 March, 1981:
567). Large-scale farmers are even more likely than previously to secure a disproportionate share of project
inputs under the ADA approach (Beckman, 1984).

Initially, the Federal Government's proposed extension
of the integrated agricultural development program to cover
all of Kaduna State encountered the resistance of the PRP
Governor. In November 1980, Governor Balarabe Musa rejected both the idea of implementing the program statewide
(originally endorsed by the military government) and the
terms of the 100 million naira I.B.R.D. loan agreement
about to be signed for financing the project expansion. A
release issued by the Governor's press secretary explained
that the Kaduna State executive differed with both the
World Bank and the Federal Government on "'some major agricultural policy directions of the project and the control
and remuneration of foreign staff.'" Regarding policy
differences, "the release stated that expatriates ... in
control of the Funtua Project ... encouraged massive application of chemical fertilizer by the farmers 'as though
this is all that modern farming methods mean.'" The Governor asserted that although this approach is "certainly in
the interest of big foreign corporations which derive huge
... [profits] from the sale of chemical fertilizers," it is
"not in the long term interest of farmers in the state,
because massive application of chemical fertilizers would
damage the soil and cause other serious ecological destruc-
tions likely to render the whole area into an acidic waste-
land in the long run." The statement also maintained that
the new government would "not allow the project to merely
serve a handful of large scale farmers who are basically
urban dwellers, but who have taken away most of the benefits at the expense of the small peasant farmer in the
villages" and that the Kaduna State executive preferred to
encourage cultivation of the food staples grown in the area
(Kaduna State, MIAI, 1981:Annex III; New Nigerian, 11, 19
November 1980, pp. 1, 11, 17). Finally, the Kaduna State
Governor complained about the expenditures imposed by the
Bank for expert salaries and benefits (Kaduna State, 1981:
Annex III; also see Sano, 1983:71; Falola and Ihonvbere,
1985:132). In 1981, however, a majority of the members of
the NPN-controlled state house of assembly voted to impeach
Governor Balarabe Musa. Their action cleared the way for
eventual acceptance of the terms set by the World Bank.

In another Second Republic development, the Shagari
administration announced plans to multiply the amount of
land to be placed under Bakalori-like irrigation schemes

run by eleven river basin authorities (New Nigerian, 22
May 1980, p.1). These projects absorbed approximately 75
per cent of the federal government's total capital budget
appropriation for agriculture in 1980 and 1981 (Sano,
1983:39) and received even greater emphasis in the budget
proposal for 1984 which the President announced two days
prior to the military coup which terminated the Second
Republic (The Guardian (Lagos), 30 December 1983, pp. 1-2).

Furthermore, the Fourth Plan emphasized the creation
of large-scale mechanized agricultural enterprises.
Through joint ventures with state governments and private
"technical partners both within and outside the country,"
the National Grains Production Company planned to establish
a 4,000 hectare mechanized grain farm in each of the 19
states at a total cost of 2.1 billion naira in public funds
(Sano, 1983:42). The five large-scale state farms in
existence by mid 1980 were located in Niger, Kaduna, Kano,
Bauchi, and Oyo states (New Nigerian, 15 August 1980, p.1).
With the aid of International Finance Corporation loans and
expanded national agricultural credit programs that principally benefitted former military officers and civil servants, the Shagari administration also promoted joint
state/private commercial estates devoted to the cultivation
of cash crops such as oil palm, cotton, sugar, rubber, and
cocoa through chemical-intensive farming techniques (New
Nigerian, 22 May, 5 August, 9 December 1980, p.1; Gusau,
1981:76, 79; Shagari, 1980:2472; Africa Research Bulletin,
31 October, 1980, p.5687; West Africa, 21 September 1987,
p. 1834; Watts, 1983:497-498).

The policies and programs advocated by the Shagari
administration aimed at fostering rapid and extensive mech-
anization of the agricultural sector. The federal government continued to provide tractors to cooperative societies
and farmer associations "at 50% subsidy," to allow duty-
free importation of farm machinery, and to offer direct
financial assistance to state governments for land-clearing
and tractor-hiring services (Gusau, 1981:76; Sano, 1983:
45). In a statement reflecting the key underlying assumptions that shaped the Fourth National Development Plan,
Nigeria's top agricultural official openly asserted that
"human labour as a source of energy on the farm has become
not only scarce and expensive but unattractive. Human
labour must therefore be replaced systematically through
the introduction of farm machinery" (Gusau, 1981:76;
Davies, 1981:69).

The expanded provision of subsidized fertilizer
supplies constituted another major component in the

government's Green Revolution Programme. Nigerians applied
over one million tons of fertilizer in 1980 (Watts, 1983:
500; "Nigeria," 1982:D7). By 1981, the federal budget
called for expenditures of 100 million naira on fertilizer
imports. Nigeria's Fourth Plan (West Africa, 16 March
1981, pp. 567-568) envisioned total purchases of 3.1 million metric tons at a cost of approximately U.S. $1 billion
to the federal and state governments between 1981 and 1985.
However, the ability of the Nigerian state to maintain and
service the extensive mechanized and import-dependent
agrarian network it introduced has become increasingly
problematic in the face of fiscal crisis (Beckman, 1984;
Andrae and Beckman, 1986:229).

In virtually every facet of its Green Revolution Programme, the Shagari administration opened new doors for
foreign involvement and profit making. The government con-
sciously emphasized large-scale privately run enterprises
in an effort to attract private foreign investment. Moreover, it transferred agricultural production and processing
from Schedule II to III of the Nigerian Enterprises Promotion Decree. This action meant that "foreigners can now
own up to 60 per cent of the equity in an agricultural
enterprise" (West Africa, 16 March 1981, p. 565, emphasis
in original; also Falola and Ihonvbere, 1985:130). The
Fourth Plan further assured potential investors that the
"fiscal incentives already provided by government for companies wishing to go into large-scale agricultural production, e.g., income tax relief for pioneer enterprises, ...
will be maintained and improved upon ..." (also see
Akinola, 1987:230). The increased capital outlays on
modern technological inputs called for by the Plan ensure
that Nigeria will continue to be an important market for
fertilizers, agricultural machinery, insecticides, herbicides, and seeds manufactured abroad as well as a major employer of expatriate technical assistance (Beckman, 1984).

Western agri-business concerns were eager to capital-
ize on Nigeria's belated interest in the green revolution
(Sano, 1983:67). Concomitantly, successive U.S. administrations, bent on reducing the trade deficit brought about
by purchases of Nigerian crude oil, assisted U.S. corporations interested in capturing an expanded share of
Nigeria's growing market for imported agricultural commodities and inputs. Former Vice President Walter Mondale's
July 1980 Lagos meetings with top Nigerian government officials and subsequent discussions held between Presidents
Carter and Shagari in Washington in October led to the
signing of bilateral agreements aimed at increasing U.S.

public and private sector involvement in Nigeria's Green
Revolution Programme. The memoranda of understanding
signed by the two vice presidents in July specifically
called for cooperative governmental efforts to expand agricultural trade and encourage and facilitate participation
by U.S. private business firms in joint farming ventures in
Nigeria (Hecht, 1981).

As a direct outgrowth of these agreements, the two
sides set up a Joint Agricultural Consultative Committee
(JACC) to promote and monitor U.S. agribusiness ventures in
Nigeria. The JACC consists of corporate executives from 72
large U.S. companies and Nigerian representatives drawn
from the private and public sectors. The U.S. members of
the Committee represent giant agri-business and financial
concerns, including farm-equipment manufacturers (FMC Corporation, Allis-Chalmers, Ford Motor Company); fertilizer,
pesticide, and seed producers (Pfizer, Occidental Petroleum, Whittaker); food-processing operations (Carnation,
Pillsbury, Ralston Purina); and financial institutions
(Chase Manhattan, First National Bank of Chicago) (see
Hecht, 1981; Young, 1981; Africa News, 29 March 1982).

The composition of the JACC left little doubt about
the intention of the U.S. government and business communities to push project purchases of high-technology agricultural inputs and heavy equipment. This approach proved
particularly appealing to the Reagan administration, which
viewed deeper involvement by the U.S. private sector in the
"development" process in general and in international agri-
business activities in particular as one of the most important ingredients in an approach to foreign aid that would
simultaneously save African economies from disaster and
benefit American business interests (Crocker, 1981a:3;
Hartwell, 1981). In 1981, therefore, Assistant Secretary
for African Affairs Chester Crocker (1981b:2-3) specifically hailed the JACC as the administration's prime model
"of the contributions and benefits of private sector involvement in Africa to which we are giving encouragement"
and expressed the hope that this novel experiment would
provide the basis "for developing similar relationships
with more African countries." The export of U.S. agricultural technology occupied center stage in this policy
approach.[7] To the delight of the corporations represented
on the JACC, the Shagari administration also encouraged the
adoption and application of U.S. technological inputs in
the agricultural sector. Nigeria's Minister of Agriculture
specifically called for greater United States involvement
in supplying Nigeria with farm machinery, pesticides, and

improved seeds at an early meeting of the JACC (Hecht,
1981).

Okello Oculi (1979:67) points out that increasing
Nigerian purchases of foreign-manufactured agricultural
equipment (along with other inputs as well as food imports,
one might add) must be attributed in part to "the emergence
of a new capitalist class of importers whose wealth depends
on the continuation of the phenomenon and who seek to influence policy makers accordingly." These beneficiaries
include powerful civil servants (see Andrae and Beckman,
1986:218-219; Toyo, 1986:241). This situation suggests
that radical changes in agricultural policy are unlikely to
be forthcoming.[8] Indeed, the Buhari regime abolished import duties on agricultural machinery and proposed to allow
non-Nigerians to own up to 80 per cent of large commercial
farming enterprises (Bangura, 1986a:53). The Babangida
administration eliminated Nigeria's six commodity marketing
boards and emphasized the construction of rural feeder
roads.[9] However, it also relaxed the expatriate controlling interest limit on large-scale production schemes
(Turner, 1986:38n), promoted export-oriented ventures, and
otherwise essentially carried on the centralized agricultural policies critically analyzed in this chapter (see
Newsweek, 7 September 1987; Nigeria, Political Bureau,
1987:56-57; Akinola, 1987:226-227).

Nigerian entrepreneurs still require a great deal of
inducement before they will make major investments in
farming (Toyo, 1986:241; Falola and Ihonvbere, 1985:138).
U.S. corporations, moreover, have shown more interest in
expanding exports of North American technological inputs
and food products than in investments in Nigerian agricultural enterprises in spite of the government's willingness
to provide foreign agri-business concerns with tax holi-
days, tariff exemptions, and special credit facilities (see
Hecht, 1981; Young, 1981; Watts, 1983:493; Sano, 1983:69;
Osoba, 1979:68; Andrae and Beckman, 1986:214). At the end
of 1981, for instance, the Plateau State government awarded
International Harvester an 18 million naira contract for
the provision of more than 500 tractors as well as other
heavy farm machinery (Sano, 1983:67; also see de Onis,
1981:9). The influence of multinational corporations over
food import policy decisions in Nigeria is illustrated by
the dominant role of Flour Mills of Nigeria, a subsidiary
of a U.S.-based shipping and wheat trading company, in
decisions affecting the domestic wheat market (Andrae and
Beckman, 1986:218).

CONCLUSIONS AND LESSONS

In Africa's current structural adjustment phase,
governments have been eager to introduce measures favoring
large-scale capitalist agricultural undertakings
(Mkandawire, 1988:18, 35). The principal results have been
escalating indebtedness and growing dependence upon
capital-intensive and export-oriented projects.[10] Large-
scale agricultural schemes of both capitalist and state
enterprise varieties typically yield unimpressive production results (see Nellis, 1986:24; Bernal, 1988:92) and
benefit a tiny fraction of the local population. The most
common outcome of the adoption of export-led agricultural
production policies is "enhanced food insecurity at both
the personal and the national level" (Loxley, 1985:135).

In Nigeria and elsewhere in Africa, externally
designed and financed agricultural production methods pose
serious threats to the long-term carrying capacity of the
land. Instead of embarking on conservation-based rural
development, focusing on the basic needs of the small-scale
producers who are responsible for cultivating more than 90
per cent of the country's food crops, and building on local
inputs and existing food production techniques, successive
Nigerian governments turned to foreign agri-business corporations and consultants that specialize in capital-
intensive, high-technology export commodities. In general,
the energy- and capital-intensive projects which Western
"experts" have conceived and promoted for export to the
Third World emphasize exploitative rather than conserva-
tionist approaches to agricultural development. The destructive environmental consequences associated with the
uncritical adoption of Western technology (Bisrat Aklilu,
1980:399) and the growing substitution of external agricultural models and foreign packages for indigenous, ecologi-
cally adapted practices are too frequently overlooked by
scholars and policy makers. Specifically, the expansion of
World Bank-supported agricultural projects has required the
overseas purchase of increasing supplies of green revolution inputs that accelerate domestic environmental destruction (cf. Le Prestre, 1981:10-33), enhance the economic
standing of a privileged group of large-scale and "progressive" farmers, force small-scale cultivators and herders
onto marginal lands or into the ranks of the landless, and
fail to generate sustained growth in local production and
consumption of staple food crops.

All of the outcomes associated with integrated rural
development projects in Nigeria, including the progressive

and absentee farmer biases, the failure of the "trickle
down" approach to improve peasant living standards, and the
financial and administrative impossibility of replicating
the conditions introduced on integrated package projects on
a wide scale, also have been experienced on similar Bank-
designed and sponsored agricultural development schemes
elsewhere in Africa (Stahl, 1974:94, 116, 122-129, 133-138,
152; Ruttan, 1974/75:15; 1984:348; Lele, 1975:203-204;
Payer, 1980:31-44; 1983:ch.8; Stryker, 1979:330-332; Lappé,
et al., 1980:72-74, 80, 88; Williams, 1981a:20-28; Leonard,
1973:3, 5ff; Hyden, 1983:93; Koehn, 1986:27; also see
Norman, et al., 1979:89, 102, 107; Thiesenhusen, 1978:
167-172). Hyden (1983:94-95; 1980:212, 232) also points to
the relative absence of popular involvement in and commitment to pre-packaged projects, to the misguided assumption
that rural officials can influence peasant decisions, and
to the administrative complexity of integrated rural development schemes as factors contributing to the failure of
such ventures (also see Migdal, 1974:210-211). Ruttan
(1984:395) further faults this approach for failing to
strengthen local government institutions and for relying
instead on "imposing centrally mandated programmes on
communities ...."[11]

Throughout Africa, government officials, businessmen,
and expatriate advisors have dominated agricultural policy
making and controlled the allocation of resources at the
project level (Yahaya, 1979:20-24, 27; Hay, Koehn, and Koehn,
1980:18, 20, 23; Palmer-Jones, 1984). The results have
been inappropriate innovations, distrust of the rural labor
force and neglect of its ecological knowledge, failure to
appreciate the role of women, support for the least needy
segment of the farming community, and the concentration of
inputs and capital accumulation in the hands of wealthy
merchants and money lenders, transporters, public servants,
traditional authorities, and politicians (Richards, 1985:
17; Le Brun, 1979:201; Fantu Cheru, 1987b:2). The Nigerian
experience indicates that sustainable, conservation-based
strategies aimed at increasing food production for domestic
consumption and bringing about the more equitable distribution of resources merit greater attention in Africa (see
Dahlberg, 1979:90, 172-192, 212-227; Pirages, 1978:
268-269).

The most promising approaches emphasize peasant empowerment and popular control over agricultural planning,
research, investment decision making, and project execution,[12] and build upon the local practices and insights of
Africa's peasant farmers and pastoralists (Richards, 1983:

56-58; 1985:157-158; George, 1987:14, 16; Hellinger, et al., 1983:2-3, 32; Korten, 1980:499). African agricultural inventions and cultivation techniques tend to be diversified, adapted to local ecological conditions, and sensitive to the critical need to preserve fragile natural resources (Gill, 1975:112-113, 116; Dahlberg, 1979:227; Richards, 1985:60-62; Andrae and Beckman, 1985:108, 110; Franke and Chasin, 1980:237; Huntington, Ackroyd, and Deng, 1981: 49-53; Ter Kuile, 1983:55; Hyden, 1980:219; Norman et al., 1979:63, 75, 104).[13] Likewise, reliance upon versatile and resilient local seed varieties reduces dependency and vulnerability (Hyden, 1986:20).

The specific dimensions of a sustainable agriculture program will draw upon indigenous insights and adapt appropriate external achievements. There will be considerable variation in approach both internally and from country to country (Hyden, 1986:42). In promoting self-reliant food security systems, attention must be devoted to rural employment opportunities and to the needs of female food producers (George, 1987:vi, 14, 18). Certainly, effective community organization is a key ingredient in the rural development process. Poor rural inhabitants, operating through formal local government structures and/or voluntary non-governmental organizations, must be positioned to shape the identification of priorities and to influence the government's service-providing agencies (Ruttan, 1984:398; Esman and Uphoff, 1984:26, 39).[14]

NOTES

1. The FMG sold fertilizer at 25 per cent of cost through 1979, when it reduced the subsidy to 50 per cent (D'Silva and Raza, 1980:12). When one includes state-supported transportation and distribution services, the effective subsidy for fertilizers may amount to 90 per cent according to Beckman (1984).

In 1977, the government added mechanized land clearing as one of the OFN's priority activities. Oculi (1979: 70-71, 68) notes that "the Federal Government in 1977/78 alone 'disbursed ... about 100 million naira for clearing 15,000 hectares of land for the use of both large and small scale farmers.' This constituted government subsidization at the rate of over 600 naira for clearing each hectare and the likelihood is that most of the land cleared went to officials and local leaders." Later, "Operation Feed the

Nation" promoted poultry farming, partly through the
heavily subsidized importation of corn (see New Nigerian,
16 July 1979, p. 9).

2. Most of the programs and inputs emphasized by the
FMG are consistent with advice tendered in the late 1960s
by a team of U.S. agricultural development "experts"
(Eicher and Johnson, 1970:386-389). However, the U.S.
agricultural economists advocated that the Nigerian government concentrate on adopting policies aimed at the "immediate expansion of export and import substitution crops
...." Specifically, "primary attention should be given to
cocoa, oil palm, groundnuts, and cotton ..." (Eicher and
Johnson, 1970:385-386; also see Crocker, 1983:4; Oculi,
1979:64). World Bank-financed agricultural development
projects originally focused on cotton, groundnuts, and
hybrid 'yellow' maize (Beckman,1984; Clough and Williams,
1983). Beckman (1982b:47) notes that the World Bank becomes involved in projects that "seek to generate a commercial food surplus" rather than "raise the general level of
food consumption among the masses ...." Also see the work
of John W. Mellor (e.g., 1985), one of the most explicit
and influential contemporary advocates of the green revolution and other production-centered strategies.

3. Impresit Bakalori later engaged an independent
anthropologist in sub-contracted project appraisal. A. F.
Robertson (1984:127) discusses the hazards of involving
anthropologists in this capacity as a substitute for
political process.

4. Disputes over the payment of compensation have
characterized major resettlement exercises in Nigeria (see
Adegboye, 1977:42; Aliyu and Koehn, 1980:2, 9; Aliyu, Koehn,
et al., 1979:127-140; Kaduna State, 1981, V:3-5, 52-53, 58;
XII:4-5).

5. The campaign platforms adopted by each of the five
political parties officially certified to contest the 1979
elections all promote this objective, as their published
manifestos (New Nigerian, 19 April 1979, p.3; 3 July 1979,
pp.I-II) demonstrate. However, the approaches advocated
varied both in emphasis and in degree of specificity. The
NPN favored "large scale agricultural and irrigation projects," the UPN an "integrated rural development" program,
and the PRP "state/company owned plantations." Both the
GNPP and the NPP more generally stressed the need to
encourage large-scale mechanized farming.

6. The new civilian governments allotted about 11 per
cent of their combined capital expenditures to agriculture

in the 1980 fiscal year, compared with the 6-7 per cent
budgeted by the FMG for this sector between 1975 and 1979
(Gusau, 1981:79).

7. Hyden (1886:9) warns that the current crisis in
Africa is less a consequence of physical conditions than it
is attributable to Western technological arrogance.

8. Also see Andrae and Beckman (1986:215, 217) on the
increasing political power exercised by flour mill
operators and bakery owners who are dependent upon massive
wheat imports.

9. In 1987, this government created the National
Directorate of Food, Roads, and Rural Infrastructure. The
Directorate operates with a ₦700 million budget. In addition to rural roads, it is responsible for providing water,
improving fisheries, establishing model villages, and
distributing seeds (see the rather critical review in West
Africa, 17-23 April 1989, p. 589).

Based on an historical study of Nigeria's experience,
Gavin Williams (1985:13) concludes that state marketing
boards are "exploitative institutions. De jure state monopolies on the marketing of crops impose high costs on
producers, on government budgets and on consumers. They
create de facto monopolies for favoured and protected
traders and the opportunities for profitable collusion
between businessmen and officials, civil, police and
military."

10. African rural economies tend to be the most
externally oriented, dependent, and vulnerable in the
world. The prospect that African countries can "export"
themselves out of the current crisis is remote given the
limited capacity of the world market to absorb many of its
unprocessed primary products and the declining terms of
trade fetched by its commodities (Hyden, 1986:17-18;
Akinola, 1987:225-226). Moreover, export crops increasing
usurp fertile land traditionally devoted to local food
staples (Fantu Cheru, 1987b:2). In Sudan, for instance,
the Blue Nile Agriculture Corporation has forced farmers
to emphasize irrigated cotton cultivation at the expense
of self sufficiency in food crop production (Bernal, 1988:
94-102).

11. In his study of the implementation of central
government agricultural extension service activities in
Zaire and Burkina Faso, Richard Vengroff (1982:14, 17, 22)
is critical of the "inordinate amount of time" which field
agents devote to purely administrative tasks and of their
negative perception of villagers as "ignorant and unwilling
to change."

12. Ethiopia's newly formed peasant associations exercised most of these responsibilities in the initial post-revolution period (see Koehn, 1979:58-63).

13. According to Shenton and Watts (1979:55), "agronomists working in northern Nigeria have come to realize that ancient, indigenous planting methods, inter-cropping and crop rotation patterns are well suited to the conditions of rainfall variability which predominate in Hausaland, as well as to the character of local soils" (also see Richards, 1983:41; 1985:16, 70-71).

14. In striking contrast to this approach, Goran Hyden (1980:31) contends that the peasantry must be subjected to even greater exploitation and subordination "to the demands of the ruling classes" in order to appropriate more surplus production for society at large. The contrary position taken in this chapter is that the "uncaptured" nature of the peasantry and its autonomy vis-à-vis other classes (Hyden, 1980:32-33) provides an opportunity for self reliance that must be protected and nurtured through community empowerment. For a critique of Hyden's coercion-oriented approach and Robert Bates' incentive-based agricultural development strategy, see Bernal (1988:89-92, 104-105). Bernal points to the overriding need for "structural changes in access to power and resources ...."

RECOMMENDED READING

The most highly recommended general source on appropriate agricultural policy is Kenneth A. Dahlberg, Beyond the Green Revolution: The Ecology and Politics of Global Agricultural Development (New York: Plenum Press, 1979). One also should consult Paul Richards' extensive review essay "Ecological Change and the Politics of African Land Use," African Studies Review 26 (June 1983):1-72, and his Indigenous Agricultural Revolution; Ecology and Food Production in West Africa (London:Hutchinson, 1985). A thorough case study of World Bank involvement in agricultural projects is Walden Bello, David Kinley, and Elaine Elinson, Development Debacle: The World Bank in the Philippines (San Francisco: Institute for Food and Development Policy, 1982). Also see Cheryl Payer, "The World Bank and the Small Farmers," Journal of Peace Research 16, No. 4 (1979):293-312. Useful studies that accommodate an establishment perspective are Uma Lele, The Design of Rural Development; Lessons from Africa

(Baltimore: Johns Hopkins University Press, 1975) and
David K. Leonard, Reaching the Peasant Farmer; Organization
Theory and Practice in Kenya (Chicago: University of
Chicago Press, 1977).   For a radical perspective, with
several chapters on Nigeria, interested readers should turn
to Peter Lawrence (ed.), World Recession and the Food
Crisis in Africa (London: James Currey, 1986).   Serious
readers also will want to be familiar with Susan George's
Ill Fares the Land; Essays on Food, Hunger, and Power
(Washington, D.C.:   Institute for Policy Studies, 1984).

   On agricultural policy and rural underdevelopment in
Nigeria, see the work of Gavin Williams, Tina Wallace,
Bjorn Beckman, and Hans-Otto Sano cited in this chapter.
One of the best studies, using wheat as its focus, is
Gunilla Andrae and Bjorn Beckman, The Wheat Trap; Bread and
Underdevelopment in Nigeria (London: Zed Books, 1985).   A
detailed, micro-level study of farming systems in
the north, with a special focus on three Zaria villages, is
provided in David W. Norman, Emmy B. Simmons, and Henry M.
Hays, Farming Systems in the Nigerian Savanna; Research
Strategies for Development (Boulder: Westview, 1982).

4

Economic Policy: Mismanagement of Foreign Debt and Structural Adjustment

Public administrators and other policy makers in
Africa are currently occupied with two particularly
pressing obligations: foreign debt repayment and structural adjustment measures. The interrelated concerns which
accompany debt and adjustment do not only present complex
and time-consuming management challenges. The choices made
in dealing with these encompassing issues of economic
policy also are of major consequence in terms of the
distribution of scarce resources and the nature of the
state.

The political economy of international debt and structural adjustment in Africa is the subject of Chapter 4.
The problems involved are immense and the choices are difficult. This chapter identifies the constraining factors,
the hard choices facing African economic policy makers, and
the most promising alternatives. The performance record to
date suggests that mismanagement is the most appropriate
terminology to employ in the subtitle for the discussion
which follows.

While total African debt does not approach the Latin
American figure, it amounts to a heavy burden on some of
the world's poorest economies. From 1974 until the end of
1986, Sub-Saharan African countries increased their foreign
credit obligations from $15 billion to $102 billion.[1] This
sum amounted to 313 per cent of total exports and 70 per
cent of total GNP (Callaghy, 1988:11). A crucial indicator
of a country's economic health is its debt service ratio
(DSR), or scheduled annual payments of international debt
principal and interest as a percentage of the value of
foreign currency received in exchange for exported goods
and services. Africa's current average DSR of 31 per cent
is not projected to decline for the rest of this century

(Callaghy, 1988:12).[2] The extent of the foreign exchange
commitment required by loan repayments has placed many
countries in the position where, in the absence of reschuling, they must decide whether to continue debt servicing
while eliminating imports and public expenditures needed
for economic development, or fall in arrears.

In the context of deepening indebtedness and declining
economic production, the IMF and the World Bank have
advanced structural adjustment as a cure for Africa's ills.
In exchange for new Fund and Bank loans and/or agreements
which are indispensable for rescheduling existing debts and
securing additional loans from private banks,African
governments must first adopt and pledge to implement structural adjustment programs (Callaghy, 1988:13; Brau, 1986:
171).[3] As the IMF's ineffective stabilization schemes
became increasingly unpopular on the continent, the World
Bank stepped in. To countries that reach an accomodation
with the International Monetary Fund, the Bank offers
structural adjustment loans (SALs) with its own macroeconomic conditionalities that are superimposed on and
intended to consolidate those insisted upon by the IMF
(Singh, 1986:113; Loxley, 1986:100; 1985:100; Helleiner,
1983:9; Lancaster, 1988:33).[4] Kenya, Malawi, Senegal,
Tanzania, Ivory Coast, and Nigeria are among the recipients
of SALs. The final structural adjustment package typically
requires steep <u>currency devaluations</u>, <u>wage restraints</u>,
<u>budget reduction</u> measures, <u>trade liberalization</u>, and
<u>privatization</u> of certain public sector activities (Hodges,
1988:22; Browne, 1988:8; Lehman, 1988:8; Serageldin,
1988:54-55; Loxley, 1986:100). The governments of 25
African countries adopted sweeping structural adjustment
programs in the 1980s. Finally, the structural adjustment
approach carries the implicit promise that internal policy
reform measures will encourage new foreign investment and,
thereby, generate economic growth.

The "additional loans with structural adjustment"
prescription has not gone unchallenged. Some critics warn
that the IMF/World Bank approach functions as a trap
designed to ensure further exploitation of the continent by
foreign capitalists. From the dependency perspective,
Africa's underdevelopment can be traced primarily to
intervention by exogenous forces. According to this
viewpoint, which is shared by many African leaders, the
critical need is less for internal reforms than for
external changes (see Browne, 1988:5). The list of
requirements includes debt relief/forgiveness, increased
concessional loans and economic assistance, grants, and

improved terms of trade for primary commodities (Callaghy, 1988:15).

Does structural adjustment offer a feasible short- and/or long-term answer to Africa's economic crisis? Is it, alternatively, aimed at deepening and prolonging dependency, pushing market-oriented economic strategies (Singh, 1986:112), and expanding world capitalism? Is it realistic in the first place to expect that the prescribed structural adjustment measures can be implemented by African public administrators? These rank among the most important questions currently confronting policy makers in Africa.

The following case study is quite revealing in terms of these crucial issues of national economic policy. In the 1980s, Nigeria initially embarked upon its own strategy of coping with foreign-exchange constraints and resisted IMF conditionalities. Later, a new government adopted one of Africa's most far-reaching structural adjustment programs. Nigeria's radically different approaches to the management of this aspect of national economic policy provide the basis for a comparative analysis which yields important lessons for other Sub-Saharan countries.

FOREIGN EXCHANGE POLICY MAKING IN NIGERIA

The military coup d'état which overthrew the Second Republic on New Year's Eve 1983 took place against a background of enormous debts. Nigeria's international financial obligations at the time of the military takeover had reached an estimated $17 billion. The country faced a serious shortage of foreign exchange in the wake of a decade-long spending spree on imported goods and the global oil glut which plagued its petroleum-dependent economy.[5] Manufacturing firms, unable to secure foreign exchange, retrenched increasing numbers of workers owing to shortages of spare parts and processing ingredients (Olukoshi and Abdulraheem, 1985:95; Falola and Ihonvbere, 1988:106).

In this political context, the military regimes which succeeded the Shagari administration faced perceived pressing needs to renegotiate payments due on existing international debts and to secure additional infusions of foreign capital in the form of further loans and invest- ments. Although there are strong constraints on state autonomy in the circumstances Nigeria encountered, federal government policy makers recognized that they possessed some room for independent action regarding the vitally

important issues involved. There are, nevertheless,
significant differences in the approaches adopted by the
two post-coup regimes. This chapter critically examines
the actions pursued by the Buhari and Babangida administrations with respect to foreign exchange, management of the
country's international debt, and structural adjustment.
Then, we analyze the impact of the adopted policies in
terms of Nigeria's political economy. The concluding
sections relate the Nigerian experience to developments
elsewhere on the continent.

The story begins when the Western governments which
insured Nigeria's credits and the International Monetary
Fund insisted that this West African country devalue the
naira by 30 per cent, terminate subsidies on the domestic
sale of petroleum products, sharply reduce state expenditures, sell off state enterprises to private owners, and
liberalize trade restrictions (West Africa, 23 April 1984).
Nigeria's military leaders are acutely aware of the dire
economic and political consequences which tend to accompany
this type of monetary and economic policy package (see
Financial Times, 25 February 1985, p. S 1; Nelson, 1984:
1005; Mittelman, 1987:58). As Gerald Helleiner (1983:
19-20) notes, "with real incomes, particularly urban ones,
already low and having fallen so far, and with the political fragility characteristic of most African states, there
is not much political room for further sharp cutbacks in
levels of consumption, employment, or the provision of
services." Nevertheless, both regimes put foward their own
foreign exchange reform program aimed at qualifying Nigeria
for (1) the rescheduling of extant debts and (2) IMF award
of a new loan in the amount of nearly $3 billion.

In this chapter, we critically evaluate the Buhari
administration's direct exchange-control policy and the
structural-adjustment reforms pursued by the successor
Babangida government in terms of debt liquidation and
economic revitalization. Did Nigeria's three-year standoff
with the International Monetary Fund inadvertently demonstrate that Africa's most populous nation and other Third
World countries that are similarly situated in the global
capitalist economy can avoid deeper international indebtedness and, with the aid of their own strict foreign exchange
controls, boldly embark on a path that fosters healthy
self-reliance?

The short-lived military regime headed by General
Muhammudu Buhari sought to end the debt crisis, curtail
corruption (especially in the acquisition and use of

foreign exchange), and place Nigeria on the path to
self-reliant economic development. In pursuit of the two
short-term goals and the long-term objective of self-
reliance, the Buhari regime introduced a number of
economic-policy measures. Several policies aimed in
general to conserve or ration foreign exchange. More
specifically, they attempted to bring under government
control all foreign exchange accruing to the citizen as a
result of services rendered at home or as a result of
activities, investments, and gifts abroad (see Ojo and
Koehn, 1986:10-14).

The measures adopted, here termed the "1984 Foreign
Exchange Control Regulations," proved to be quite extreme
in terms of their impact on the ability of the ordinary
citizen to engage in international financial transactions.
They principally involve the Exchange Control (Anti-
Sabotage) Decree (No. 7) of 1984 and supplementary regulations. One of these regulations restricts how much
Nigerian currency a traveller may take out or bring into
the country. Another requires immediate official exchange
into naira of any imported foreign currencies or traveller's checks. Yet another reduces maximum salary remit-
tances by expatriates from fifty to twenty-five per cent.
Closure of all foreign bank accounts by public officials --
political appointees, civil servants, university teachers,
administrators in the parastatals --is required by the
regulations, as is the payment of 100 naira as levy on all
travel outside Africa which originates in Nigeria.

The Buhari regime also reduced the total allocation of
foreign exchange to commercial and merchant banks for
apportionment among their various customers to $4.85 bil-
lion in 1984. This amount comprised about 30 per cent of
the foreign exchange needed to keep the economy operating
at the 1983 level (U.S., D.O.C., 1985:9). The government
distributed the total sum to the banks according to their
size and required each bank to allocate its share according
to an end-use formula which reserved 58 per cent for industrial raw materials, spare parts, and equipment, 18 per
cent for food imports, 12 per cent for other consumer
goods, and 12 per cent for invisibles such as foreign
travel, education, medical services, and repatriation of
expatriate salaries (West Africa, 20 February 1984, p.
395).

The overthrow of the Buhari government in 1985 led to
fundamental changes in Nigeria's monetary policy. Major
General Ibrahim Babangida raised the issue to center stage
by immediately launching a nation-wide debate on whether or

not Nigeria should accept and implement conditions laid
down by the International Monetary Fund in order to secure
additional credit from that organization and obtain debt
rescheduling. In the midst of the world oil glut and con-
tinuously deteriorating foreign exchange earnings, participants in the debate overwhelmingly rejected both the loan
and IMF conditionalities.

Nevertheless, the Babangida regime proceeded to introduce the chief components of the IMF plan at the same time
that it publicly rejected the most resented requirements
and kept the Fund in the background by arranging for the
World Bank to monitor compliance. The government adopted
an ambitious structural-adjustment program in June 1986.
In exchange, Nigeria received access to new loans and support for the rescheduling of its existing debts (Tallroth,
1987:20). The decisive step in implementing structural
adjustment occurred on 26 September 1986, when the Second-
tier Foreign Exchange Market (SFEM) took effect and the
value of the naira plunged dramatically.

Impact of the Buhari Foreign Exchange Policy

The issue of devaluation is particularly volatile in
Nigeria. Instead of official and abrupt change, therefore,
most administrations have preferred to allow the value of
the naira to slide. In 1980, the official rate for the
naira came to $1.80. Over the next four years, the federal
government quietly and gradually reduced its value to $1.12
by altering the mix in the basket of currencies to which
the naira is pegged. The Buhari regime further contended
that strictly enforced foreign-exchange regulations are
more likely than devaluation is to achieve the stated
objective of revitalizing the local economy -- provided
that the new loan would be forthcoming and existing debts
rescheduled.

Nigeria's foreign exchange control regulations had an
immediate impact on its balance of payments situation and
on the domestic economy. First, the 1984 regulations
caused problems for those individuals needing foreign
currency and shortages for the society at large. In
addition, the foreign-exchange regulations, together with
changed procedures whereby all imported goods came under
specific licenses, further reduced imports. Austerity
measures and closed lines of credit on account of rising
debt already had led to a reduction of imports from the
all-time high of about $17 billion in 1981 to $13 billion

in 1983. During the first year of the new regulations, in
spite of some carryover 1983 licenses, the import bill fell
to $11.7 billion -- a decline of 10 per cent from 1983.
For 1985, the government imposed a $3.5 billion ceiling on
imports. The new ceiling, a 70 per cent reduction from
1984, led to shortages of many products and sizeable price
increases for numerous household commodities (West Africa,
20 February 1984, p. 395; 14 January, 1985; U.S., D.O.C.,
1985:3-4, 9).

The sharp reduction in imports affected the economy
and the populace in other ways. Many manufacturers faced
shortages of raw materials and spare parts as a result of
receiving only 10 to 15 per cent of their requested imports
after waiting months for the licenses. Some closed down;
others operated at between 30 and 50 per cent capacity
(U.S., D.O.C., 1985:3-4). This accentuated commodity
shortages and the price spiral.

Increased unemployment turned out to be another short-
term consequence of the 1984 foreign-exchange regulations.
As factories closed or operated at lower levels of production, thousands of workers found themselves laid off. On
its own initiative, but coincident with conditions for the
multi-billion dollar IMF loan under negotiation, the Buhari
regime reduced government expenditures by withdrawing subsidies for health and education services at all levels and
embarked on retrenchment in the public sector (West Africa,
15 October 1984, p. 2069; Olukoshi and Abdulraheem,
1985:96). With the exception of a few departments and
parastatals, the government mandated 15 per cent personnel
reductions in the federal and state public services.

Although the 1984 foreign-exchange regulations
involved short-run hardships and sacrifices, they also
exerted a salutary impact on the economy that augured well
for self-reliant economic growth and development in
Nigeria. The performance of certain major firms after the
new foreign exchange rules came into force provided one
cause for optimism (Ojo and Koehn, 1986:18-19).[6] The
light at the end of the debt tunnel is the production and
processing of raw materials domestically and the
installation of spare-parts factories. It is by using
local resources and fabriciating one's own spare parts that
the transformation toward self-reliant development begins.
In an encouraging development, therefore, Nigerian firms
started to rely on domestic raw materials or experimented
with locally available products.

Furthermore, the Buhari administration's foreign-
exchange regulations helped to cushion Nigeria's debt

crisis. The country's overseas debt at the beginning of
1984 included approximately $10 billion in medium- and
long-term loans held by the government and about $7 billion
in overdue short-term credits (plus interest) for goods and
services already supplied. Uninsured obligations to
overseas export firms constituted roughly $5 billion of the
total foreign trade debt; the national governments (mainly
OECD countries) of the creditor companies guaranteed the
balance (see Economist, 3 May 1986, pp. S 6-7). Under
Buhari, Nigeria sought both to reschedule these loans and
to secure an additional capital infusion of between $2.5
and $3 billion from the IMF. Strategists maintained that
without the IMF loan, Nigeria would be unable to meet
rescheduled repayment terms on its short-term loans.

At first, negotiations for rescheduling these debts
bogged down largely because government officials in the
creditor states (led by Britain's Export Credit Guarantee
Department) insisted that Nigeria reach agreement with the
IMF on pre-conditions labelled an "economic recovery
program" (West Africa, 20 February 1984, pp. 373-375; 23
April 1984, pp. 866-867). They also insisted that the
negotiated terms of the agreement apply to the uninsured
creditors. Eventually, however, the principal uninsured
creditors, among them the Mobil Oil Corporation and ITT,
gave in. This allowed Nigeria to reschedule about $3.5
billion in debts without the necessity of first reaching
agreement with the Fund and opened the way for rescheduling
the uninsured debt owed to other companies.

The insured $2 billion debt remained subject to
reaching prior agreement with the IMF. On 5 October 1984,
both sides negotiated a face-saving "secret debt accord" in
Paris.[7] The terms of this agreement were identical to
earlier settlements with uninsured creditors.[8] The accord
provided that principal on these loans would be stretched
over six years, with repayment to begin after two and a
half years. Interest payments commenced on 1 January 1985,
with the interest rate set at one percentage point above
the rate that London banks charge each other (Wall Street
Journal, 24 October 1984).

The deadlock still existed with the IMF over the
latter's more demanding conditions for the $2.5-$3 billion
loan (Bangura, 1986b:31). Consequently, Nigeria did not
receive the additional foreign exchange which the Buhari
administration had been counting upon.[9] Nevertheless, at
the rate of $300-$400 million per month, the government
continued to repay both the rescheduled bona fide

short-term obligations and its non-rescheduled medium- and
long-term debt to Western financial institutions.[10]
    By the end of February 1985, the country's external
liabilities had decreased to about $15.1 billion from $16.1
billion in December 1984 and $17 billion at the end of
1983. Foreign-exchange reserves totalled $1,365 million in
February 1985, compared with $1,285 million the previous
December and less than $1,000 million at the time of the
coup in December 1983 (West Africa, 6 May 1985, p. 905 and
13 May 1985, p. 958). In essence, Nigeria had implemented
its own stabilization program without relying on the IMF,
or being bailed out by other external sources of capital.
    Nigeria held firm to this economic policy position
until the overthrow of the Buhari regime in 1985.[11] The
government resisted global pressures to abort its experiment. It convinced the politically active strata that
their sufferings and inconveniences were temporary and
necessary for a better future, and that the situation would
be worse under the IMF's prescription. In Finance Minister
Onaolapo Soleye's words, "'if you go for the Fund, you can
expect more and more stringent tightening of the economy'"
(West Africa, 15 October 1984, p. 2069).

Developments Under the Babangida Policy

    The wide support which the Buhari regime's position
enjoyed (West Africa, 4 June 1984, p. 1152) and the
intensity with which Nigerians resented the IMF's patroniz-
ing attitude and unsuitable conditions for securing a loan
surfaced early in the successor Babangida administration.
In one of his first actions, General Ibrahim Babangida
promised to resume negotiations with the Fund over the
proposed loan (see Blackburn, 1986:18) and to move from
"'austerity alone to austerity with adjustment'" (Tallroth,
1987:20). He personally favored accepting the conditionalities and loan. In early moves which manifested these
inclinations, the new Head of State appointed a former
official of the IMF, Dr. Kalu Kalu, as Finance Minister and
stacked the newly established "Committee on the IMF Loan"
with appointees who favored accepting the money.[12]
    Expecting that key political actors in Nigeria would
be unwilling to tolerate for long the hardships which the
extant foreign exchange-based approach entailed, the new
Head of State initiated a vigorous national debate on the
issue in various public fora and on the pages of the
newspapers. However, Nigerians overwhelmingly repudiated

the IMF conditionalities and loan (see, for instance,
National Concord, 20 and 21 September 1985, pp. 6, 1;
Bangura, 1986a:56). They indicated a willingness to
tolerate hardship for as long as necessary in order to
restore the economy by self-reliance. In the face of the
nearly unanimous public opinion expressed in the debate it
had inaugurated, the government officially abandoned loan
negotiations with the International Monetary Fund on
13 December 1985 (Washington Post, 14 December 1985).

While respecting the public's rejection of an
agreement that would approve of the IMF's terms, the
Babangida regime, nonetheless, proceeded to adopt most of
the Fund's prescriptions for revamping the economy. It
partially withdrew the state subsidy for the domestic sale
of petroleum.[13] This action nearly doubled the pump price.
The government also dismissed more public servants and
trimmed wages and salaries by between 2.5 and 20 per cent.
Later, it reduced the import prohibition list from 74 to 16
items (West Africa, 12 October 1987, p. 2023). Finally,
the Babangida administration allowed the official value of
the naira to slide further (African Guardian, 16 January
1986, p. 19; Economist, 3 May 1986, p. S 12). By mid 1986,
the naira only drew 85 U.S. cents.

The sudden tumbling of petroleum prices to under $10
per barrel during the first half of 1986 put further
pressure on Nigeria to negotiate debt rescheduling. The
country's western creditors continued to insist upon IMF
guarantees (see Washington Post, 14 January 1986). Thus,
the oil glut forced Nigeria back to the IMF's doorstep and
to additional conditionalities. The Fund demanded the
privatization of government-owned enterprises in conformity
with its long-standing policy vis-a-vis debtor Third World
countries and the Reagan administration's campaign to
transform Africa into free-market economies. The Nigerian
government, broke as a result of events in the world-wide
petroleum market, obliged the IMF. It dissolved some
companies, began a program of gradual divestment in others,
and removed subsidies to monopoly parastatals which, in
turn, milked their captive customers (Nigeria, Federal
Government, 1987:10).

On the other hand, in a challenge to Nigeria's
international creditors, General Babangida unilaterally
established a ceiling on the amount of foreign-exchange
earnings which would be devoted to servicing the country's
external debt beginning in 1986. Most estimates had
indicated that Nigeria's debt repayments would consume
nearly 60 per cent of total foreign-exchange earnings in

each of the next five years. The Babangida regime set the
ceiling on servicing medium- and long-term loans at half of
this level; i.e., at 30 per cent (Washington Post, 5
January 1986 and 14 January 1986; African Guardian, 16
January 1986, pp. 13, 15). The low price of petroleum on
the world market during most of 1986 dramatically undercut
Nigeria's capacity to repay its foreign debt in any
event. According to Terisa Turner (1986:38), the
government serviced 28.9 per cent of the total debt of
$18.4 billion in 1986 and projected a debt service ratio of
21.6 per cent in 1987.

Under such circumstances, rescheduling is imperative
for creditor and debtor alike. Nigeria proved far from
powerless in this situation. In the first place, de facto
"rescheduling" occurs in the absence of a formal agreement
should a country be unable or unwilling to meet its
obligations. This is precisely what happened in 1986, when
Nigeria unilaterally declared a moratorium on debt
repayment for nine months (Christian Science Monitor, 7
October 1986, p. 13). Moreover, the extremely precarious
state of the country's economy highlights the shared
interest of all parties in Nigeria's economic survival and
revival. In the words of one analyst, "Nigeria can only
pay her external debts if she is economically alive and
kicking" (African Guardian, 16 January 1986, pp. 18, 15).[14]
Thus, by 1987, the President of the World Bank, R. Barber
Conable, referred to Nigeria's "creeping default" strategy
of limiting debt service interest payments to 30 per cent
of the country's export earnings as "'reasonable.'" He
also remarked that "'the bank acknowledges that if a
country's debt service ratio exceeds 25 per cent, it
becomes very difficult to get enough additional resources
so that there can be growth'" (Thisweek, 7 September 1987,
p. 35).

Devaluation and SFEM

Devaluation of the naira constituted the most
far-reaching IMF condition. Since the Nigerian public
resoundingly rejected outright devaluation, the Babangida
regime could not officially accept this condition.
Instead, it opted to achieve the same end by creating a
Second-tier Foreign Exchange Market (SFEM) in which
domestic banks are free to buy and sell foreign currency at
rates determined by market forces of supply and demand.
Most public sector and all private transactions would be

channelled through the new market.[15] The scheme opened on 26 September 1986, with the naira exchanging for about 22 U.S. cents (₦4.62=$1). In one week, the naira had declined in value by 70 per cent against the dollar.[16]

The SFEM alternative to outright devaluation proved acceptable in principle to the IMF, to Nigeria's international creditors, and to most Nigerian policy makers. From the beginning, the Fund and the creditors viewed SFEM's existence as the functional equivalent of an outright devaluation which would deflate the value of the naira far in excess of what the IMF originally asked for. They also saw this step as introducing freer trade since it would make the stringent import controls and the 30 per cent generalized import levy unnecessary.

Nigerians accepted SFEM, in principle, as the lesser of two evils. They foresaw an end to the import levy, import licensing, direct foreign-exchange allocation, and the accompanying bureaucratic bottlenecks and corruption. Furthermore, SFEM essentially replaced the Exchange Control (Anti-Sabotage) Decree of 1984. The latter's iron-fisted approach to foreign-currency management had made it nearly impossible for people to remit small amounts abroad, had encouraged a thriving black market as well as international drug trafficking, and had resulted in long jail terms for minor foreign-exchange offenses.

Some proponents even proclaimed that SFEM would knock a hole in Nigeria's notorious import-based consumption pattern by raising the exchange rate to such a high level as to discourage frivolous importations as well as to force import-dependent industries to seek alternative domestic sources (Guardian, 13 July 1986, p. 9). These advocates of the new approach anticipated that a fundamental structural reorganization of the Nigerian economy would occur provided that the citizenry could be persuaded to tolerate the prevailing harshness of life for long enough.

Immediate Impact of Structural Adjustment

Nigeria's experience with structural adjustment has been influenced by external pressure, internal debate, and the dramatic drop in the price of a barrel of crude. Citizens and policy makers have unquestionably gained greater awareness of the negative domestic consequences associated with IMF "remedies." Nevertheless, the Babangida government had set virtually all of the Fund's conditions in place or in motion within Nigeria by the end of 1986. Acting alone and confronting low prices for crude oil on the world market and a receptive new military regime, the Nigerian public proved to be no match for the unified counter attack mounted by the country's creditors. The post-coup measures introduced by Nigeria's latest military regime are more accurately described as capitulation to IMF and creditor demands than as independent economic policy making.[17] They effectively reversed the previous government's efforts to replace Fund conditions and loans with strict foreign-exchange controls.

The Babangida regime sold SFEM as a milder, market-imposed rather than government-mandated "alternative" to official devaluation, the main IMF conditionality, and to the 1984 regulations. The short-term success of the scheme rests upon the government's ability to cushion the inevitable inflationary spiral, plug all possible loopholes that could lead to abuse of the market, reduce to the barest minimum the incidence of corruption, curtail public spending, and hold down wages through state intervention in the labor market.

The results of the Babangida regime's structural adjustment shock treatment have been far less salutory for the country's economy than the predictions suggested. In the wake of SFEM, imported inputs became so expensive that industries operated at less than one-third of capacity. Unemployment remained high, and the cost of manufactured goods (both domestic and foreign) increased several fold (National Concord, 24 September 1987, p. 3; Tallroth, 1987:22). New foreign private investments outside of the oil sector have not been forthcoming and are not on the horizon (New York Times, 23 November 1987, p. 6). Other concerns center on the ubiquitous corruption that undermines the formal objectives of the scheme and on the uneven social impact of structural adjustment.

The new policy allowed the foreign-exchange market to be cornered by a few financially powerful organizations and individuals, including commercial banks -- the officially designated dealers. Given that commercial banks are among the most corrupt institutions in Nigeria, their pivotal role in SFEM is suspect to begin with (see West Africa, 1-7 May 1989, p. 678; Nigeria, Political Bureau, 1987:58; also see Makgetla, 1986:420 on Zambia). In an effort to allay this fear, curtail speculative buying, and protect financially less-powerful organizations, the Central Bank decided to sell foreign exchange to the commercial banks at the rate they individually bid, while permitting them to resell at a specified level marginally above the buying

rate. Even so, SFEM unleashed pent-up demands for foreign
exchange by large firms hitherto compelled by import
licensing and foreign-exchange allocations to operate below
capacity and by multinational corporations which had huge
sums tied down in unremitted profits. Some commercial
banks colluded with these companies to make SFEM a conduit
for siphoning off foreign exchange, thereby facilitating
massive capital flight and the elimination of small- and
medium-scale enterprises which lack the financial strength
and connections that count most in a market system. It is
not surprising, therefore, to find that commercial banks
are making money from SFEM. The First Bank of Nigeria,
Union Bank, and United Bank for Africa all declared even
more massive profits in 1986 than they had in the two prior
years (see Business Concord, 4 July 1986, p. 2 and 12 May
1987, p. 22).

Another basis for skepticism lies in the demonstrated
incapacity of Nigerian governments to implement policies
that defy corruption and in the susceptibility of the
Nigerian marketplace to speculation and unscrupulous
manipulation (see Guardian, 1 August 1986, p. 9). This
pessimistic perspective stems from the corrupt practices
engaged in by the makers and executors of public policy
themselves and the prevailing social values and business
standards which accommodate such behavior.[18]

Structural adjustment programs tend to be mismanaged
in ways that are attributed to lack of administrative
discipline and competence (see Frisch, 1988:71) as well
as to pervasive corruption. Budget expenditure controls
have presented a particularly difficult condition to adhere
to in practice (Nelson, 1984:991-992; Callaghy, 1988:14).
According to the World Bank economist assigned to Nigeria,
implementing fiscal restraint has been the most problematic
part of the structural-adjustment program for that country
(Tallroth, 1987:21). Western analysts are particularly
distressed by the 58 per cent increase in capital expenditures included in the federal government budget for 1988
and by the mammoth outlays on construction of a steel
manufacturing complex and on the new federal capital of
Abuja. The IMF remains dissatisfied over the revenue loss
resulting from the government's current subsidy (roughly
80 per cent) on the domestic sale of petroleum (Washington
Post, 13 June 1988; Africa Analysis, 15 April 1988, p. 5;
also see New York Times, 27 September 1988, on Zaire's
budget deficit). Across Africa, governments have managed
to engage in public sector personnel cuts and/or wage
reductions (Frisch, 1988:68; Fromont, 1988:94; Nsingo,

1988:82). However, these actions have not been accompanied by demonstrated improvements in performance or increases in administrative capacity and efficiency (see Nelson, 1984:994).

Adoption of a structural-adjustment program did produce some movement on the external-debt front. The introduction of SFEM paved the way for new arrangements with Nigeria's external creditors. In November 1986, the government reached a rescheduling accord with representatives of its commercial bank creditors that included a ten-year postponement in complete repayment of the $1.5 billion in principal originally due between 1 April 1986 and 31 December 1987 and a four-year grace period that only required interest payments. As a result of this debt rescheduling arrangement, Nigeria received $320 million in new foreign-trade loans from commercial banks[19] and a $452 million foreign-trade promotion and export-development loan from the World Bank (Wall Street Journal, 16 October 1986; New York Times, 22 November 1986).

Within two months, moreover, the Bank had promised to grant Nigeria $4.3 billon in project loans over the next three years (West Africa, 12 January 1987, p. 47; Thisweek, 7 September 1987, pp. 34-37; Christian Science Monitor, 17 November 1987, p. 14). In January 1987, the International Monetary Fund approved a stand-by arrangement of $830 million in support of Nigeria's structural-adjustment program. Content to allow the World Bank to stand in for the discredited IMF in Nigeria (see Payer, 1986:659-676; Cline, 1987:43-45), the Babangida regime declared its intention not to draw on these stand-by funds (Tallroth, 1987:20).

CONCLUSIONS AND LESSONS

There are several advantages to the Buhari administration's short-lived tight foreign exchange control strategy over the IMF's prescription for economic prosperity. The principal benefits of foreign-exchange strictures over SFEM in terms of promoting national economic self-reliance are (1) the former can be selectively employed to control access by type (e.g., replacement parts and/or new industrial machinery can be exempted), while the latter distributes foreign currency primarily according to individual or corporate purchasing power on the open market; (2) the first approach is

predicated upon domestic resourcefulness, whereas SFEM
facilitates and promotes foreign penetration via corporate
investment and new loans; and (3) the independent foreign
exchange control strategy involves no obligations or
pressures to implement other odious conditions contained in
the typical IMF package or in the World Bank's structural-
adjustment program. Under SFEM, in sum, Nigeria faces
little prospect of rectifying the entrenched structural
problem of dependency which afflicts its ailing economy.

There is no longer any question that the regeneration
of African economies is retarded by structural adjustment.
In the first place, structural-adjustment programs fail to
address the underlying constraints on small-farmer
production in Africa. In particular, they do nothing to
rectify the inequitable distribution of basic inputs or to
reverse the promotion of inappropriate agricultural
"development" schemes (see Chapter 3).

Furthermore, the IMF's devaluation requirement is
defended on the grounds that it makes local products
cheaper, leads to increased exports, and thus prompts
higher production. At least one prominent spokesman for
the Nigerian business community, Dr. Michael Omolayole,
Chairman of Lever Brothers, supported devaluation on the
grounds that it would make locally manufactured goods
competitive on the world market (Guardian, 21 August
1984). The first problem with this argument is that it
only works for an industrialized, export-oriented economy
that possesses the capacity to penetrate highly competitive
foreign markets (Loxley, 1986:96; 1985:121). Like most
other African countries, however, Nigeria does not possess
an industrialized economy and exports no manufactured
products. Its chief export, petroleum, already was priced
in dollars. Devaluation of the naira brought no benefits
to Nigeria in terms of increased sales of crude oil. Most
of the country's main potential agricultural exports,
commodities such as rubber, palm kernal, and palm oil, take
years to cultivate and, therefore, are not likely to show a
substantial response immediately to reductions in the
naira's value, or even to the more beneficial abolition of
the country's exploitative marketing boards. Moreover, if
all debtor Third World countries simultaneously succeed in
following the standard IMF prescription and expand their
exportation of the same or similiar crops, the result would
be "a fall in commodity prices ..." and the perpetuation of
dependency (Singh, 1986:110; Usman, 1986:98; Koester, et
al., 1988:1,4; Makgetla, 1986:418; Saitoti, 1986:28-29;
Hyden, 1986:18).

Accepting the Fund's devaluation shock therapy means that industrial machinery and spare parts will have to be imported at far greater expense by countries intent on developing their productive capacity and breaking out of their dependent economic position (Ojo, 1985:167; Okongwu, 1987:1962; Singh, 1986:111-112; Hodges, 1988:24). In order to avoid starvation and to feed rapidly growing populations at required nutritional standards, African countries also are forced to continue to import a major proportion of their food needs regardless of the increased cost. Sustained self-sufficiency in food production is not on the horizon for Nigeria and many other African countries (Shaw and Fasehun, 1980:561; Africa Analysis, 11 July 1986, p. 4).[20] Indeed, higher prices for cash crops induce farmers to shift out of food-crop production (Hyden, 1983: 197; Fantu Cheru, 1987b:2).

One result of devaluation throughout Africa has been higher inflationary expectations and spiraling domestic prices, especially for food. For instance, Nigeria continues to provide a seller's market for imports owing to dependency and oligopoly. The principal beneficiaries of structural adjustment have been the foreign firms that trade with Nigeria or seek to remit profits from the country, some local traders, and the banks. Rural cash-crop producers also have received higher prices for their commodities, while small-scale, food-crop producers remain disgruntled.[21] In general, devaluation "leads to a redistribution of income away from those who produce for the domestic market to those who are involved in production for exports ..." (Singh, 1986:107). It also reduces real wages and overall consumption. In Nigeria, annual per capita income fell from $800 to $375 under General Babangida's economic policies, prompting the World Bank to downgrade the country's classification from "middle" to "low" income (Brooke, 1988:D4). In sum, the working class, the 5 million unemployed job seekers, and poor urban dwellers as well as those dependent upon fixed incomes have assumed the heaviest losses imposed by structural adjustment, rising food prices, and the unprecedented collapse of the standard of living (Blackburn, 1986:20; Nigeria, Political Bureau, 1987:37; Bangura, 1986b:35; also see Nelson, 1984:995-996; Parfitt and Riley, 1986:522; Fantu Cheru, 1987b:53; Fromont, 1988:94).

The IMF demand that a loan recipient liberalize its external trade policy has been aptly described by a former finance minister in Nigeria as "'an invitation to commit suicide'" (West Africa, 22 October 1984, p. 2113).

Although luxury spending is a common problem (see Makgetla, 1986:403), Nigeria's propensity to import foreign manufactures may be the highest in Africa. Relaxation of import restrictions allows the renewed importation of luxury products such as automobiles, refrigerators, VCRs, television and stereo sets, lace and polished cotton cloth, fashionable shoes, etc. -- in place of industrial machinery. Domestic import-substitution industries, whose products Nigerians shun in favor of "the original" (identical, but foreign-made), suffer further decline and produce even higher rates of unemployment (Okongwu, 1987: 1962; Parfitt and Riley, 1986:525; Mittelman and Will, 1987:66; Mawakani, 1986:110). This outcome serves to depress wages -- a fortuitous consequence for foreign and domestic capitalists (Mittleman and Will, 1987:56, 68; Usman, 1986:97).

In contrast, Nigeria's 1984 foreign-exchange regulations, coupled with restrictive controls on imports, ensured that essential raw materials, equipment, spare parts, and food consumed scarce foreign exchange. In this sense, the approach initiated by the Buhari regime promised to constitute a more effective and equitable method of restraining imports and made more economic and political sense for Nigeria than did the IMF's shock-treatment emphasis on drastic devaluation of the local currency (Economist, 3 May 1986, pp. S 8,12). Elsewhere in Africa as well, "closer examination of the external deficit suggests the need, not to abolish government control, but to make it more rigorous and consistent in directing foreign exchange to meet development needs" (Makgetla, 1986:415; 1988:26-27).

In conclusion, the Buhari administration's foreign exchange control strategy contained some of the elements needed to set Nigeria on a self-reliant economic path. It is difficult to envision such potential in the Babangida policy. SFEM and structural adjustment invite increased direct foreign investment. New loans and rescheduling, designed in part to finance food and other imports (Christian Science Monitor, 7 October 1986, p. 13), show little promise of leading Nigeria away from a state of perpetual debt crisis.

Other Experiences With Structural Adjustment

The available evidence from other African countries is consistent with the Nigerian experience. For instance,

Kenya's orthodox structural adjustment program has produced
escalating international indebtness without generating
economic growth. By 1987, Kenya had to devote more than 40
per cent of its export earnings to servicing its $6 billion
debt. At the same time, the government implemented severe
cuts in social-service expenditures and refused to consider
matters of economic redistribution. As a result, the
principal burden of adjustment fell on the poor (Lehman,
1988:7-9, 14-15).

After seven years of IMF involvement in Sudan's
economy and multiple loan reschedulings, the country needed
all its export earnings to pay the interest on its debt and
had been declared ineligible for further assistance by the
IMF for falling more than one year in arrears in its
payments. The social impact of structural adjustment in
Sudan has been "catastrophic" (Fantu Cheru, 1987:5-6).
Zaire closely adhered to an IMF program from 1983 to 1986.
During that time, the cost of debt service grew from 10 per
cent to 50 per cent of its budget and development
expenditures declined dramatically (New York Times, 27
September 1988).

At one time, the World Bank singled out Zambia as a
particularly promising performer (Clausen, 1985:8-9).
However, as Neva Makgetla (1986:396-411, 415-421) points
out, the standard IMF remedy applied there "conflicted with
overall development goals" and "incorporated virtually no
attempt to minimize the cost to the poor majority of
Zambia's population." The outcome of increased reliance
upon market forces included reduced wages for poor
laborers, "a persistent capital outflow, a bias towards
production [and importation] of luxuries, and a failure to
develop the peasant and small-scale manufacturing sectors."
At the same time, heavy borrowing from international
creditors at high interest rates led to rising net capital
outflows and growing dependence on multinational
corporations (Makgetla, 1988:15-17). Zambia's DSR reached
105 per cent in 1986 (Nsingo, 1988:80). When implementation of drastic food-price increases led to riots and
strikes, President Kenneth Kaunda cancelled the reform
measures and changed his team of economic advisors
(Lancaster, 1988:33). After ten years of IMF stabilization
and World Bank structural adjustment, the Zambian
government rejected both in 1987, asserting that these
economic policy prescriptions had wrecked the national
economy without providing debt relief.[22]

In place of structural adjustment, Zambia decided to
pursue a self-initiated program "based on development of

our own resources" (Kaunda, 1988:76). The government
resumed the exercise of state control over foreign-exchange
allocation, severely restricted most imports except for
"machinery required for increasing selective capacity
utilisation" (Nsingo, 1988:84), raised the corporate income
tax, and set a 10 per cent DSR ceiling "until we gain
strength enough to be able to pay back what we owe"
(Kaunda, 1988:76; Rule, 1987). Most of these measures are
similar to those pursued earlier by the Buhari regime in
Nigeria. The World Bank retaliated by suspending a $400
million loan commitment, the IMF cancelled its agreement,
and additional lending by foreign creditors came to an
abrupt halt (Makgetla, 1988:18-19; Lancaster, 1988:33).[23]
Nevertheless, both the rate of inflation and Zambia's
balance of payments deficit declined and the utilization of
manufacturing capacity increased (Nsingo, 1988:84; Maine,
1988).

Zimbabwe offers one of the few examples of an African
country which has successfully resisted structural
adjustment. The government has refused to allow dramatic
devaluation of its currency, has vigorously defended its
unwillingness to limit the expansion of health and
education projects, has increased wages while freezing
prices,[24] and has rejected other IMF and World Bank
conditionalities. On its own, Zimbabwe has restricted
imports, strictly controlled foreign exchange, and employed
"tools of state economic management," including dividend
and remittance controls, "in an attempt to balance economic
growth with redistribution" (Lehman, 1988:18, 20-23). By
steadfastly repaying its loans to foreign creditors and
refusing to reschedule its external debt, Zimbabwe has
avoided escalating indebtedness and the need to adopt IMF
conditionalities. The drawback of this strategy is a
severe shortage of foreign exchange for the domestic
economy (Lehman, 1988:21-23).

By the end of the 1980s, the International Monetary
Fund and the World Bank were hard pressed to find reliable
African examples of economic recovery and sustainable
development associated with embarking on structural
adjustment (Fantu Cheru, 1987:4-5; Lancaster, 1988:32;
Callaghy, 1988:14).[25] As the Director-General for
Development at the European Economic Commission points out,
the 25 Sub-Saharan countries working on structural-
adjustment programs with IBRD and IMF help "have managed
little in terms of economic performance so far" (Frisch,
1988:68; also see Loxley, 1985:119). New investments have

not been forthcoming, maintenance of existing facilities is
not assured, and urban poverty is spreading rapidly in the
wake of inflation, dwindling per capita income, and
mounting unemployment (Frisch, 1988:68, 70; also see
Callaghy, 1988:11; Killick and Martin, 1989:3-5). The
International Labor Organization estimates that per capita
incomes in Africa were 7 per cent lower in 1987 than in
1978, that urban unemployment rose from 10 per cent in the
1970s to 30 per cent in 1985, and that Africa might be
experiencing the beginning of a large-scale
de-industrialization process (Fromont, 1988:94). In sum,
"after a decade of increasing IMF involvement, Africa has
fallen deeper into crisis" (Hardy, 1986:466).

The IMF and Economic Nationalism

It is clear from both the detailed case study and the
general discussion presented in this chapter that the
International Monetary Fund's prescriptions have not been
associated with sustained national economic growth and
equitable distribution of material goods and services in
Africa.[26] Indeed, the most noticable outcomes have been
widespread "adjustment fatigue," antipathy toward the IMF,
and domestic political turmoil (see Frisch, 1988:69, 71;
Fantu Cheru, 1987:5; Callaghy, 1988:15). Why is it that
the Fund insists upon the implementation of structural
reforms which only lead to the further deterioration of
conditions in adopting Third World countries? One answer
to this question can be found in the organization's raison
d'etre. The IMF is designed to maintain the restructured
international trade system and to stabilize the
international monetary system. This dual task is
ostensibly economic, but in essence political. Since 1945,
the Western powers which have dominated the global economy
have viewed the IMF as a political instrument for the
propagation of capitalism. External capital is to be
stimulated to flow into "developing" economies through
trade liberalization and the creation of a favorable
climate for direct foreign investment (Spero, 1981; Harris,
1986:88-95).[27]
The IMF promotes the flow of private foreign capital
by insisting on certain types of internal reform programs
and policy measures that rely on free enterprise economic
principles. Its typical stabilization program involves
both economic and political requirements designed to

curtail anti-capitalist propensities and economic
independence. The economic components, usually including
favored treatment of foreign private investment and
devaluation of the local currency as well as liberalization
of trade, are aimed at "de-nationalizing" the recipient
economy and making it dependent, via transnational
corporations, on Western capitalist institutions. The
political components, typically including reductions in
government expenditure and intervention in the economy,
elimination of state subsidies, reorganization of public
enterprises, trimming of bureaucracy, and termination of
barter trade (frequently with the Soviet Union), are
intended to "de-socialize" the economy (Hutchful, 1981;
Mittelman and Will, 1987:55).

The underlying economic policy issue at stake in these
international negotiations concerns the economic path which
African and other Third World countries will follow. The
Fund and the other external creditors demanded that Nigeria
open its doors to further penetration by foreign capital,
particularly transnational agribusiness corporations. In
the blunt terms of The Economist (3 May 1986, p. S 8),
foreign exporters "have made huge fortunes out of Nigeria,
and would like to make more." The World Bank also
continues to supply Nigeria with loans designed to
"stimulate growth in private industry" and to promote
inappropriate and discredited agricultural production
projects (see Daily Times, 20 September 1985, p. 12; Africa
Analysis, 11 July 1986, p. 4). The "market forces" which
such institutions extol are, in Bala Usman's terms (1986:
114-115, 110) "decisively external and structured to serve
the transfer of the wealth of ... [Nigeria] abroad." The
policy reforms exported by the IMF and the World Bank serve
to entrench African countries in their current dependent
roles as providers of primary products and importers of
manufactured items (see Usman, 1986:98; Browne, 1988:9;
Loxley, 1985:127).

Ironically, its essentially political ends, to be
achieved by economic means, create dilemmas for the IMF.
It must insist upon programs that it knows are injurious to
the domestic economy of Third World countries that have
accumulated vast outstanding debts to Western financial
institutions. One IMF official acknowledged that Nigeria's
devaluation controversy "'can raise a lot of problems for
the government,'" and that "'it is a dilemma, but there is
no other alternative'" (Washington Post, 9 March 1984).
In fact, there are other options. One of them, the foreign

exchange control based approach, undermines the IMF's ability to dominate a country's economic policies in the short run and, in the long run, threatens to enhance self-reliant development and independence from Western transnational corporations.

The prospect of successful economic adjustment outside the IMF framework clearly is not in the interest of the Fund, the Bank, and the Western powers (see West Africa, 22 October 1984, pp. 2113-2114). Such an outcome would pose a major threat to the IMF's authority elsewhere in the Third World and reduce the leverage of Western institutions concerned with enforcing a dependent "solution" for debt crises. This explains the pressures placed upon Nigeria and other African countries to accept a package of reforms which exacerbate the existing economic crisis and perpetuate dependence (see Guardian, 27 August 1984; New Nigerian, 21 September 1985, p. 5; West Africa, 22 October 1984; Nigerian Tribune, 29 March 1986, p. 1; Gruhn, 1983: 45).[28]

Structural Adjustment and National Economic Policy Making

Structural adjustment has had a major impact on national economic policy making in recipient countries. In desperate straits due to diminished access to foreign exchange and pressing debt-rescheduling obligations, most African governments proved unable to resist an "unprece-dented seizure by the donor community of the levers of policy ..." (Browne, 1988:8; Lancaster, 1987:226). This situation developed in part via alliances which IMF execu-tives cultivated with Ministry of Finance and Central Bank technocrats responsible for carrying out and monitoring policy reforms (Lancaster, 1988:33; Loxley, 1986:100).[29] Nationals working in these key economic agencies frequently defer to IMF decision-making (see Lehman, 1988:12 on Kenya). In other cases, such as Zaire, "the IMF and the World Bank went as far as placing their own teams of expatriates into the Central Bank, the finance and planning ministries, and the customs office" (Callaghy, 1987:107). The Babangida regime's enthusiasm for structural adjustment can be better understood when one considers that "currently economists from Washington, D.C. virtually run Nigeria's Central Bank and finance ministry" (Turner, 1986:34-35). Creditors also have pressured African governments to engage foreign banks as financial-policy consultants (Makgetla,

1988:18) and to allow direct external involvement in economic policy making (Gould, 1980:59, 122). Callaghy (1987:107) reports that "nowhere else in the world has IMF, World Bank, and ... private bank advice played such a major and public role."

One consequence of the external penetration of central government financial institutions is loss of flexibility and independence in economic policy making. This phenomenon goes a long way toward explaining why African governments continue to buy into inappropriate (Frisch, 1988:70), unpopular, and discredited structural-adjustment programs. It is noteworthy, therefore, that President Kaunda of Zambia changed his team of economic advisors just prior to charting a course independent of the IMF, "reappointing several of the party faithful in the place of individuals who had supported the reforms" (Lancaster, 1988:33).

There also are major internal administrative costs that accompany compliance with structural-adjustment programs and dealing with the complexities of debt rescheduling and relief. Scores of public-policy analysts must be trained for and assigned to the demanding tasks associated with striving to meet Fund/Bank conditionalities without alienating domestic social forces (Shaw, 1988:36). Repeated debt rescheduling consumes scarce managerial resources. Economic policy makers must "spend months on end merely securing agreement to reschedule a single year's foreign debt payments" (Hawkins, 1987:27; Helleiner, 1983:15). Concomitantly, there is a danger that the reduced public expenditures on educating one's citizens required under structural adjustment could further retard the ability of indigenous public administrators to resist external manipulation and exploitation (see Usman, 1986:97).

The implementation of structural adjustment often produces serious political consequences as well. In the face of protests against the standard IMF economic policy-reform prescriptions, labor unrest, and popular opposition to the negative impact of conditionalities on the standard of living, African governments have been forced to invoke increasingly repressive measures (Loxley, 1986:97; Bangura, 1986b:26; Newswatch, 31 July 1989, pp. 19-21). In Nigeria, this tendency raises chilling prospects for the transition to civilian rule. In spite of student-led riots in several cities, General Babangida continued in mid 1989 to insist that "'there is no

alternative'" to structural adjustment (West Africa, 12-18
June 1989, p. 950; Nigeria, Federal Government, 1987:
12-13). As the result of prior rescheduling actions,
moreover, Nigeria's debtservice burden will increase
dramatically at the time of the scheduled return to
civilian rule in 1992 (Agbese, 1988b:21-25). A democratic,
civilian administration will find it more difficult than an
authoritarian military regime to sustain the repressive
actions and institutions required to execute structural-
adjustment programs. Rather than move in a repressive
direction, therefore, the new regime will be tempted to
ignore or abandon externally propagated conditionalities.

Domestic Economic Consequences

The Nigerian case confirms that the Western financial
establishment will firmly resist major departures from
currency devaluation and the other components of structural
adjustment. The added experiences of Zambia and Zimbabwe
also make clear that any government which embarks upon an
independent strict foreign exchange control strategy must
be prepared to forego debt rescheduling and additional
loans.[30] In any event, accepting IMF conditionalities,
negotiating new loans, and rescheduling old ones are more
likely to constrain than to facilitate progress toward
national economic recovery and self-reliant development.
Borrowing at market or near-market interest rates is
particularly injudicious for countries which rely on the
export of one or more primary commodities (Hardy,
1986:474).

To begin with, new loans saddle African states with
additional foreign debts which must be repaid. Under
General Ibrahim Babangida, for instance, Nigeria's total
international indebtedness grew from $15 billion in 1985 to
$27 billion in 1988. The regime encountered renewed
difficulty rescheduling the $6 billion in annual interest
payments it now owed on this external debt. In light of
this huge burden and the continued lack of interest which
foreign investors demonstrated in Nigeria, the Babangida
administration launched a debt-equity conversion scheme and
announced that it would consider rolling back the country's
long-standing indigenization regulations (Brooke, 1988:D4;
Washington Post, 13 June 1988; Wall Street Journal, 19
January 1988; Kalu, 1987:26).

New lending to Africa became increasingly short-term

and non-concessional in the mid 1980s (Hardy, 1986:468).[31]
Rescheduling buys time, but at the cost of more onerous
repayment terms (see Hardy, 1986:462-464; Parfitt and
Riley, 1986:523; C.E.C., 1988:92). By 1986, "charges on
debts incurred at original average interest rates of 6 percent had increased to an average 10 percent" in the wake of
repeated refinancing (Hardy, 1986:462). Refinancing debt
increases future burdens because arrears are capitalized at
market rates (see Callaghy, 1988:13-15; Brau, 1986:172).
After eight reschedulings between 1977 and 1988 on external
loans eventually totaling $5 billion, Zaire had amassed $2
billion in new obligations largely attributable to
capitalization of interest, rescheduling fees, and late
payment penalties (New York Times, 27 September 1988). In
the face of similar developments across the continent,
Africa's total debt-servicing burden escalated from $10
billion per year in 1984-86 to $15 billion per annum
between 1987 and 1989 (Frisch, 1988:68). The situation is
rapidly approaching the point where the poverty-stricken
continent of Africa will be a net exporter of capital to
Europe and the United States (Hardy, 1986:472, 467;
Lancaster, 1987:221; Adedeji, 1988:53; New York Times, 27
September 1988).

In conclusion, debt servicing itself is increasingly
responsible for the expansion of poverty and economic
decline in Africa (Parfitt and Riley, 1986:520; Hardy,
1986:468). Payments on external debts are consuming a
growing proportion of Africa's diminished export earnings
and declining net capital inflows, in spite of "the fact
that almost the only debts currently being serviced are
those to the World Bank and the IMF" (Hardy, 1986:469,
467). As a consequence of the need to meet debt-service
obligations, fewer and fewer resources are available for
domestic investments and public expenditures on health,
education, infrastructure, and nutrition programs.[32] Thus,
the Fund's emphasis on regular debt-service payments
(Mawakani, 1986:110) exacerbates rather than relieves
problems of underdevelopment in Africa.

ALTERNATIVES TO STRUCTURAL ADJUSTMENT

In order to break out of their reactive mode, African
public policy makers must design promising alternative
approaches of their own (Helleiner, 1986:10). Several
potential alternatives to structural adjustment already

have been discussed in this chapter. The Buhari regime's
tight foreign exchange control strategy, coupled with the
Babangida administration's ceiling on debt repayments,
offers one possible option that deserves further
consideration and sustained implementation. The Government
of Zimbabwe's determination to avoid both deeper indebtedness and IMF conditionalities and to maintain public
expenditures on essential domestic services provides
another model worth exploring. An effective, progressive
system of domestic taxation must constitute a central
component in such efforts. To date, the incomes of wealthy
citizens in Nigeria and elsewhere on the continent "are
scarcely taxed" and property taxes are not a major tool of
revenue collection (Forrest, 1986:25).

In addition, John Loxley (1985:134-136, 140-141)
advocates a food-security strategy targeted to benefit
low-income citizens. This approach encompasses expenditures on basic needs, land redistribution, limited
subsidies on selected food items consumed by the poor,
increases in the minimum wage, and an emphasis on expanding
budget revenues. Loxley argues that promoting the
production of traditional staples would exert a progressive
distributional impact in terms of meeting basic needs,
would heighten self-reliance, reduce import bills and
inflationary pressures, and encourage growth of the peasant
economy.

There is growing recognition that the primary need of
most African countries is for additional capital resources
(i.e., long-term net resource inflows) which can be devoted
to domestic investment and social services and, thereby,
promote sustainable economic growth that benefits the
poorest classes (Hardy, 1986:469). Even the proponents of
structural adjustment admit the need for substantial
infusions of new resources and decry the fact that external
assistance and foreign investment has not been forthcoming
on the scale pledged and required (Adedeji, 1988b:53;
Helleiner, 1986:4, 9; Browne, 1988:7; Lancaster, 1987:221;
Brau, 1986:169-170).[33] Far less agreement exists, however,
over the strategy for attracting new capital to Africa.
Past experience and careful analysis does reveal the stark
limitations of the "export-agriculture" strategy which has
been popular for so long among Western donors and
creditors. Africa's export potential for the immediate
future is insufficient to service existing debts, no less
to finance long-term development (Hardy, 1986:470; Loxley,
1985:127, 136). Indeed, with world prices for many raw

materials at their lowest levels in thirty years, "the income from African exports fell from US$ 64 billion in 1985 to US$ 45 billion in 1986" (Fromont, 1988:94). The export-led approach will not work without substantially improved terms of trade for primary commodities and vastly increased regional and intra-continental economic interaction and cooperation -- two oft-discussed measures which still are not present on the African horizon.[34]

The place to begin is with debt forgiveness and debt relief. Without question, "the most effective solution to Africa's debt problem would be to cancel its obligations" (Hardy, 1986:470). In the African case, relief must include payments owed to the IMF and the World Bank (Callaghy, 1988:16; Brau, 1986:170).

The seven largest industrial countries have recently shown serious interest in a debt relief package that includes the option of cancellation (Laishley, 1988:1, 16).[35] Some governments have already converted the bulk of their outstanding development assistance loans to low-income countries into grants (Brau, 1986:173).[36] The weight of the evidence presented in this chapter demonstrates that it is a mistake to tie eligibility for any debt-relief measures to adoption of a structural-adjustment program.

Debt forgiveness/relief releases funds that are urgently in demand for domestic development (Nsingo, 1988: 84). However, a pressing need remains for the infusion of additional resources. In President Kaunda's (1988:76) words, Africa requires "new money on better terms." For the foreseeable future, the form should be loans on concessional terms plus grants of development assistance in amounts sufficient to meet the mutually agreed target levels of 0.7 per cent of GNP in aid for all developing countries and 0.15 per cent for the poorest ones (Hardy, 1986:472; Sanusi, 1986:65-66).

NOTES

1. The external debts incurred by public enterprises constitute a major part of the continent's total obligations (Nellis, 1986: vii, 12). A relatively high proportion of the total African debt is owed to the International Monetary Fund and the World Bank (about 20%) and a large proportion of its private debt is guaranteed by Western government agencies and, therefore, negotiated under the "Paris Club" mechanism rather than by the banks themselves (the "London Club") (see Callaghy, 1988:12). With the

decline in official lending in the mid 1970s, African
governments increasingly contracted loans from private
sources which carried floating interest rates with shorter
maturities. The average interest rate rose from 5.6 per
cent in 1975 to 10.4 per cent in 1981 (Fantu Cheru, 1987:
3-4). The value of loans issued on a concessional basis
also has declined in absolute and relative terms (Callaghy,
1988:12).

    2. Some countries (e.g., Sudan, Somalia, Zambia, and
Tanzania) have a DSR that far exceeds this average figure.
Nigeria stood at 36 per cent in 1986-87 (Callaghy, 1988:11)
and Zaire at about 50 per cent in 1988 (New York Times, 27
September 1988). Twenty-two Sub-Saharan countries have
projected DSRs for 1988-90 that exceed 30 per cent
(Callaghy, 1988:11, 18).

    3. Carol Lancaster (1988:32) also reports that
assistance from U.S.A.I.D. to 12 African countries has been
tied in recent years to adoption of policy reforms (also
see Helleiner, 1983:9; Perlez, 1989:4). The countries
include Senegal, Malawi, Zaire, Zambia, and Tanzania.

    4. World Bank officials share the Fund's basic
market-centered orientation toward "development" issues and
pursue similar goals. Staff of the two institutions usually promote the same policy-reform recommendations. In
countries accepting IMF structural adjustment facility
loans, "both institutions use joint documents that have
been negotiated with the Governments concerned -- the
policy framework papers (PFP) which combine stabilization
measures and reforms of structures" (Boidin, 1988:52; also
see Serageldin, 1988:54; Loxley, 1985:128).

The Bank's blueprint for the conditions attached to
SALs is the 1981 publication entitled Accelerated Development in Sub-Saharan Africa: An Agenda for Action and popularly known as the Berg Report. The Berg Report underplays
external factors, criticizes overvalued currencies and
government pricing practices (along with state intervention
in the domestic economy), and advocates a wider role for
private enterprise and a focus on large successful farmers,
export promotion, and regions with high growth potential.
The World Bank's follow-up to the Berg Report is the World
Development Report 1983. The 1983 document gives greater
attention to external shocks and deemphasizes progressive
farming schemes, but otherwise is consistent with the
earlier volume. David Murray (1983:292-295) criticizes
both the unsubstantiated support for privatization and the
technocratic emphasis found in the 1983 Report (also see
Loxley, 1986:97-98; 1985:129).

5. On the economic policy measures introduced by the Shagari administration to deal with Nigeria's foreign-exchange crisis, including the Economic Stabilisation Act of 1982, see Bangura (1986a:51-52).

6. Many of these firms experienced a decline in both turnover and profits in 1986 (Newswatch, 14 September 1987, p. 57).

7. In 1987, the Paris Club more openly agreed to a Brazilian rescheduling without a formal IMF program (Cline, 1987:44).

8. Eduard Brau (1986:172) points out, however, that "both principal and interest payments due are rescheduled in the Paris Club framework, whereas banks reschedule only principal payments."

9. Including another $2 billion in tied funds from the World Bank and Western commercial banks (Washington Post, 5 January 1986).

10. West Africa (23 April 1984, p. 866 and 4 June 1984, p. 1153) estimated debt servicing at between $2.5 and $5 billion annually. The 1985 national budget earmarked 44 per cent of estimated foreign exchange earnings for this purpose.

11. There is wide speculation that General Buhari's intransigence regarding the IMF's conditions played an important part in the coup which overthrew his regime (see Bangura, 1986b:31-32).

12. In the words of one Political Bureau member (Oyovbaire, 1987:17), it is "clear to observers that the present [Babangida] Administration is not a revolutionary regime in the sense of an articulated radical alternative world view and strategies for a structural and normative break of the existing predominantly capitalist system and inauguration of a completely new one." Yusuf Bangura (1986b:32) argues more bluntly that the Babangida regime is the first in Nigeria's history which is "ready to reason along the lines of the IMF in its details...."

13. Terisa Turner (1986:40) reports that the World Bank pressured the Nigerian government to cut subsidies on local oil products as a condition for issuing a $200 million loan for a gas pipeline to the Egbin power plant in Lagos. After the government reduced the subsidy in January 1986, the Bank added a new demand that foreign companies be allowed to share in the project as a condition for release of the funds.

14. Parfitt and Riley (1986:526-527) also note that "creditors have much more to lose there than in most other African states, and this gives Lagos some leverage that

other governments do not possess, especially since Nigeria still represents one of the largest markets in black Africa ...."

15. Initially, debt repayment and contributions to international organizations were pegged at the official rate of exchange and handled through the first-tier market (the Central Bank). A single Foreign Exchange Market (FEM) took effect on July 2, 1987 (see Okongwu, 1987:1961-1962).

16. By 1989, the exchange rate required 7.7 naira to the dollar (Daily Champion, 25 July 1989).

17. The Babangida administration remains quite sensitive about charges that it capitulated to IMF pressures. In its official reply to the Political Bureau's Report, the Government went out of its way to "correct the erroneous impression" that its policies are "IMF-induced." Instead, the White Paper defends SFEM and the privatization and commercialization policies as Government-initiated and implemented decisions "in pursuance of the much-needed structural adjustment of the economy" (Nigeria, Federal Government, 1987:9-10).

18. For instance, the Economist (3 May 1986, p. S 12) maintained that "everyone knew that a two-tier exchange rate would open the door to massive cheating by a business class that has become expert at getting around regulations" (also see Nigeria, Political Bureau, 1987:213; Aina, 1982: 76).

19. Disputes continued to delay implementation of the accord and actual receipt of the new credits (estimated at $360 million) from Paris and London Club creditors throughout 1987 (Okongwu, 1987:1961; Newswatch, 14 September 1987, p. 55; Christian Science Monitor, 17 November 1987, p. 14).

20. Nigeria experienced serious food shortages in 1988 (Washington Post, 13 June 1988).

21. Hyden (1980:24) points out that an agricultural development strategy that relies upon higher prices for producers "presupposes that the market is a significant factor influencing peasant behaviour" -- an assumption which is "highly questionable in a situation where peasants are only marginally incorporated into the capitalist economy" (also see Frisch, 1988:69). The most likely results of this approach, then, are that large-scale farmers and those involved in the market economy will monopolize the benefits and the gap in income between these producers and the peasantry will grow (also see Bernal, 1988). Free-market enthusiasts also are unhappy with the World Bank's emphasis on streamlining government marketing boards and raising farm-gate prices (see Wall Street Journal, 13 April 1988).

These efforts have been less effective in practice in
Africa than anticipated (Browne, 1988:6-7).
   22. When the government broke with the IMF in May
1987, Zambia owed the Fund more than $800 million in
arrears (Callaghy, 1988:14).
   23. In addition to Zambia and Sudan, Liberia, Sierra
Leone, and Somalia also have been declared ineligible for
further loans after falling in arrears for over a year to
the IMF (Washington Times, 15 September 1988). Failure to
repay the Fund on time also "means no rescheduling and
probably no [additional] private credit either" (Callaghy,
1988:13; Hardy, 1986:462-465). Thus, the prevailing system
makes debt relief virtually impossible under such circumstances (C.E.C, 1986:91).
   24. The typical IMF reform program, in contrast,
involves wage restraints and price increases (see Makgetla,
1986:402).
   25. In April 1989, the United Nations' Economic
Commission for Africa (ECA) pointedly disputed the statistical analysis employed by the Bank in its March 1989
positive assessment of Africa's SAP experience (see West
Africa, 15-21 May 1989, p. 791).
   26. At least one internal World Bank report entitled
"Beyond Adjustment: Toward Sustainable Growth With Equity
in Sub-Saharan Africa" also reportedly concedes that increased attention needs to be devoted to sustainable and
self-reliant economic growth, income distribution, and environmental protection (see Christian Science Monitor, 13
April 1989).
   27. The World Bank has the same primary mission
(Payer, 1986:667).
   28. On Tanzania's experience, see Fantu Cheru (1987a:
16-37).
   29. Via its sectoral-loan programs, the World Bank
has developed even wider alliances within the public bureaucracies of most African countries (Lancaster, 1988:33).
   30. This point is clearly articulated by Nils
Tallroth (1987:22), World Bank economist working on
Nigeria. He states that "should the [Babangida] Government
lapse in its pursuit of reform and return to the previous
policy regime of controls, external financing would be less
forthcoming" (also see Perlez, 1989:4).
   31. Between 1980 and 1985, IMF lending in Africa
involved short- and medium-term credit (repayment within
3-5 years) at market rates of interest (Lancaster, 1987:
223; Helleiner, 1983:14, 22). In 1987, the Paris Club
granted Mozambique, Somalia, Zaire, Uganda, and Mauritania

reschedulings with longer than normal repayment terms
(15-20 years) and grace periods (6-10 years). However,
there has been no progress in interest rate relief on non-
concessional loans (Callaghy, 1988: 16; C.E.C, 1986:95;
Killick and Martin, 1989:5).

32. For instance, per capita expenditures on social
services decreased between 44 per cent and 62 per cent from
the early to mid 1980s in Madagascar, Senegal, and Somalia
(Fromont, 1988:95).

Amidst reports of increasing malnutrition among chil-
dren, declining school enrollments, and growing unemployment, participants at an international conference in
Khartoum issued a declaration in March 1988 that criticized
existing structural-adjustment programs for neglecting
urgent human needs (Africa Recovery 2 June 1988:1, 22; also
see Fromont, 1988:95; Usman, 1986:97; Helleiner, 1986:3).

33. Loxley (1985:141, 134) rejects the assumption
that "foreign private investment is either necessary or
desirable in the forms and amounts in which it is likely to
appear." He advocates front-end loading of external
assistance in order that it will be "easier for an economy
to absorb the strains of policy adjustment ...."

34. See Judith Hurley (1986:12-13) on the importance
of delinking with multinational lending institutions,
establishing a unified debtors' cartel, and developing
South-South trading relationships (also see Ojo, 1985:
167-168; Nsingo, 1988:84; Nyirabu, 1986:33; Ndegwa, 1986:48;
Sanusi, 1986:67). Ajit Singh (1986:110) advocates commodity agreements that will bring about adequate prices.
Loxley (1985:138) views sudden delinking from the world
economy as impractical, but recommends that African governments commit themselves to annual contributions for projects that foster regional economic cooperation. Consistent
with the recommendations of the Lagos Plan of Action, C. M.
Nyirabu (1986:33, 38-40) emphasizes regionally based manufacturing that is characterized by low direct and indirect
recurrent import content.

35. Fantu Cheru (1987:8) would add repayment in local
currency and reimbursement in kind to the allowable mix in
the debt-relief package. He would like to see African governments pay off their debt by funding environmentally
sound projects and by pledging to preserve their own
national heritage through conservation and reforestation.

In 1989, the Bush administration advanced the "Brady
plan." This approach calls for voluntary bank reductions
in the value of the debt owed by countries pursuing policies that encourage private investment in exchange for IMF

and World Bank guarantees of repayment on the reduced-value
loans (see Kilborn, 1989:1, 28; New York Times, 24 July
1989, pp. 1, 4).

36. President Bush announced in July 1989 that the
U.S. would cancel its aid debt in Africa (New York Times,
7 July 1989). Nevertheless, the proportion of U.S. foreign
assistance devoted to Sub-Saharan Africa has declined and
the Reagan administration had opted to pursue the option
(C) from the Toronto package which increased long-term
debt-service burdens rather than provided debt relief
(Killick and Martin, 1989:3-6).

RECOMMENDED READING

For a critical and perceptive view of the International Monetary Fund and the World Bank, see Cheryl Payer, The
Debt Trap: The IMF and the Third World (New York: Monthly
Review Press, 1974) and her more recent "The World Bank: A
New Role in the Debt Crisis?" Third World Quarterly 8
(April 1986):659-676. Another frequently consulted source
is Frances M. Lappe, Joseph Collins, and David Kinley, Aid
as Obstacle (San Francisco: Institute for Food and Devel-
opment Policy, 1980). Recommended general background works
are Joan E. Spero, The Politics of International Economic
Relations, 3rd edition (New York: St. Martin's Press,
1985) and Jeffrey A. Frieden and David A. Lake (eds.),
International Political Economy: Perspectives on Global
Power and Wealth (New York: St. Martin's Press, 1987).

Sources on Nigeria related to the issues treated in
this chapter include Claude Ake (ed.), Political Economy of
Nigeria (London: Longman, 1985); and Thomas J. Biersteker,
Multinationals, the State, and Control of the Nigerian
Economy (Princeton: Princeton University Press, 1987).

Recent studies of the impact of IMF and World Bank
prescriptions in other African countries are particularly
helpful for comparative insights. See the papers presented
at the Conference on "Africa; The IMF and the World Bank"
held at City University, London, 7-10 September, 1987;
Fantu Cheru, et al., From Debt to Development; Alternatives
to the International Debt Crisis (Washington, D.C.: Insti-
tute for Policy Studies, 1986); Chandra S. Hardy, "Africa's
Debt: Structural Adjustment With Stability" in Robert J.
Berg and Jennifer S. Whitaker (eds.), Strategies for
African Development (Berkeley: University of California
Press, 1986), pp. 461-475; Gerald K. Helleiner (ed.),
Africa and the International Monetary Fund (Washington,
D.C.: IMF, 1986).

5

State Land Allocation

Increasingly, sub-national governments in Africa constitute a prized avenue for access to lucrative sources of
capital accumulation. These sources include credit, contracts, fertilizer and other agricultural inputs, rent
payments, privatized enterprises, and land (Smith, 1985:
195). The subject of this chapter is land.

Control over land is an important factor affecting the
distribution of wealth as well as the structure of political power -- especially when suitable land is in short
supply and provides the most critical ingredient in the
prevailing mode of production. Disputes over land are a
major factor in the struggle for development. In rural
areas, the outcome typically affects matters of title and
terms of holding, the size of the landless class, cultivation patterns (including the scale of operations), and the
incentive structure (see Ega, 1987:425). In urban centers,
the outcome usually directly affects housing patterns and
conditions.[1]

Throughout Africa, public policy makers are preoccupied with issues of land tenure and land use (see, for
instance, Cohen and Koehn, 1977, on Ethiopia, and Seymour,
1975, on Zambia). Much of the "action" with respect to
land in Africa takes place at the sub-national level and
continues to be influenced by traditional customs and
authorities. This presents challenging assignments for
state and local administrative officials charged with
implementing land-use and land-allocation regulations.

Nigeria's Land Use Decree of 1978 consolidated the
central position of the state in the two activities which
most profoundly influence rural and urban land: acquisition and allocation. This chapter primarily deals with
state government land allocation. It provides a revealing

case study of bureaucratic policy making through the implementation process.[2] Although this type of allocation study
is relatively rare, it is valuable in explicating how public administrators use control over their community's pre-
cious land resources to entrench the privileged position of
the ruling class at the expense of the rural and urban
poor.

LAND ACQUISITION AND ALLOCATION IN NIGERIA

Nigerian state and local governments currently
exercise extensive powers over urban and rural land. The
1978 Land Use Decree "empowers the local land allocation
committees to expropriate almost any land within their
areas of control and to allocate it for either public or
private use" (Francis, 1984:7, 14). State government
authorities have made frequent use of their strategic
position and broadly defined powers under the 1978 Decree
to expropriate urban and rural land -- often without
adequate compensation to the poor and powerless people who
live on and survive off of the affected areas.[3] State
control over rural land is intended, in part, to promote
large-scale, private agricultural enterprises by making it
easier for entrepreneurs to overcome the barriers to
accumulating land holdings presented by traditional tenure
systems (Smith, 1985:195).[4] A great deal has been written
on the subjects of expropriation and compensation (see, for
example, Kaduna State, 1981:6-39; Wallace, 1980:2-7, 11;
Frishman, forthcoming; Francis, 1984:14; Ibraheem,
1981:22-23; Christelow, 1987:246; Ega, 1987:425, 429;
Olayemi, 1979:351-352; and Nigeria, Political Bureau,
1987:54).

Less is known, however, about the end use of state-
acquired lands. In Nigeria, the state level controls the
distribution of statutory land use rights, while local
governments grant and certify customary rights of occupancy.[5] For what public and private purposes have lands
been expropriated and people displaced and dispossesed in
Nigeria? Is the agony and suffering caused by state
acquisition justified by some higher ("developmental") use
to which expropriated land is put?

The focus here is on the disposition of lands under
state government authority. This means that we will be
concentrating on statutory rights. Since the Decree only
requires that state governments become involved in
allocating rural land-use rights when the area at stake

exceeds 500 hectares for agricultural purposes or 5,000 hectares for grazing (Francis, 1984:7), we will mainly be investigating urban and peri-urban distribution patterns.[6] The results are important because statutory titles over land constitute a valued source of capital accumulation in contemporary Nigeria (Koehn, 1983b:461-462).[7] In order to maintain and solidify the dominant strategic position of the ruling class in the national political economy, it is vital that the bureaucracy control and manipulate peripheral land allocations. Data on the distribution of statutory rights of occupancy provide valuable evidence regarding the outcome of this effort.

In Nigeria, state governments allocate statutory certificates of occupancy (C of O). Individuals and firms apply for C of O with respect to (1) land held under customary rights, (2) unclaimed lots in state layouts, and (3) plots currently occupied as the result of purchase or invasion. The statutory C of O provides a state-recognized right to use an urban or rural plot of land. Equivalent legal protection is not afforded by customary rights, squatting, or purchase on the secondary market (see Frishman, 1977:391; forthcoming; Goonesekere, 1980:33, 36; Hamma, 1975:59). Thus, the C of O is a necessary ticket for admission to large-scale industrial and agricultural undertakings as well as for entree to the urban real estate business. Billy Dudley (1982:270-271) refers to the latter, which encompasses the rental of city dwelling units as well as the sale of land and housing, as "one of the most profitable forms of private investment in contemporary Nigeria" (also see Barnes, 1975:2; 1982:27, 14).[8]

Moreover, major institutional sources of domestic credit treat statutory titles as necessary collateral against various types of loans -- including bank mortgages and commercial and agricultural credit (Goonesekere, 1980:17-18, 26, 33, 42; Famoriyo, 1979:10). Thus, holders of statutory rights possess privileged access to domestic money markets and are able to secure loans at favorable terms for investment in private capital accumulation (Hamma, 1975:77). Those who do not obtain a state-issued C of O are essentially shut out from such opportunities. For some people, therefore, denial of statutory rights amounts to a guarantee of continued poverty and exploitation.

The allocation of statutory titles in Nigeria is regulated by federal laws and controlled by state administrative actions. The principal policy measures are the Land Tenure Law (1962) and the Land Use Decree (No. 6 of 1978). In sweeping terms, section 1 of the 1978 Decree

stipulates that "all land comprised in the territory of
each state in the federation are hereby vested in the
Military Governor of that state and such land shall be held
in trust and administered for the use and common benefit of
all Nigerians in accordance with the provisions of this
Decree." Individuals and organizations that desire to
obtain a new residential, commercial, or industrial plot in
government layouts, as well as those seeking to convert
customary rights of occupancy into statutory titles, must
petition the state for a legal grant of land-use rights.
The Decree sets limits on the size of holdings in the case
of undeveloped urban and rural land acquired after 29 March
1978, but there is no ceiling on the size of developed
holdings and one individual may simultaneously possess
rights to undeveloped land up to the upper limits in any
number of states (Graf, 1986:120-121). As a result of
provisions enhancing the ability of state governments to
acquire land for "public" purposes and to control the
allocation of C of O, the Land Use Decree constitutes a
powerful and potentially volatile vehicle for redistribu-
tive policy implementation.

Although the importance of statutory land-use titles
is widely recognized, the implementation of state
allocation policy has not been subject to careful scrutiny
and detailed documentation by students of development
planning and administration. The research study discussed
in this chapter investigated land-allocation processes and
outcomes in two states. The findings provide the basis for
empirically grounded conclusions regarding the way in which
state gatekeepers shape the structure of local economic
opportunity through the distribution of rights to utilize a
crucial indigenous resource that is increasingly in scarce
supply throughout Nigeria (Ega, 1984:96-97; Frishman,
forthcoming). We are specifically interested in determining the extent to which the Decree's policy mandate that
land shall be "administered for the use and common benefit
of all Nigerians" affected allocation practices in the
initial period following its promulgation and subsequent
incorporation into the 1979 Constitution. One cannot
assume that the formal provisions embodied in the Decree
have resulted in changes in the manner of administrative
implemention. A persuasive assessment of the Decree's
impact must be based upon careful analysis of "'who are the
beneficiaries?'" (I. Okpala, 1979:16).

The next parts of the chapter are devoted to detailed
scrutiny of the implementation of land-allocation policy by
state-level public administrators in Nigeria. Specifically,

we will examine the ways in which the bureaucracy controls
the award of statutory land-use rights in two northern
states of Nigeria (Kano and Bauchi) and identify policy-
implementation biases through detailed analysis of process
and beneficiaries. The question of primary concern is:
who benefits from the interpretations that state gatekeepers apply in determining access to land-use rights and
in the award of statutory titles?

THE LAND-ALLOCATION PROCESS

    In addition to exploring the ways in which state
administrative gatekeepers control access to statutory
rights of occupancy through policy interpretation and
application, attention is devoted in this section to
uncovering the principal factors which enable applicants to
satisfy mandated procedures and qualifications. Then,
the implications of process-study findings for the
distribution of land-use rights among different classes
will be considered.
    In Nigeria, decisions on the award of statutory rights
of occupancy over specific parcels of land are made by a
handful of state government officials. Section 5 of the
Land Use Decree vests final authority to grant a statutory
C of O over urban and rural land in the military administrator. The elected civilian governors who replaced military administrators in all 19 states assumed this
responsibility from 1 October 1979 until it reverted to
military men following the 31 December 1983 overthrow of
the Second Republic. High-ranking administrative and
professional officers in the Ministry of Lands (Works) and
Survey and the state Urban Development Board typically
perform central roles in the application-review process.
The 1978 Decree also required the governor to establish a
state Land Use and Allocation Committee to provide advice
"on any matter connected with the management of land" in
designated urban areas (sections 2, 3).

Bauchi State

    The principal gatekeepers in the process of allocating
statutory rights of occupancy in Bauchi are officials in
the Land and Survey Division of the Ministry of Works and
Survey, staff of the State Development Board (Town Planning
Division), the Commissioner for Works and Survey, and the

Governor. The Land Use and Allocation Committee also became centrally involved in the urban land-allocation process following its establishment in 1978 under the terms of the Land Use Decree. The Military Administrator designated the membership of the Bauchi Land Use and Allocation Committee in April 1978 and it met for the first time on 26 May of that year. Members are the Commissioner for Works and Survey (Chairman), Attorney General, Chief Estate Officer, Surveyor-General, Chief Commercial Officer of the Ministry of Trade, Industry, and Cooperatives, Chief Executive of the State Development Board, the local government supervisory councillors in charge of land matters in the four designated urban areas (Bauchi, Gombe, Azare, and Misau), a representative from each of the Emirate Councils concerned, and the Senior Town Planning Officer in the Ministry of Works and Survey (who serves as secretary). The State Development Board has been represented by its Principal Town Planning Officer and the four Emirate Councils by their secretaries.[9]

All applicants for statutory land-use titles over plots they do not presently occupy (including unclaimed lots in state layouts) and for the conversion of customary to statutory rights of occupancy apply first to the Ministry of Works and Survey. Applications are reviewed by the Chief Estate Officer, the Surveyor General, and the Town Planning Officers. Local governments and traditional authorities must certify the possession of customary rights of occupancy in conversion cases. In the case of rural land, the Commissioner for Works and Survey then makes a recommendation directly to the Governor.

Local governments may grant customary rights of occupancy over land falling within the old city and over rural land found within their boundaries. However, the Ministry for Works and Survey has ruled that only statutory titles may be granted over land falling outside the old city but within the urban planning zone of the four local governments concerned -- even though customary holdings do exist in these areas (Senior Surveyor, MOWS, interview, 12 July 1979). Applications for land lying within the four urban planning zones are reviewed by the Town Planning Division of the State Development Board to ensure that their proposed use is in conformance with the town land use plan and the Board's development proposals. Urban land applications are next scrutinized by the Land Use and Allocation Committee for compliance with provisions of the Land Use Decree. The Committee submits its recommendations for award or rejection to the Governor for final action on each application for a statutory C of O over land within an urban-planning zone (interviews with the Senior Surveyor, MOWS, 12 July 1979 and the Chief Town Planning Officer, SDB, 13 July 1979).

Kano State

Different officials have performed decisive gatekeeping roles in the state land-allocation process in Kano. In 1970, the Military Governor assumed the Commissioner of Works' authority to allocate plots within Kano township (Hamma, 1975:71). The Military Governor's office would receive C of O applications in the first instance, determine the specific plot to be awarded, and then notify both the applicant and the Land Office of the Ministry of Works that a plot allocation had been made by "His Excellency." From July 1975 through September 1979, the Military Governor (Administrator) retained the post of Commissioner for Lands and Survey and controlled the allocation process from that strategic vantage point. Decisions on the award of statutory rights of occupancy continued to be made by the Military Administrator, together with a small group of high-level administrative officers in his office and the Ministry. The Urban Development Board acted in an advisory capacity, but the officials who controlled the process would ignore its recommendations when it suited their purposes to do so. The Military Administrator of Kano State established a Land Use and Allocation Committee following promulgation of the Land Use Decree in 1978. However, this Committee did not play an active part in the land-allocation process and soon dissolved.[10]

Within a short interval following the return to civilian rule on 1 October 1979, the newly elected governor of Kano State announced that no new land applications would be entertained or statutory rights awarded pending the completion of an inquiry into past practices and revision of the policy guiding allocation decisions. In late May of 1980, the Governor's Cabinet resolved that a new Land Use and Allocation Committee should be appointed and land allocations unfrozen. At that time, an estimated 10,000 C of O requests submitted prior to the freeze had not been acted upon by the state (Commissioner, Kano State Ministry of Lands and Survey, interview, 11 June 1980).

Procedures and Criteria

There are a number of ways in which land-allocation practices in Kano and Bauchi states have effectively denied the poor access to statutory rights of occupancy. Sule Hamma (1975:72-78) reports that, in the early 1970s, influential people successfully petitioned the Military Governor of Kano State directly for award of the choicest residential and industrial plots found "on the maps in his office." Paul Lubeck (1979:39) maintains that government officials grant land-use rights to individuals in exchange for monetary compensation which is beyond the means of most rural and urban residents (also see Kaduna State, 1982:40; Salau, 1980:52).

In any event, official administrative requirements imposed on the land-allocation process in both states preclude the vast majority of the population from gaining access to statutory rights of occupancy. One factor alone virtually excludes all but the wealthiest members of society from securing a statutory C of O. This is the requirement that applicants demonstrate the financial capacity to complete improvements on a plot within a stipulated period of time (usually 2-3 years).[11] In order to satisfy Kano State officials that the land applied for would be developed in an appropriate and timely fashion, one needed (in 1973) to possess or be able to obtain 10,000-20,000 naira for residential accommodations, ₦50,000 for commercial establishments, and ₦100,000 for industrial firms (Frishman, 1977:294, 387). Such interpretations have been invoked by state gatekeepers to discourage prospective applicants and as the primary justification for refusing to consider numerous C of O requests (Hamma, 1975:52). Furthermore, low-income residents are prevented from securing mortgage loans from the Nigerian Building Society, state housing corporations, and commercial banks by a combination of factors. The main disqualifying conditions are low monthly income, inability to afford the required down payment and other charges, and lack of bureaucratic contacts and procedural understanding (Onibokun, 1976:22; Stren, 1988a:108).

No changes occurred in the financial capacity requirement in the post-Decree period. Indeed, at its 21 March 1979 meeting, the Chairman of the Bauchi State Land Use and Allocation Committee (i.e., the Commissioner of Works and Survey) stressed that "plots should be allocated to suitable persons, i.e., to those who can really develop them" (minutes of the meeting, File No. MOW/LAN/PAP/S.77/

226, p.4). At a later meeting, the Committee did not
recommend 27 applications for residential plots in Bauchi
Local Government. In 21 of these cases, it cited "doubtful
intention and ability to develop plot" as the reason for
refusing to recommend allocation of a statutory C of O
(minutes of the meeting of the Land Use and Allocation
Committee held on 27 December 1979, Appendix C). In Kano
State, after deciding that the freeze on new allocations
should be removed, the Cabinet adopted two criteria for the
award of future statutory rights of occupancy: the applicant (1) must be 21 years of age (required under the Land
Use Decree) and (2) must demonstrate financial capacity to
develop the plot with a suitable structure for the type of
land use and nature of the area (interview, Commissioner
for Lands and Survey, 11 June 1980).

The criteria which state bureaucratic officials rely
upon in determining an applicant's ability to complete
improvements are of crucial importance. Process gatekeepers typically base their assessment on two considerations: occupation and salary/income. In Bauchi State, it
is widely known that applicants who are not senior civil
servants or prominent businessmen are likely to be disqual-
ified. In applying the "ability to develop" standard,
private businessmen qualify to receive an award if they are
expected to possess sufficient wealth to develop the type
of plot they have applied for and/or if they are judged
likely to secure a bank loan. A civil servant seeking a
residential plot must be eligible to secure a government
housing loan; i.e., be at GL 06 or higher (interview,
Senior Surveyor, MOWS, 12 July 1979; D. Okpala, 1979:
33-34). These minimum qualifications clearly exclude the
bulk of the rural and urban populace from the award of
statutory rights of occupancy (see Salau, 1980:52-53; D.
Okpala, 1979:32-34; Kaduna State, 1981:43).

In the case of urban plots, other administratively
imposed financial requirements present a formidable
economic barrier for most residents. The government's
insistence that all new layouts be provided with expensive
services (water, electricity, roads, drainage) inflates the
total cost of a C of O. In Bauchi, the amount of funds
required for most residential and commercial plots in
government layouts came to 600-700 naira in 1979, and
industrial plots cost 2000 naira.[12] The 125 naira deposit
which must accompany a residential plot application by
itself exceeded the discretionary income available to many
urban inhabitants. In Kano, similar regulations have
excluded the "majority of the population ... from obtaining

plots even if enough were available" and have tended to
restrict plots in new government layouts to "a higher
income class" (Frishman, 1977:387, 394; also see D. Okpala,
1979:29-33 for similar findings about the Lagos urban
area). In 1979, the permanent secretary of the Bauchi
State Ministry of Works and Survey (1979:3) noted that the
charges levied for a statutory C of O have made it
"difficult for an ordinary man to get a plot in an urban
area ...; thus defeating the purpose of the decree."

In summary, the implementation of statutory land-
allocation policy in Kano and Bauchi states has been
structured both formally and informally in a fashion that
is biased in favor of those who are wealthy and well-
connected. The states did not remove the principal
administrative barriers to access by poor rural and urban
residents in the post Land Use Decree period. The next
sections of this chapter explore the extent to which these
process biases are reflected in the kinds of statutory C of
O applications acted upon and awarded by Kano and Bauchi
State officials.

GENERAL OUTCOMES

Entries recorded in registries maintained by the two
states enable us to identify some general statutory land-
allocation patterns and trends. From the time Kano became
a state in 1967 through 31 December 1976, the government
approved 2,274 statutory C of O. It granted 6,048
additional land-use requests between 1 January 1976 and 31
January 1980, the majority (3,557) in the interval
following promulgation of the Land Use Decree on 31 March
1978 and prior to Governor Abubakar Rimi's freeze on new
allocations which went into effect on 1 February 1980.[13]
By 1980, then, fewer than 9,000 persons held statutory
titles to rural and urban land in Kano State. The total
adult population of the state in that year is estimated to
have exceeded 3.5 million.[14] These figures starkly reveal
the narrow scope of state government land allocations.[15]

Moreover, the small proportion of statutory rights of
occupancy awarded over urban land relative to demand has
forced many people into the secondary (private) market,
particularly in the densely populated Kano metropolitan
area. As a consequence, unregulated land-sale prices have
risen far beyond the reach of the poor.[16] Two further
results are the exploitation of increasing numbers of
unsubsidized tenants and squatters and a widening of the

gap between landlords and the landless (see Frishman,
1977:335, 393-396; forthcoming; Francis, 1984:15; Mohammed,
n.d.; Dar al-handasah, I, 1978:B.6; Lubeck, 1979:39;
1987:101-102; Aronson, 1978:253-265; Kaduna State, 1981:43;
Sada, n.d.:74; Barnes, 1982:14).

The Kano State Ministry of Lands and Survey recorded a
total of 652 applications in its industrial registry
between 5 November 1976 and 13 February 1980.[17] Out of
this total, 418 (64%) are for industrial plots. The rest
are primarily commercial plot applications. Private firms
submitted 322 of the registered industrial plot applications (77%), and individuals submitted 96 (23%). Among the
company applications, roughly a dozen land requests had
been submitted by or on behalf of transnational corporations, although there are undoubtedly others which cannot
be identified on the basis of registry analysis.[18]
Applicants based in Kano State submitted 350 industrial
applications (84% of the total); those in Lagos, Kaduna,
and other places submitted 53, 6, and 9 requests,
respectively. Relative to Bauchi State, at least, the
number of plot requests filed by local enterprises
indicates that indigenous entrepreneurs had undertaken a
substantial amount of new or expanded manufacturing
activity in the important Kano industrial center (also see
Biersteker, 1980:34).

Between 1971 and 1975, the Northeast State Ministry of
Works and Survey registered 618 applications for land
within the present boundaries of Bauchi State. Land
registration activity increased dramatically following
state creation. Ministry of Works and Survey records
indicate that it awarded a total of 3,064 statutory C of O
to non-governmental applicants through 30 June 1980 (Chief
Estate Officer, Lands Office, Bauchi State Ministry of
Lands, Housing, and Environment, 30 June 1980). Specifically, the state government granted statutory titles over
2,044 residential plots (67% of the total); 806 commercial
plots (26%); 88 industrial plots (3%); 20 plots for farming
purposes (1%); and 106 plots for religious use (churches,
mosques) (3%). The total number of C of O awarded amounts
to a tiny fraction of the adult population of Bauchi State,
estimated at more than 1.5 million persons in 1980 (see
Dudley, 1982:199, 201).

Based upon evidence collected in Gongola State, Umar
Ibraheem (1981:222-223) suggests that the frequency of
public complaints about expropriated land will be higher in
areas where government agencies are actively involved in
building new office structures and staff quarters,

operating agricultural and public housing projects, and constructing schools and hospitals. In Bauchi, government agencies clearly were active in this respect during the initial state-formation period under consideration. Agency submissions constituted a relatively high proportion (16%) of the total number of applications registered between May 1976 and July 1979. The Kano State government granted C of O to only 69 public agencies between 1976 and 1979. Over the same time frame, it awarded statutory rights of occupancy to 572 private organizations.

It is interesting to note that 76 individuals holding military titles secured allocations in Kano State in the last year prior to the return to civilian rule; the total over the previous <u>three</u> years is 48. Furthermore, counting the number of files allotted to selected individuals in the registry confirms that influential figures in Kano have been issued multiple plot allocations by the state.   One Kano family of businessmen succeeded in accumulating statutory rights of occupancy over 100 plots between 1967 and 1979.[19]   Other well-known Kano names appeared between 3 and 30 times in the registry. In 1975, the newly appointed Military Governor of Kano State revoked a number of the C of O over plots in the Airport Road New Layout which the previous Governor had granted. After determining that holders had not been awarded another C of O, the new Governor reallocated these plots on 18 September 1975. Nevertheless, the state government continued to issue multiple plot allocations. One investigation reported in the Kano files found that an employee of the Ministry of Lands and Survey had been allocated two plots in Takuntawa Layout by the former Commissioner in the same month (September 1979).

BENEFICIARIES: ACCESS

In analyzing beneficiaries, it is necessary to possess reliable background information on those granted access to the state land-allocation process. Investigation of a random sample of the files acted upon provides detailed, generalizable data concerning the nature of individual applications and applicant attributes. I attempted to examine every tenth file considered by the Kano and Bauchi ministries in 1976 and 1979 in order to draw a reliable sample for each type of land-use request.[20] With the removal of missing cases, the sample drawn amounts to about 8 per cent of all applications acted upon in both states for each year under study. Kano State applications (N=217) constitute 68 per cent of the total sample, with cases from Bauchi State (N=102) comprising the remaining 32 per cent. Applications for residential plots constitute three-fourths of the sampled files. There are 35 commercial or indus-trial land use requests (11%) and 35 conversions for farming purposes (11%). The rest of the sample consists of applications involving government offices or projects (4%). As expected, a decided urban bias exists in the C of O applications acted upon by the Kano and Bauchi state governments. About 90 per cent of the petitioners in the total sample specifically applied for plots that are located in urban areas or towns.[21]

We have observed that in their policy-implementation role, state officials have introduced requirements and applied criteria which are likely to prevent poor rural and urban residents from gaining entry to the land-allocation process. Although the new policy embodied in the Land Use Decree of 1978 stipulated that state governments should administer land in the interests of all Nigerians, the administrative regulations which previously restricted access to statutory rights of occupancy remained in effect. We would expect, therefore, that data on the socio-economic backgrounds of the individuals whose applications for C of O have been acted upon will confirm that the rural and urban laboring classes have been denied admission to the distribution process both prior and subsequent to promulgation of the Decree.[22]

In the first place, analysis by current place of residence shows that urban applicants possess superior access to the state land-allocation process. In Kano State, for instance, residents of the capital city LG submitted three-fourths of the applications requesting conversion of customary to statutory rights for agricul-tural purposes. These applicants are not farmers; they include businesspersons, contractors, speculators, and others seeking to use the state allocation process to obtain secure agricultural land-use titles both in the rural periphery and in the high-demand outskirts of the rapidly expanding capital city (also see Beckman, 1982:13). The application files study also suggests that firms and individuals based in Kano control a major part of the expanding commercial sector in the northern states. Kano city applicants presented 2 of the 3 industrial plot requests and nearly all of the new commercial plot appli-cations. Furthermore, the second largest proportion of commercial applications acted upon in Bauchi State came from petitioners based in Kano city.

Occupation of Applicants

Table 5.1 presents the study's findings with respect to the occupational backgrounds of the individual applicants granted access to the land-allocation process in

TABLE 5.1
Applicant's Self-designated Occupation (N=265)

Occupation	No.	% Total
Self-employed businessperson	59	22.3%
State ministry or parastatal employee	58	21.9
Self-employed trader	56	21.1
Sr. mgr/owner, large private firm/bank	18	6.8
Local govt./Emirate Council employee/ councillor, District Head	13	4.9
Federal ministry or parastatal employee	10	3.8
Contractor	8	3.0
Police officer/commissioner	5	1.9
Army/air force officer	5	1.9
Employee, large private firm/bank	4	1.5
Physician	3	1.1
University staff	3	1.1
Court judge	3	1.1
Soldier/airman	3	1.1
Other (2 applicants each)[a]	10	3.8
Other (1 applicant each)[b]	7	2.6

[a]Farmer (farmer-trader), driver, student, teacher, state commissioner/attorney general.
[b]Housekeeper, housewife, journalist, prince, engineer, lawyer, Military Administrator.

1976 and 1979. Self-employed businesspersons, traders, and contractors submitted nearly half of the sampled forms on

which an occupation is designated. Senior managers and
owners of large private corporations, including multinational firms and banks, presented another 7 per cent of
the applications acted upon by Kano and Bauchi states. An
additional 2 per cent of the land requests can be traced to
employees of such concerns. State government officials
(including a former Military Administrator) head the list
of public sector applicants (23%). They are followed by
local government officials (5%), military and police
officers (4%), federal ministry and parastatal employees
(4%), university staff and other teachers (2%), judges
(1%), and military men who do not hold officer rank (1%).
Farmers and drivers submitted less than 2 per cent of the
petitions entertained by both states and not a single
application from a laborer is encountered in the sample.

Organizational Affiliation of Applicants

Ministry application forms allow petitioners to list a
specific organizational affiliation as their current
address.[23] The most frequently named organizations are
reported in Table 5.2. This information reveals that
individuals associated with administrative agencies
involved in reviewing C of O applications possess
privileged access to the state land-allocation process.
The Kano State Ministry of Lands and Survey heads the list;
9 of the land requests found in the sample of applications
acted upon during the years under study are attributed to
personnel in this ministry. Fourteen other petitioners
report affiliation with a state government agency directly
involved in the land-allocation process (Ministry of Works
and Survey, Development Board, Cabinet Office). Members of
the Nigerian army and police also proved particularly
successful in securing access to land-use rights in 1976
and 1979. Moreover, public administrators connected with
influential state ministries not directly involved in land
allocation, including finance and establishments, clearly
have not been excluded from the process. It is probably
not coincidental that the one private organization placing
3 petitioners on the list (Bank of the North, Ltd.) has
issued nearly half of the authorized C of O mortgages
reported in the sample. Individuals holding government
positions accounted for 54 per cent of the mortgages
obtained by those awarded statutory C of O. The value of
most bank mortgages exceeds 60,000 naira.

158

TABLE 5.2
Specific Organization Applicant Reports Affiliation With[a]

Organization	Number of Applicants Affiliated
Kano State Ministry of Lands and Survey	9
Nigerian Army	7
Nigerian Police	5
Kano State Ministry of Works and Housing	5
Bauchi State Water Board	5
Bauchi State Ministry of Finance	4
Bank of the North, Ltd.	3
Kano State Urban Development Board	3
Bauchi State Ministry of Works and Survey	3
Kano State Ministry of Establishments	3
Kano State Ministry of Agriculture and Natural Resources	3
Government House, Kano	3
Kano LG/LGA	3
NTV	3

[a]Only those organizations mentioned by 3 or more of the applicants in the sample are reported in this table.

Income of Applicants

In the sample of files selected for analysis in this study, 121 petitioners reported their annual salary or income. The information provided (see Table 5.3) confirms that poor rural and urban residents are unable to gain admission to the state land-allocation process. Although the income received by most families living in Kano and Bauchi is less than the minimum wage (₦1,200 per annum in 1980), neither state government acted upon a single C of O request by an applicant with self-reported earnings below ₦1,300. Moreover, administrative gatekeepers entertained only 15 applications (12% of the reporting sample) presented by petitioners with incomes between 1,300 and 2,500 naira. In contrast, 40 per cent of the submissions acted upon came from individuals or organizations reporting annual earnings in excess of ₦20,000.

Table 5.3
Applications Acted Upon: By Applicant's Reported
Annual Salary/Income (N=121)

Reported Salary/Income	No. Applications Acted Upon	% Total
₦ 1,300 to 2,500	15	12.4%
₦ 3,000 to 5,880	33	27.3
₦ 6,000 to 15,000	33	27.3
₦ 20,000 to 80,000	32	26.4
₦ 100,000 to 14,000,000	8	6.6

Impact of the Land Use Decree

The impact of the 1978 Land Use Decree on the
implementation of allocation policy can be assessed by
comparing the attributes of those granted access to the
process in pre-and post-Decree intervals. The data
collected show that some shifts occurred in the socio-
economic characteristics of applicants found in the 1979
sample. However, the findings do not indicate that the
Decree has led to admission of the rural and urban poor to
the state land-allocation process.

Individuals employed in private occupations
experienced a slight decline in access subsequent to
promulgation of the Land Use Decree (from 64% in 1976 to
58% of the files acted upon in 1979). The proportion of
entertained applications submitted by persons working for
public agencies rose from 36 to 42 per cent over the same
period. Petitioners reporting affiliation with domestic
business firms recorded the sharpest decrease (7%) in
C of O files considered by the Kano and Bauchi state
governments, while employees of state agencies involved in
the land-allocation process registered the largest increase
(11% of those treated in 1979 versus 2% in 1976). Those
working in other state government ministries and parastatals, however, experienced declining access (from 20% to
15%). Applicants connected with the armed forces and
federal government agencies slightly improved their share
of the requests acted upon by the two states (from 3% to 6%
in both cases).

Finally, a majority (57%) of the income-reporting applicants who gained entry to the state land-allocation system in 1976 earned in excess of 20,000 naira per annum. The proportion of treated applications which petitioners in this income bracket submitted fell to 23 per cent in 1979. Over the same time frame, applicants earning between 3,000 and 15,000 naira increased their share of the sampled files from 30 to 65 per cent. Individuals at the lowest income level (1,300 to 2,500 naria), however, experienced less success in gaining access to the allocation process in 1979 (12% of the files acted upon) than they had in 1976 (14%).

BENEFICIARIES: AWARDS

The application files treated by the Kano and Bauchi state governments have been categorized as approved (C of O awarded or about to be awarded) or rejected (C of O refused or likely to be refused). The value of access to the land-allocation process is apparent from the overall success rate for the available sample of files acted upon. Gate-keepers in the two states awarded statutory C of O in 80 per cent of the cases they accepted for consideration; only 11 per cent of the applications acted upon had been clearly rejected by 1980. One type of C of O request had not been approved by a state in the majority of cases. The Kano State government refused to approve three-fourths of the new commercial plot applications it entertained.

Further confirmation regarding the importance of access to the process is available from the minutes of four meetings of the Bauchi State Land Use and Allocation Committee held between the time it first convened on 28 September 1978 and 23 April 1980.[24] The Committee recommended approval for 89 per cent of the 1,083 C of O applications presented following Ministry of Works and Survey evaluation; it rejected (recommended against approval) only 57 files (5%). Members opted to defer any action in the remaining cases. All of the rejections involved residential or residential-commercial plots; the Committee did not issue a single recommendation for final disapproval with respect to the strictly commercial (78), industrial (23), and agricultural (39) land requests it reviewed. With the possible exception of industrial plot requests, where lack of Ministry of Trade and Industry clearance is frequently cited as a reason for deferring action, these findings reveal that securing consideration of one's application in Bauchi State has virtually guaranteed receipt of a valuable and potentially highly profitable C of O when an urban or rural business activity constitutes the proposed land use. In the case of new residential plot applications, the Bauchi State Land Use and Allocation Committee rejected those submitted by individuals who failed to pass its version of the "means" test. Nearly all of its recommendations for disapproval questioned the "ability" of the applicant to develop the plot or cited the applicant's possession of an "undeveloped" plot as the basis for rejecting the current request.

The limited information about statutory C of O recipients made public as a result of official investigations into allegations of malpractice reveals only that politically influential individuals have acquired extensive land-use rights throughout Nigeria (see Collins, 1977:141; D. Okpala, 1979:34-35; Hamma, 1975:72; Sada, n.d.:76; Kaduna State, 1981:40-43). The information provided in this chapter allows more precise distinctions to be drawn among the beneficiaries of state land-grant policies. Specifically, we can compare the occupational, organizational affiliation, and income characteristics of successful and unsuccessful C of O applicants. The overall objective here is to identify any policy-implementation biases that operate after one has been granted admission to the land-allocation process.

Occupation of Awardees

Some interesting variations exist in the C of O approval rates recorded by individuals in the sample possessing different occupational backgrounds. The lowest rates of success in securing state approval for entertained land use requests occur among applicants (principally in Kano State) who identify themselves as farmers (50%), soldiers or airmen (67%), and self-employed traders (70%). These percentages stand in sharp contrast to the higher than average approval rates attained by state ministry employees (86%), officers in the armed forces and police (90%), local government officials (92%), and owners/senior managers of large banks and private business firms (94%).

Organizational Affiliation of Awardees

Table 5.4 reports on the proportion of treated C of O requests awarded to applicants affiliated with different types of public and private organizations. These results show that petitioners connected with domestic business firms record the lowest success rates in the entire sample

TABLE 5.4
Status of Applications for C of O; By Applicant's Reported Organizational Affiliation (N=306)

| | - Application Status - | | | |
| Applicant's Reported Organizational Affiliation | Awarded; likely to be awarded | | Rejected; grant unlikely | |
	No.	%	No.	%
State Govt. parastatal, hospital, university, etc.	17	100.0%	0	0.0%
Federal govt. agencies	14	100.0	0	0.0
Kano/Bauchi LG, Emirate	8	100.0	0	0.0
Bank or insurance company	7	100.0	0	0.0
Other state govt. ministries	33	91.7	3	8.3
Other LG, Emirate	8	88.9	1	11.1
Armed Forces or police	12	85.7	2	14.3
Multinational corporation	5	83.3	1	16.7
State govt. land allocation agencies (MLS, UDB, Gov.'s office)	19	82.6	4	17.4
None; none reported	107	77.0	32	23.0
Domestic business firm	24	75.0	8	25.0

(75%). Indeed, even those who fail to report any organizational affiliation are more likely to secure C of O (77%). Association with a small private company clearly does not result in favored treatment when state land-allocation decision makers act upon applications accepted for consideration. On the other hand, applicants with ties to banks, large insurance companies, and multinational corporations have a high proportion (92%) of their C of O

petitions approved by state authorities. Petitioners
affiliated with state government land-allocation agencies
and the military or police prove to be somewhat less
successful at the award/rejection stage of the process (83%
and 86%, respectively) in comparison to those connected
with other state ministries (92%), local governments (94%),
state parastatals (100%), and federal government agencies
(100%).

Income of Awardees

Analysis by income levels indicates that over half
(56%) of the C of O awards issued by the Kano and Bauchi
state governments in 1976 and 1979 (in which income information is reported) have been secured by individuals or
companies with annual earnings in excess of 6,000 naira.
However, this income group also submitted 74 per cent of
the application files which have been considered but

TABLE 5.5
Status of Applications for C of O; By Applicant's Income (N=121)

| | - Application Status - | | | |
| | Awarded; likely to be awarded | | Rejected; grant unlikely | |
Applicant's Income Level	No.	%	No.	%
From ₦ 1,300-2,500	13	86.7%	2	13.3%
From ₦ 3,000-5,000	28	84.8	5	15.2
From ₦ 6,000-15,000	28	84.8	5	15.2
From ₦ 20,000-80,000	20	62.5	12	37.5
From ₦ 100,000-14,000,000	5	62.5	3	37.5

rejected by state land-allocation authorities. Table 5.5
reveals that petitioners in the two highest income brackets
recorded the lowest application success rate (63%). A
large number of these rejections involved applications for
commercial (but not industrial) plots in Kano submitted by
private domestic enterprises.

In summary, the findings of this study with respect to application status suggest that factors which facilitate entry to the state land-allocation process (occupation as a self-employed businessperson, affiliation with a government agency involved in the process, and wealth) do not provide a guarantee that one will be successful in obtaining a C of O award. Nevertheless, the biases which have the strongest effect in determining who is excluded from statutory awards operate at the initial access stage.

Within the small pool of those allowed entry, a limited degree of competition for the award of statutory certificates occurs among elements in the dominant class. For instance, the new permanent secretary in the Kano State Ministry of Lands and Survey conceded, in a note to the Commissioner dated 12 March 1980, that the July 1979 reduction in C of O granted to organizations "was certainly made with a view to satisfying some pressure groups since most of the revoked plots were reallocated to [other] individuals." In addition to variations in the amount of influence possessed by individuals who are eligible for admission to the land-allocation process, outcomes are affected by particular local concerns. At its 15 April 1980 meeting, the Bauchi State Land Use and Allocation Committee adopted an explicit policy that "indigenes of Bauchi State should be given preference over non indigenes" in the allocation of plots. The Committee then refused to recommend that C of O be awarded to 13 non-indigene applicants (including 3 from Oyo State, 2 each from Borno and Bendel, and 1 Lebanese resident); it deferred ten of these files "until applications by indigenes are cleared." Finally, land-use planning considerations and technical criteria are not completely overlooked by administrative gatekeepers (see Koehn, 1983:476).

Impact of the Land Use Decree

Available data on the earnings reported by those granted C of O in Kano and Bauchi during the pre- and post-Decree periods (see Table 5.6) reflect the trends identified in the earlier discussion of access to the land-allocation process. It also is noteworthy that all of the C of O grantees with incomes of ₦20,000 or more are private sector applicants, whereas approximately 70 per cent of those awarded statutory titles who report annual incomes in the 3,000 to 15,000 naira range are employed by government agencies. The ₦3,000 to ₦5,880 category

TABLE 5.6
Period C of O Awarded; By Grantee's Reported Annual Income (N=93)

Grantee's Reported Annual Income	- Period C of O Awarded -			
	-Pre Decree-		-Post Decree-	
	No.	%	No.	%
From ₦ 1,300 to 2,500	3	12.5%	10	14.5%
From ₦ 3,000 to 5,880	3	12.5	25	36.2
From ₦ 6,000 to 15,000	4	16.7	23	33.3
From ₦ 20,000 to 80,000	11	45.8	9	13.0
From ₦ 100,000 to 14,000,000	3	12.5	2	2.9
Total	24	100.0	69	99.9

corresponded roughly to GL 07-10 in the public service
(mainly executive and technical officers), while the ₦6,000
to ₦15,000 income range encompassed public servants who
occupied GL 11-17 positions (administrative and professional officers). Both of these income groups improved
their C of O award position relative to high-income private
sector applicants following the Decree.
      Few shifts occurred in the distribution of awards by
specific occupational category. The only notable changes
are for federal government officials (increase from 1%
before the Decree to 6% in the 1978-80 period) and for
owners and senior managers of banks and large private
enterprises (decrease from 10% to 6%). In the case of
state government personnel, those affiliated with agencies
involved in the land-allocation process expanded their
share from 2 per cent of all pre-Decree grants to 11 per
cent of the C of O awarded in the post-Decree interval. An
equivalent decline took place in the proportion of awards
secured by their counterparts working in other state
ministries and parastatals. Overall, public sector
applicants increased their share of the statutory rights of
occupancy issued by Kano and Bauchi states to 51 per cent
(from 41%) after the Decree went into effect. Most of
their gains apparently came at the expense of petitioners
associated with domestic business firms. The latter's

share of all C of O awards diminished from 14 per cent prior to promulgation of the Land Use Decree to 5 per cent between 1978 and 1980.

The findings reported in this study reveal the continued presence of class bias in the application of allocation policy in the immediate post-Decree period. There are some interesting changes, however, in process beneficiaries. The available data on income and organizational affiliation of successful applicants indicate that intermediate and senior administrators in state land-allocation agencies, the armed forces, and federal government positions secured an expanded share of the statutory rights of occupancy issued by Kano and Bauchi officials charged with implementing the Land Use Decree.

CAPITAL-ACCUMULATION STRATEGIES

In the sample selected for analysis, 5 per cent of the C of O recipients had officially transferred their titles via notarized, government-authorized sales and an additional 6 per cent had legally mortgaged their certificates of occupancy by 1980.[25] Fully 90 per cent of the statutory titles transferred or mortgaged involved plots allocated for residential purposes. Most (70%) of the individuals who sold their rights of occupancy are employed in public-sector occupations. Individuals, rather than firms, purchased the vast majority of the C of O which had been reassigned. The reported sale price ranged from 4,000 to 50,000 naira, with purchasers securing a majority of the plots for less than 13,000 naira.

Slightly more than half of those who mortgaged their titles held public-service positions. The value of the low-interest mortgages obtained by those in the sample ranges from ₦30,000 to ₦200,000; most amount to more than ₦60,000.[26] These findings indicate that, for grantees with access to lending institutions, mortgaging a C of O provides a superior avenue for accumulating capital relative to plot transfer (sale).

Four cases drawn from the Kano files illustrate how the mortgage of a C of O over an urban residential plot is used as a means of private capital accumulation in Nigeria. In the first, an employee of the Kano Cooperative Bank offers, in May of 1979, to purchase the right of occupancy over a plot in Gyadi-Gyadi Layout which the original awardee had secured in 1977. In August 1979, the Ministry of Lands and Survey approves the transfer. A notorized deed of sale for ₦50,000 is signed by both parties to the transaction in November. Three months later, the new holder applies for permission to mortgage the property covered by the C of O to his employer. The Ministry approves his request in March of 1980 and the employee obtains an ₦80,000 mortgage from the Kano Cooperative Bank.

The second case involves a residential plot in the Airport Road New Layout awarded in 1976 to a member of the Bayero University staff. In August 1977, he asks for consent to mortgage his C of O to the Bank of the North for a short-term loan of 7,000 naira. Two months later, he applies for permission to sublease four flats on the plot to a construction company for an annual rent of ₦12,000 (with the first year's rent fully paid in advance). In May of 1980, he reports that his previous mortgage has "lapsed" and applies for permission to mortgage the property covered by his C of O to the Bank of the North for ₦80,000. The Ministry quickly approves each request. This file is particularly informative. Through possession of a statutory C of O and access to a lending institution, the university staff member managed within four years to erect rental units with the aid of an initial mortgage and to secure a second low-interest bank loan of ₦80,000 by presenting as collateral property from which he continued to collect ₦12,000 annually in rent income.

In a third case, the Ministry allows one of its own officials (a GL 09 civil servant) to mortgage his plot in the Kundila Housing Estate to the Bank of the North for ₦75,000 in October 1979. This intermediate-level employee of the Ministry of Lands and Survey had previously managed to purchase a house valued at ₦65,000 on the Kundila plot.

Finally, a District Head requests that his customary right of occupancy over a 27-acre farm two kilometers outside of Wudil town be converted into a statutory holding. The Ministry of Lands and Survey grants him a C of O in 1978. In September of 1979, he applies for permission to mortgage the C of O to the Arab Bank of Nigeria for ₦35,000. The Ministry approves this request in January 1980.

The four cases described in some detail here indicate the value of receiving a statutory land-use allocation from the state. In order to obtain a sizeable mortgage at favorable terms, one must possess a C of O. Personal connections play an important part in determining which holders of statutory rights of occupancy will be issued a mortgage or loan by the banks. In the absence of the required connections, a grantee may settle on the sale of

his/her right of occupancy for a smaller sum to another
individual who is positioned to secure a mortgage upon presentation of a C of O. Following a land-sale transaction,
state officials again perform a crucial gatekeeping role.
The Ministry must record its approval of any "transfer" in
order for the purchaser to receive secure, statutory title
to the land. The files only report the purchaser's name
when the state government authorizes the reassignment of a
C of O. In the Kano sample, a handful of businessmen
secured a large share of the authorized transfers. Since
the legal limitations governing new C of O allocations do
not apply, the secondary transaction arena offers wealthy
individuals and those who are fronting for corporations an
unrestricted "backdoor" opportunity to accumulate statutory
rights over multiple plots of valuable urban land.

It also is noteworthy that the Bank of the North
issued nearly half of the authorized mortgages in the
sample of files selected for study and that state
government officials involved in the land-allocation
process are among the recipients of those capital outlays.
Officials of this particular bank have fared especially
well in terms of access to statutory rights of occupancy.
Among the sampled files, state gatekeepers acted upon
requests submitted by Bank of the North employees more
frequently than they did for staff of any other private
business firm or bank (see Koehn, 1983:474). This occurred
even though Kano State had issued 10 residential C of O to
the Bank of the North in 1976 specifically for the purpose
of building housing quarters for its staff.[27] These
findings suggest that a particularly close symbiotic
relationship has evolved between influential elements in
the state bureaucracy and officials of the Bank of the
North, whereby the latter grant the former relatively easy
access to mortgage capital in exchange for the award of
statutory land-use rights.

CONCLUSIONS AND LESSONS

Both process and beneficiary analysis support the
conclusion that poor rural and urban residents have been
effectively barred from the land-allocation system in Kano
and Bauchi states (also see D. Okpala, 1979:40 on Lagos).
Further confirmation for this finding comes from a survey
of 1,028 randomly selected household heads in Makurdi,
Benue State, conducted by Richard Stren. Stren's study
(1988a:119-122) revealed that urban residents, with the

exception of wealthy landlords, are convinced that it is "very difficult" to secure a residential plot due to the complexity of the land-allocation process, corruption, and the "influence of the rich in the allocation process."

The results of this chapter's critical examination of policy-implementation outcomes also elucidate the specific and shifting nature of the biases governing administrative treatment of statutory land-use applications in the immediate pre and post Land Use Decree periods. While competition existed for the limited number of C of O distributed by the state, allocations by Kano and Bauchi gatekeepers were broadly inclusive of all elements in the ruling class. The most strategically positioned members of the state bureaucracy (i.e., officials affiliated with land-allocation agencies) received a disproportionate and growing share of the statutory rights awarded by the two governments.

The files also point to several means by which the land-allocation process is used to promote capital accumulation in Nigeria and suggest that an informal understanding has evolved regarding the primary preserve of key elements in the ruling class. Access to statutory land-use rights for agricultural undertakings has been open to urban applicants engaged in commercial activities and to government agencies and officials (also see Beckman, 1982:13; Ega, 1987:431). New commercial and industrial lots generally have been reserved for wealthy individuals and large private firms. Public officials, for their part, predominate among those who have received, sold, and mortgaged titles to plots located in the most desirable urban residential areas.

Finally, the findings of this study lend support to the skeptics who challenged provisions of the Land Use Decree on the grounds that its principal impact in the Nigerian political-economic context would be to enhance the ability of state authorities "to assemble land for the elite ..." (I. Okpala, 1979:17, 20-21; also see Udo, 1977:9). Detailed analysis of post-Decree applications and awards in Kano and Bauchi indicates that prevailing class biases had not been altered following the Decree.[28] The rural and urban poor continued to be denied access to statutory titles. In Bauchi, the State Land Use and Allocation Committee applied a new "means test" in rejecting residential plot applications. In Kano, wealthy businessmen utilized "backdoor" methods (e.g., the official registration of purchased land) to acquire multiple statutory rights of occupancy. The land-allocation process

had at most been extended to incorporate additional
intermediate-level civil servants into the small circle of
beneficiaries of state control over urban and rural land.
Senior government officers, particularly those connected
with state land-allocation agencies and the armed forces,
also recorded gains in securing statutory C of O over residential plots in the immediate post-Decree period. In the
absence of fundamental changes in Nigeria's political
economy which are reflected in administration of the land-
allocation process, particularly at vital access points,
the prospects appear remote that state governments
will distribute land "for the use and common benefit of all
Nigerians."

Are public administrators in Nigeria primarily engaged
in promoting development, underdevelopment, or self-
aggrandizement? The results of this focused study on land
allocation do not provide a comprehensive and conclusive
answer. However, the findings presented here lend support
to those who maintain that state officials function as a
"parasitic class" (Anise, 1980:23) and to those who have
emphasized their pivotal comprador role in the commercial
triangle (Williams, 1976a:32-33, 37; Collins, 1977:131-143;
Joseph 1978:229-230). We will return to this issue in the
final chapter of the book.

Land and Urban Housing

It is clear, in any event, that the poor have suffered
as a result of prevailing land-allocation practices. The
difficulty of acquiring urban residential plots, for
instance, has contributed to increasing rental charges. At
the same time, the public housing units constructed by the
Nigerian government tend to be too few and too costly for
low-income city dwellers and to be located at substantial
distances away from places of employment (Taylor, 1987:439;
Onibokun, 1976:22; Salau, 1980:54; Barnes, 1982:7). As a
result of process biases, a few high-level government officials and wealthy families ended up in possession of many
of these units (Salau, 1980:54; Barnes, 1982:7-8; Onibokun,
1989). The consequences of land and housing policy implementation have been overcrowding, the growth of illegal
settlements, and compounded problems of service delivery
for concerned public officials (see, for instance, Olayemi,
1980:350; Salau, 1980:49).

Available research findings suggest that quite similar
developments have occurred in other African cities. The

state has not provided effective assistance in locating
secure plots for low-income inhabitants or facilitated land
registration and house construction. The result has been
the spontaneous growth of illegal settlements which are not
serviced by federal, state, and local governments (Cooper,
1983:31; Stren, 1982:80, 89; Barnes, 1982:10-11; Temple
and Temple, 1980:249). As Richard Stren (1982:89) points
out, "inadequate sanitation, water supplies, refuse
removal, and community facilities in these unplanned areas
can only lead to a worsening of the life chances for those
who cannot take advantage of the bureaucratic apparatus."

In Africa, "squatting arises from lack of choice,
often due to bureaucratic decisions which ignore the needs
and wants of ordinary people" (Peil, 1976:165). In both
Tanzania and Senegal, officially well-intentioned plot-
allocation programs never benefitted the urban poor. The
reasons include the scarcity of low-cost loans, the
complexity and cost of the application process, lack of
personal contacts within the bureaucracy, and corruption
and bias at the implementation stage (Stren, 1982:80-81,
85-86; 1988:111, 116; White, 1985:516-519, 524-526). The
wealthy, educated, and well-connected in Tanzania have
benefitted disproportionately from the prevailing land-
allocation system (Stren, 1982:89). In Khartoum, civil
servants have gained preferential access to the best city
sites because of biases in the plot-allocation process
(Stren, 1988a:123-126).

Similar biases are built into other processes that
determine housing allocation. In Tanzania, low-income plot
holders have encountered difficulty securing loans from the
Tanzanian Housing Bank and permission to build from government officials. The process is cumbersome and protracted
and the costs associated with obtaining a mortgage,
including incidental fees and the services of draftsmen,
would consume one month's salary (Stren, 1982:86-88;
1988a:112-115). Elsewhere, governments assigned responsibility for low-cost housing construction to parastatals.
The centralized, state-centered approach has consumed
substantial public revenues in Ivory Coast, Senegal, and
Kenya. However, housing corporations have compiled poor
performance records throughout the continent. The number
of units constructed has not begun to meet the housing need
in rapidly expanding urban areas and the few who secure
highly subsidized allocations tend to be influential
middle- and high-income residents who often sublet their
awarded dwelling unit at the market rate (Stren, 1988a:
107-110; Temple and Temple, 1980:224, 231-235, 247, 249;

Peil and Sada, 1984:295, 299, 302-303; Cohen and Koehn, 1980:155-156, 200).

Since the colonial period, however, the state has been successful in providing government housing for civil servants. In anglophone Africa, government housing frequently consists of luxury units constructed in highly segregated and uniquely well-serviced enclaves (Barnes, 1982:6-7, 11-12, 26; Peil and Sada, 1984:294, 301-302). Such schemes are a valuable asset for public administrators at all levels. Richard Hodder-Williams (1984:171) uses the example of the permanent secretary to illustrate this point. Armed with government housing and a prestigious position in the bureaucracy:

> "a permanent secretary can borrow money to buy not only a small home for himself but also another property and thus let out, at high rents to international companies like the airlines, two houses. The profits from these enterprises can then be invested in further housing, probably at the lower end of the market where the demand for homes is so very high, and so make the bureaucrat ... a major rentier in the capital city."

In sum, African governments have accomplished little in terms of housing the urban poor. The implementation of housing policy has favored civil servants and relatively wealthy residents (Cooper, 1983:30; Barnes, 1982:5, 26). Throughout the continent, "with respect to the bureaucratic allocation of housing and land, two crucial resources in today's rapidly growing African cities, both procedures and outcomes are generally biased in favour of the dominant social groups" (Stren, 1988a:125; Temple and Temple, 1980:224-225, 235, 242, 246-247, 249; Peil and Sada, 1984:302).

ALTERNATIVE APPROACHES

Several alternative approaches promise to achieve more than prevailing public policies have accomplished in terms of addressing the land and housing needs of the poor in Africa. One strategy is to grant legal recognition to existing settlements and broaden access to inexpensive plots. Secure tenure offers settlers a powerful incentive to invest in housing improvements and paves the way for the delivery of vital public services. For instance, Seth

Asiama (1984:170-83) shows how the provision of inexpensive
plots of land in the Medina area of Accra constituted the
crucial factor in allowing low-income migrants and
resettled residents to develop their own accommodations.

Government support for self-help, slum-upgrading
activities is a more realistic and equitable option than
site-and-services schemes given the massive scale of the
housing shortage in most African urban centers (see Peil,
1976:165; Peil and Sada, 1984:296-298, 313-316, 323;
Onibokun, 1989; White, 1985:519).[29] Public administrators
must encourage the spread of affordable dwelling units primarily constructed with local materials as well as the
conservation of traditional dwellings. Such housing will
necessarily be basic and rudimentary given that government
subsidies will not be available on the scale required for
major improvement and that low-income households typically
are not prepared to pay a large proportion of their income
for shelter (Stren, 1988a:117). The availability of low-
interest credit is an essential ingredient for the success
of this type of undertaking. Progressive lending on a
group basis helps to guarantee repayment and program
sustainability (see Hossain, 1988:2-3). At the same time,
the underlying problem of poverty induced by lack of
employment opportunities and low wages must be addressed
(Seymour, 1975:75; Peil and Sada, 1984:323) and provisions
must be made for future servicing of upgraded neighbor-
hoods. Piped water and effective human waste disposal
systems are vital public health considerations.

Lawrence Ega (1987:432-433) provides helpful recommendations aimed at improving the land-holding situation
experienced by poor rural residents. First, he advocates
that African governments refuse to support large-scale
projects that require the massive displacement of farmers.
Ega also supports legal recognition of customary rights and
transactions in land, the establishment of minimum and
upper limits on the size of individual holdings of agricultural and grazing land, and the adoption of land-use
conservation measures. In places where landlessness is
widespread, plans for the redistribution of possessory
rights also should receive careful consideration.

Finally, the findings on state land allocation in
Africa presented in this chapter reveal that major changes
are required in processes as well as in policies in order
for the rural and urban poor to benefit from government
programs. The bureaucratic procedures regulating land,
housing, and credit allocation, in particular, must be made
less complex (see Stren, 1982:81-82) and redesigned out of

consideration for the needs and capacity of low-income
citizens. Specifically, barriers to access by the poor
must be eliminated and biases in implementation overcome.
Convincing sub-national public bureaucracies to respond in
this fashion will not be easy. Such an outcome will only
be realized through the empowerment of peasants and the
urban masses and widespread community participation in
public policy making by previously excluded classes (see
Stren, 1988a:126; Ega, 1987:433).[30]

NOTES

1. Stren (1985:60) refers to urban housing in Africa
as "a relatively under-researched but crucial focus of the
class struggle ...."
2. On the importance of implementation studies, see
Biersteker (1987:296-297). It is particularly revealing to
analyze policy implementation at the level of citizen
interaction with public officials (see Peters, 1989:193).
3. On the other hand, landlords have profitted from
over-compensation by the state for cheaply and hastily
constructed buildings on land designated for expropriation
(Agbese, 1988a:278).
4. Expropriation also allows the state to make land
available (e.g., by leasing) to foreign concerns for large-
scale farming when this is precluded by customary tenure
(see Akinola, 1987:230; Mkandawire, 1988:30).
5. Traditional authorities also continue to influence
outcomes in some areas.
6. Francis (1984:15) reports that "by the end of 1979,
few certificates had been issued [in Oyo State] for land
outside the main urban centers, and most of those which had
were in respect of land already developed before March
1978."
7. According to Sara Berry's (1983:270) analysis of
contemporary Yoruba society, land and housing are especially
valued investments as a result of the interaction of
descent-based relationships and the process of class
formation (also see Barnes, 1979:59, 67; Aronson,
1978:253-267).
8. For instance, the Nigerian military establishment
pays private landlords exhorbitant sums (up to five years
in advance) in order to rent choice urban housing for officers in the armed forces (Agbese, 1988a: 278; also see
Salau, 1980:52).
9. Minutes of the Land Use and Allocation Committee
meeting held on 26 May 1978, pp. 1, 2; Senior Surveyor,

Bauchi State Ministry of Works and Survey, interview, 12 July 1978.

10. Similarly, few Land Use and Allocation Committees had met by late 1979 in Oyo State (Francis, 1984:15).

11. On the colonial origins and intentions of this condition, see Goonesekere (1980:12-15). Similar requirements are enforced in Ghana by the Lands Commission (Asiama, 1984:173-174).

12. The breakdown of 1979 charges is: (1) application processing fee (₦50); (2) survey fee (₦75); (3) reimbursement for government compensation in acquiring the land paid at a rate of ₦2,000 per hectare (up to ₦1000 for the maximum half-hectare residential or commercial plot; ₦2000 for the maximum one-hectare industrial plot). Conversions from customary to statutory rights of occupancy carried a higher survey fee (₦100), but involved no compensation for land acquisition. Interview, Senior Surveyor, MOWS, 13 July 1979.

13. Source: Right of Occupancy Register, Kano State, Ministry of Lands and Survey (as of May 1980). This register mainly records applications for statutory C of O that have been awarded by the Ministry. However, some C of O have subsequently been revoked; others have been rejected or are still pending final action. We still refer to the registry figures as though all files have been approved since the actual status of the application could not be ascertained without consulting each file.

14. This population estimate is based upon figures agreed to in 1977 for the purpose of allocating seats in the Federal House of Representatives (see Dudley, 1982:199, 201).

15. Indeed, in the Makurdi LGA, capital of Benue State, the state granted only 162 plots for residential purposes (Stren, 1988a: 119).

16. In the face of uncontrolled market processes, "access to land is getting tighter for low-income groups in many Third World cities" (Gilbert and Healey, 1985:11).

17. Source: Register of Industrial Plot Applications, Kano State, Ministry of Lands and Survey (as of May 1980).

18. See the discussion of "fronting" found in Biersteker (1980:23-25, 32). In one case reported in the Kano files, a 58 year old self-employed trader secured a C of O over a G.R.A. plot in 1979. In 1980, he subleased the plot to a construction firm for 1,000 naira per annum over the next 25 years. Asked to comment on the "correctness" of such transactions, the Principal Land Officer in the Ministry of Lands and Survey noted (1 April 1980) that "it

is quite in order for a holder of a right of occupancy to arrange with a building company to develop the plot and occupy it for a certain number of years until the building company recovers the total amount spent on the buildings."

19. This family's holding companies rank among the largest and strongest in the country (Biersteker, 1987: 272). Family members initially promoted their business interests through strategic positions in government. Their enterprises benefitted from access to credit and agency patronage (Izah, 1987:286).

20. A grant awarded by the Joint Committee on African Studies of the Social Science Research Council and the American Council of Learned Societies assisted the research reported here. The author also acknowledges with gratitude the assistance rendered by staff of the Ministry of Lands and Survey in Kano and the Bauchi State Ministry of Works and Survey (later renamed Ministry of Lands, Housing, and Environment), as well as the field contributions made by research assistants. For details regarding the research methodology utilized in this study, see Koehn (1982:5-6, 22-23). Some results also are reported in greater statistical detail in that source (pp. 16-20).

The Kano data may have been collected in timely fashion. Diamond (1981:11) reports that demonstrators burned down the Ministry of Lands and Survey office building on 10 July 1981.

21. A primate (capital) city bias occurs in Kano State, but not in Bauchi State.

22. Although not discussed in detail here, age and gender also bear upon one's ability to have an application for statutory rights of occupancy processed in Kano and Bauchi. Petitioners under age 30 submitted only 16 per cent of the applications found in the reporting sample (N=98). Nearly three-fourths have been prepared by applicants in the 30-49 age range. Access by women to the state land-allocation process is extremely limited. The eleven female applicants comprise only 4 per cent of the sample (N=286). These findings stand in sharp contrast to the results of Barnes' (1979:64-66) study of land holders under customary rights in pre-Decree Lagos.

23. This data set includes a small number of applications submitted by private and public organizations.

24. The Committee reviewed requests for statutory rights of occupancy in four urban areas of the state -- Bauchi, Gombe, Azare, and Misau (minutes of Bauchi State Land Use and Allocation Committee meetings on 28 September 1978, 21 March 1979, 27 December 1979, and 23 April 1980).

Actions taken on requests reviewed at the 3 other meetings held by the Committee during this period are not recorded in the minutes.

25. Paul Francis (1984:15) points out that "in spite of their being outlawed by the [1978] Decree ..., sales in land continued, although documents confirming such transfers had to be fraudulently backdated if they were to appear legal under the new legislation."

26. On the "social" loans issued for the purchase of "owner-occupied" housing by the Federal Mortgage Bank, see Dudley (1982:272).

27. This (10 consecutive plots) is the most extensive residential allocation encountered in the Kano files. The Bank of the North lost two of its Hotoro GRA plots in the July 1979 revocation/reduction exercise.

28. Although not considered here, class biases in land allocation also are pronounced at the local government level (see Kaduna State, 1981:19-24; Mohammed, n.d.; Francis, 1984:12). In Kenya, LG councillors and administrative officers "routinely used council authority to allocate housing and business plots to their own advantage, until the process was removed from their control in the early 1970s" (Stamp, 1986:34).

29. For a critique of the World Bank-inspired approach to upgrading adopted in Lusaka, see Seymour (1975:74-77).

30. A useful step in this direction is the Political Bureau's (1987:118) recommended decentralization of local government responsibilities with respect to rural land in Nigeria to village committees.

Readers interested in the alternatives outlined in this chapter should carefully consider the rural and urban land reform measures introduced following the Ethiopian revolution. See Cohen and Koehn (1977:3-62).

RECOMMENDED READING

Interested readers can obtain a useful comparative perspective on some of the land acquisition and allocation issues addressed in this chapter by referring to Erik Eckholm, The Dispossesed of the Earth: Land Reform and Sustainable Development, Paper No. 30 (Washington, D.C.: Worldwatch Institute, 1979); chapters 4 and 7 of J.P. Dickenson, et al., A Geography of the Third World (London: Methuen, 1983); Alan Gilbert and Patsy Healey, The Political Economy of Land; Urban Development in an Oil

Economy (Brookfield, Vt.: Gower Publishing Company, 1985); John M. Cohen and Peter H. Koehn, "Rural and Urban Land Reform in Ethiopia," African Law Studies 14 (1977):3-62; Seth O. Asiama, "The Land Factor in Housing for Low Income Urban Settlers," Third World Planning Review 6, No. 1 (Feb. 1984):170-184; and Richard E. Stren, "State Housing Policies and Class Relations in Kenya and Tanzania," Comparative Urban Research 10, No. 2 (1985):57-75.

The best sources for data on land-allocation practices in Nigeria are Kaduna State Government, White Paper on the Report of the Land Investigation Commission, Vols. I-VII (Kaduna: Government Printer, [1981]); Peter Koehn, "State Land Allocation and Class Formation in Nigeria," Journal of Modern African Studies 21, No. 3 (1983):461-481; Ifebueme Okpala, "The Land Use Decree of 1978: If the Past Should be Prologue ...!" Journal of Administration Overseas 18 (Jan. 1979):15-21; and Sandra T. Barnes, "Migration and Land Acquisition: The New Landowners of Lagos," African Urban Studies 4 (Spring 1979):59-70. For background, see Reuben K. Udo, Report of the Land Use Panel; Minority Report on Nationalisation of Land in Nigeria (Lagos: Federal Government Printer, 1977); Land Use Decree 1978 (published in Nigerian Herald, 1 April 1978, pp. 3+); and R.K.W. Goonesekere, Land Tenure Law and Land Use Decree (Zaria: Faculty of Law, Institute of Administration, A.B.U., [1980]).

6

Development Planning

Chapters 6 and 7 of this book are concerned with
local-level policy making and administration. Throughout
Africa, local government activities exert an immediate, but
understudied, impact on the lives of the populace. The
role of local government in the policy-making process
deserves closer attention and critical scrutiny. Development planning constitutes one important local decision-
making mechanism and provides a useful focal point for
analysis.

The past decade has witnessed a major shift in the
prevailing approach to development planning in Africa.
Centralized planning has generally produced five-year
documents which are unrealistic, rigid, and ineffectual as
guides to resource allocation (Bryant and White, 1982:235;
Caiden and Wildavsky, 1980; Conyers and Hills, 1984:46-47,
51). The approach has failed, in part, because the
economists and short-term foreign experts who dominate
national government agencies are unaware of local-level
conditions, opportunities, and constraints (see Cohen,
1984:189; Hyden, 1986:3, 34-35). In the absence of positive results, macro-level strategies have been discredited
and gradually amended (see, for instance, Rondinelli,
1981:596). In place of centralized planning, African
governments are placing increased emphasis on developing
local capacity to identify, initiate, and carry out
development projects (Vengroff and Johnson, 1987:287).
Decentralized planning allows government units to respond
to diverse needs and demands (Rondinelli, 1981:596).

Although the continent's growing interest in grass-
roots involvement in national development planning is
promising in theory, local government participation cannot
be approached uncritically. Several crucial dimensions of

179

LG involvement in national planning merit careful study and critical analysis. In this regard, process studies are useful in determining whether or not the prevailing method of plan formulation enhances articulation of the needs and interests of local residents and their incorporation into the national development plan. Systematic financial analysis of plan proposals promotes understanding of the types and objectives of development projects on which particular LGs intend to devote the greatest emphasis, attention, and resources over the plan period. This chapter presents a critical process study and attempts to demonstrate the advantages of engaging in multidimensional "impact" cost analysis during the formulation stage of the planning process.

At the local government level, the blueprint approach to planning typically prevails. Planners generate "shopping lists" of standard projects for inclusion in the national plan that are accompanied by crude sectoral and areal analysis. Plan formulators tend to proceed without considering fundamental development policies and goals, basic local needs, alternative proposals, and program consequences. Cost-benefit analysis is misleading and can easily be manipulated by the political and administrative elite in most local government areas (Conyers and Hills, 1984:11-12, 85, 138, 195, 198-201; Kent and McAllister, 1985:66-67, 16-17; Denhardt, 1984:154; Gran, 1983:298-299; Chambers, 1978:210-215). In any event, standard economic indicators are far from flawless as measures of development (see Cohen and Uphoff, 1977:162-163; Robertson, 1984:124, 126).

Local development planners need to adopt more appropriate appraisal techniques and to consider a more diverse set of factors at the decisive project-selection and priority-setting stages of the planning process. One goal of this chapter is to illustrate how this can be done and what some of the advantages would be. The novel approach advanced here incorporates elements of social impact assessment and planning balance sheet analysis (Conyers and Hills, 1984:80, 139, 141-142, 148).

The specific aims of this chapter are (1) to illustrate the process of local government involvement in national development planning, (2) to evaluate the priorities attached to various types of programs by local planners, and (3) to explore alternative approaches for assessing project proposals at the LG level that can be employed at an early stage in the planning process.

The plan submissions prepared by two local governments in Nigeria provide the principal focus of study.[1] Costs of the initial capital development proposals presented by the Bauchi and Kaduna LGs for inclusion in the Fourth Plan are scrutinized in the sections below partly to determine the extent to which these plan submissions embrace a people-oriented development strategy and partly to demonstrate the value for local policy makers of moving beyond preoccupation with unidimensional sectoral considerations.

LOCAL GOVERNMENT INVOLVEMENT IN NATIONAL DEVELOPMENT PLANNING IN NIGERIA

Nigeria's Fourth National Development Plan (1981-85) for the first time included input from the local government level. The incorporation of LG development projects in the Fourth Plan is the result of a deliberate policy of encouraging grass-roots participation in the national planning process. The Second Republic Constitution enshrined this policy in Article 7, Section 3, which reads: "It shall be the duty of a local government council ... to participate in economic planning and development of the area ...." In more specific terms, the Federal Ministry of National Planning's 1979 Guidelines for the Fourth National Plan 1981-85 (pp. 22, 7) provided that:

> The surest way to get to the grass roots through planning is via the local government authorities. In the past, local governments were not regarded as significant planning and executing agencies.... As a marked departure from the past, the local government authorities should play a crucial part in the next plan. Under the general supervision of the State governments, they will initiate and prepare their own projects which will be integrated into the plan. They will also be responsible for the implementation of such projects using the resources expected to be placed at their disposal.

The policy of involving local governments in Nigeria's national development planning resulted largely from official dissatisfaction with previous planning exercises which relied upon project identification and plan preparation at the top and failed to address the actual or felt needs of local populations (see Mabogunje, 1972:4;

Muhammed, 1979:6; Abubakar, 1978:132; Uwazurike, 1987: 251).[2] This point is illustrated by the following argument advanced by an official of the Kaduna State Ministry of Economic Development:

> There is an urgent need ... for the proper integration of the local government units into the planning process in order to ensure that the local units play a significant role in the preparation and implementation of the development plans. This involvement is necessary if development is to emerge from below thereby correcting the past mistakes caused by the existing arrangements (Ajala, 1978a:2; 1978b:8).

These national and state government references make official intentions clear. This chapter is concerned with evaluating the outcome of the Nigerian experiment.

PROCESS ANALYSIS: TWO CASE STUDIES

In order to establish the context for impact-cost analysis, we first examine the process employed in 1979-80 for involving Kaduna and Bauchi LGs in the national planning exercise. While several authors have examined the structure of development planning at the national and state level in Nigeria, little has been published to date regarding the nature of local government involvement in the process.

Participation in the national plan formulation exercise at the LG level in Kaduna and Bauchi exhibited important parallels and major variations. In accordance with the Federal Government's Guidelines, local governments in both states actively took part for the first time in the preparation of project proposals for inclusion in the Fourth National Development Plan. Once approved, these proposals constituted a separate local government segment in each state's final overall submission to the National Planning Office. In most other respects, the process of plan formulation differed markedly in Kaduna and Bauchi. The major differences in procedure encountered with respect to the two capital city LGAs considered here are highlighted in the following discussion.

Kaduna Local Government

In Kaduna State, the Ministry of Economic Development undertook special efforts to explain local government roles with respect to the Fourth Plan. In addition to issuing a call circular and project proposal forms, senior field advisers from the Ministry met with various LG officials in late 1979 to underscore the importance being attached to grass-roots involvement in development planning and to describe and answer questions about the mechanics of the planning process. In January 1980, the Ministry divided the 14 Kaduna LGAs into 4 zones and assigned two senior planning officers to each zone. These officers, who spent a week in the field, worked with and assisted the local government personnel responsible for identifying and preparing project proposals, particularly the official assigned to coordinate the LG's planning effort. The planning officer's principal roles involved establishing rapport and gaining the confidence of local government administrators and councillors, assessing past performance under the 1975-80 development plan, identifying available resources and revenue sources, advising on essential data needs for local planning purposes, exploring community problems, guiding local officials in the completion of the project proposal forms issued by the Ministry, and answering any questions that arose during the planning process.[3]

During January 1980, the Kaduna Local Government undertook several measures aimed at identifying community and project needs and costing the capital development projects selected as plan proposals for submission to the state Ministry of Economic Development.[4] At an initial meeting of the Council's Finance and General Purpose Committee (FGPC), high-level LG officials asked village and district heads to compile a list of projects and services needed in their area. From this list and other project proposals[5] submitted by the local government secretary, department heads, and individual councillors, the FGPC compiled its own ranking of high-, medium-, and low-priority projects to be proposed to the Ministry of Economic Development.[6]

Following certain deletions from the list made by top local government officials due to duplication and other reasons, each department prepared written justifications for the projects falling within its area of jurisdiction on the Ministry of Economic Development's forms. The Supervisor for Works or the concerned head of department provided a rough estimate of the capital costs of each project over the duration of the plan period (incorporating a 5 per cent projected annual inflation rate). The full council did not grant formal approval to the proposed development plan for Kaduna Local Government, although officials did inform the council regarding their actions. The final proposed Kaduna plan for 1981-1985 consisted of 66 capital projects. Since the Ministry did not place a ceiling on the number or estimated cost of capital development projects, the Kaduna LG endeavored to include as many projects as it could justify in its proposed plan.

A thorough examination of these documents, which also provide the substantive basis for the financial analysis section of this chapter, yields crucial insights into the planning process at the local level in Kaduna. First, the written justifications rest entirely on general expressions of project or service "need." For example, the justification for Project No. 46 (construction of more staff quarters and furnishing the existing ones) reads: "There is an urgent need to house the officers of the local government and furnish the quarters already built. Houses for officers are essential for efficiency of the officers ...." The Kaduna LG assigned this proposal "high priority." In a few instances, a case is advanced for locating the project in a certain area based on the current availability of that service only in other districts or villages. Thus, the justification for Project No. 32 (construction of permanent and temporary market stalls at Ungowar Sarki, Malali, and Badarawa villages) states, in part, that these places have large populations, are located far away from the central market, and are presently the only villages without markets. Although many projects are rated "top," "high," or "medium" priority, no effort is made to defend the attached rating by reference to the relatively greater urgency or seriousness of the problem involved in relation to other proposals, because of greater citizen demand, or in cost-effectiveness terms.

Upon receipt of local proposals, the Ministry of Economic Development's Planning Division called upon each LG to defend its capital-development submissions in day-long sessions held at the Ministry's headquarters. The SLG, heads of departments, and Ministry for Local Government officials typically attended these sessions.[7] Based upon the outcome of these meetings, its own review of the materials submitted, and consultations with other state ministries,[8] the MED made its own revisions and recommendations with respect to each local government's development plan. The Planning Division compiled the Ministry's revisions, which included budget additions and reductions, deletion of inappropriate (i.e., non capital development) projects, and rearranging of priorities, in the form of summary sheets entitled "Kaduna State Fourth National Development Plan (Local Government) 1981-85." These sheets describe each project by title, number of units involved, total estimated capital expenditure over the plan period, and the priority attached by the Ministry and the LG to each project proposal.[9] The Ministry submitted its final recommendations on each local government's project proposals and overall development plan to the State Planning Committee.[10] Following revisions and deletions by the SPC, the Governor's Policy Impact Council reviewed local government development plans prior to forwarding them to the State House of Assembly for consideration, amendment, and approval.

Final approval authority at the state level resided with the State Executive Council and the Governor, who then transmitted all LG development plans in consolidated form to the National Planning Office in the Federal Ministry of National Planning (see the helpful chart in Adamolekun, 1983:163; Oyovbaire, 1985:217-219; Uwazurike, 1987: 181-184).[11] At the national level, the review and approval of state and local government development plan proposals also involved the Joint Planning Board, the National Economic Council, the Conference of Ministers and Commissioners for Economic Planning, the Director of Budget, and the President's Special Adviser on Economic Affairs. The National Assembly, as well as the house of assembly in several states (including Kaduna), unsuccessfully sought to establish a more influential role in the development-planning process (Adamolekun, 1983:159-166; Oyovbaire, 1985:211-216, 220).

Bauchi Local Government

Local government participation in national development planning in Bauchi State differed in important respects from the process followed in Kaduna. Due to staff shortages, for instance, the Ministry of Finance and Economic Development proved unable to provide direct assistance to LGs during the planning process. In November

of 1978, however, the Ministry sponsored a seminar on
planning for local government officials throughout the
state in collaboration with the National Planning Office.
The seminar principally aimed to prepare local public
servants for their new roles in national development
planning (see Muhammed, 1979:7; Onitiri, 1979:9).

Input for the Bauchi Local Government's Fourth Plan
proposals came from various sources. Heads of LG
departments and state ministry field agents played major
roles in shaping the 64 proposals submitted in response to
the Ministry for Local Government's circular of November
1979.[12] Touring reports forwarded by supervisory
councillors, along with notification regarding decisions
reached by hamlet, village, and district-level community
development committees on needed projects and communicated
through village-level workers and the Supervisor for
Community Development to the SLG and council, also received
some consideration in the planning process.[13] The precise
roles played by the FGPC and the full local government
council are not clear, although consideration by the latter
body is likely to have been perfunctory and limited to
debate over the location of projects. In his cover letter
to the MLG, the Secretary to the Bauchi LGA reported that
the attached proposals resulted from a series of meetings
held under his chairmanship with "all government officers."
He added that the Council "agreed" at its meeting of 31st
January 1980 that they should be submitted to the Ministry
"for consideration and approval."[14] If so, this
constituted one of the last acts of the subsequently
dissolved Bauchi Local Government Council.

Analysis of project submissions by the Bauchi Local
Government reveals striking contrasts with the practices
utilized in Kaduna. LGs in Bauchi State did not present
their plan proposals on the forms provided by the National
Planning Office. Instead, each local government submitted
a list of proposed projects described by title along with
the estimated annual capital and recurrent expenditure. No
written justification for the selection of particular
projects accompanied these proposals.[15] The LGs in Bauchi
State did not assign a priority ranking to any of their
projects. In short, they showed little interest in and
capacity for development planning. Lack of personnel
trained in project design and plan preparation at the local
level clearly contributed to this situation.[16]

Only 10 out of the 16 LGs in the state had managed to
submit their Fourth Plan proposals to the Ministry for
Local Government by April 1980.[17] In the meantime, the

Federal Government notified state authorities that it had imposed a ceiling of ₦400 million on all Bauchi State projects for the entire plan period. In contrast to their counterparts in Kaduna State, Bauchi State officials interpreted the ceiling to be <u>inclusive</u> of local government projects. The MLG, therefore, informed each local government that its combined capital and recurrent development plan expenditures for 1981-85 could not exceed ₦15 million. Most of the plans already submitted to the Ministry had exceeded the ₦15 million ceiling in estimated capital development charges alone.[18] The Ministry returned these plans to the LGs involved (including the Bauchi Local Government) in early April with instructions to revamp them in accordance with the newly imposed ceiling.[19]

Upon receiving revised proposals, the Ministry for Local Government (Inspectorate Division) completed the process of checking figures and compiled a consolidated expenditure summary for each LG. MLG and Ministry of Finance and Economic Development officials then reviewed and revised all local government plan proposals before transmitting each package to the State Planning Committee[20] for final administrative examination and approval prior to submission to the House of Assembly.

Process Weaknesses

Several problems in the process devised for involving Nigerian local governments in the national planning exercise are apparent from this discussion. In the Fourth Plan preparations, LGs in both states failed to provide meaningful and useful project justifications for most (all, in the Bauchi case) of their proposals. In spite of criticisms raised about previous local government planning practices (see Ajala, 1978b:1, 8), the Kaduna LG did not utilize cost-benefit analysis or decision matrices (see Chambers, 1978:216) in the selection and rating of the projects included in its proposed 1981-85 development plan, conduct the mandated feasibility studies or identify specific sites for projects, articulate and consider the relative merits of alternative approaches for achieving desired objectives (see Montgomery, 1979:59), or carry out citizen preference surveys.

The financial information contained in both sets of proposals typically spreads estimated capital expenditures evenly over the five-year period. Only a few project proposals give attention to accompanying recurrent costs or

provide estimates of expected revenue to be generated by
the scheme or service.[21] The failure of LGs in both states
to provide useful justifications for their Fourth Plan
proposals and to engage in meaningful project analysis
constitutes a serious barrier to the effective decentralization of planning. To be fair, lack of reliable data,
inadequate administrative capacity to undertake complex
project evaluation tasks (i.e., carry out surveys and
feasibility studies; conduct cost-benefit, cost-effectiveness, sensitivity or risk analysis; employ cost-minimiza-
tion or performance-maximization techniques; and establish
operational criteria for appraising progress),[22] and the
absence of an effective and responsive forum for articu-
lating citizen demands have made it difficult to assess and
justify capital-development projects at the local government level in Nigeria.

In addition, there is little room for effective public
or interest group participation in the project identification, design, and selection process.[23] For this reason,
among others, the 1979–1980 exercise departed markedly from
the ideal model of council involvement proposed by 'Ladipo
Adamolekun (1979a). The blanket dissolution of LG councils
which occurred in the wake of the return to civilian rule
also meant that none of the council members who approved
local development plans would be in a position to oversee
their execution. State government intervention at various
stages in the planning process and the overall lack of LG
autonomy further constrained the role of grass-roots units
of government. State government authorities also
complained about unilateral National Planning Office revision of their priorities during the final stages of the
Fourth Plan formulation process (Oyovbaire, 1985:257).
Finally, local planners and policy makers did not apply
appropriate criteria and standards when selecting,
reviewing, and assessing project proposals.[24]

As a result, the Fourth Plan proposals submitted by
the two LGs might be more aptly characterized as an
extension of regular departmental operating responsibilities than as the basis for a coherent long-range development plan. It should not be surprising, therefore, when
local government officers later evidence little plan
discipline and introduce new projects which compete for
scarce budgetary resources with those in the five-year
plan.

Avenues for Improvement

Three basic directions are available for alleviating prevailing deficiences in the evaluation of local government plan proposals. The first would require deeper involvement in plan preparation by well-trained state government planning officers. This approach possesses the twin disadvantages of violating both the spirit of Nigeria's 1976 local government reform and the objectives of the Guidelines for the Fourth Plan in that it would result in a reduction of local government autonomy and grass-roots involvement in planning. However, state government officials can perform an extremely valuable facilitating role for LGs engaged in development planning by strengthening and expanding their data-collection services. State involvement in data collection is advisable for three reasons: (1) the availability of more qualified research staff; (2) the need for standardization in data-gathering methodology and units of analysis; and (3) the cost effectiveness involved in sending a single team of researchers to work on a rotating basis in all local governments rather than establishing data collection units in every LGA.[25] To maximize the utility of this service, LG planners should be given authority to set data-collection objectives for the state government's research unit based upon common planning needs encountered at the local level, and the local government area should be established as the primary unit of statistical organization for all state research activities, reports, and planning studies.

As an alternative to deeper state government involvement, extensive efforts can be initiated to strengthen the capacity of LGs to engage in development planning on their own. Such efforts certainly deserve encouragement. However, an unrealistically high level of commitment and investment would be required to bring most local government organizations up to the point where they can successfully employ technically sophisticated planning techniques (see Caiden and Wildavsky, 1980:18-19). This assessment even applies to basic cost-benefit analysis. That technique still requires expensive investments in trained economists. Moreover, cost-benefit calculations exclude several criteria which policy makers should consider in project selection -- including equity and geographical distribution (see Roemer and Stern, 1975:viii, 3). Finally, there is

no guarantee that elected or appointed local government
councillors would understand and accept complex and frequently inappropriate planning devices in those areas where
the capacity to utilize them could be developed.

For these reasons, serious attention should be devoted
to another method of improving the evaluation of local
government development plans: impact analysis. As
developed in this book primarily on the basis of my own
thinking on the subject, this approach would require that
every proposal considered for adoption by a LG council
carry a straight-forward justification that is not only
grounded on local need, but also shows the likely or
projected impact of the project in terms of several
important and easily comprehended considerations. Whatever
this less complex method of project appraisal may lack in
terms of statistical precision is compensated for by
directness and simplicity.[26] In addition, reliance on
basic impact cost comparisons as the principal aid in
selecting appropriate projects is likely to be the most
realistic and suitable approach at the LG level. Impact
analysis requires that plan formulators ask "who benefits"
and "relative value" questions regarding proposed goals,
and facilitates critical evaluation of organizational
activities. It can be implemented immediately by local
governments lacking specialist personnel and highly
reliable data (see Conyers and Hills, 1984:83-85).[27] Since
the methodology, terminology, and figures are readily
understood, they are prone to be used in policy making by
democratically elected council members.

IMPACT COST ANALYSIS

Unfortunately, Nigerian planning authorities lumped
project proposals into broad and vague sectoral categories
during the 1980 planning exercise. Bauchi State employed
four standard classification categories: administrative,
economic, social, and regional development. These broad
sectors are ambiguously and arbitrarily defined. As a
result, the Bauchi Local Government classified ambulances
under the administrative sector, wells as "economic,"
slaughter slabs and houses under the social sector, and
markets and motor parks as "regional development" projects.
Questionable operationalization of broad sectoral headings
also occurred at the local and state level in Kaduna.[28]

In short, the sectoral categories used for aggregating
LG proposals and cost estimates in 1980 are not useful for

analytical or planning purposes. Indeed, sector is one of
the weakest and least uniformly applied planning categories; it simply "cannot be defined with sufficient
clarity nor manipulated with sufficient precision ..."
(Robertson, 1984:116-118). The sums reported for such
broad and ambiguous categories yield little information of
value to policy makers and planners seeking to assess the
likely overall impact of a development plan on local
classes, specific geographical areas, and natural
resources.[29] Local government planners need to break out
of this overly restrictive mode of analysis (see Caiden and
Wildavsky, 1980:x).

The analytical categories developed in this chapter
are employed in part to illustrate the advantages of
adopting new criteria for measuring the likely impact of
a local government's development plan. The criteria used
here allow for greater differentiation and more precise
impact assessments than do the broad categories used by
government agencies to date, while they remain simple to
apply and easy to comprehend.[30] They may be utilized
independently as alternative standards of evaluation, or in
combination in complementary fashion.[31] In either case,
all plan proposals are evaluated in several different ways
rather than relying exclusively on a single (sectoral)
breakdown. Thus, this approach takes cognizance of the
multiple policy issues and value decisions involved in
development planning and enables plan formulators to
balance or assign different weights to sectoral,
locational, socio-political, employment, basic needs, and
other important considerations[32] (see Chambers, 1978:212,
217; Murray, 1983:294-296; Curry, 1987:268-269).

In the analytic scheme employed here, an attempt is
made to define each category in relatively specific terms
in order to provide criteria which can be consistently
applied within and across local government areas. In
theory, given sufficiently tight category definitions, it
should be possible to reach concensus on where to assign a
particular project.[33] In questionable cases, planners
should apply criteria related to desired impact objectives
more strictly in order to promote accountability among LGs.
A procedure should be provided that permits inter-subjec-
tive verification or challenge of the way in which the
planner has distributed project proposals among the various
categories.

A secondary objective of this chapter is to undertake
a comparative analysis of the plan proposals submitted by
the Kaduna and Bauchi LGs according to the impact

categories recommended for evaluating local development
plans. Among the many interesting bases for comparison
between these two areas are the common status they possess
as capital city local governments and the fact that both
encompass rapidly growing urban population centers.[34] On
the whole, however, the Kaduna Local Government Area is
more densely populated, and the Bauchi LGA is smaller in
population size.[35] Kaduna also is a long-established
capital city, while Bauchi only has been a state
headquarters since 1976.

Primary source materials provide the data base for the
analysis presented in this chapter.[36] The Kaduna and
Bauchi figures for the Fourth Plan period utilize the cost
estimates which accompanied each LG's preliminary submission of a development plan. The initial grass-roots
proposals, which had not yet been modified at higher levels
of government, are the most interesting for our purposes
because they show how the two local governments approached
development planning. These plan proposals provide a
general indication of each LG's principal objectives[37]
and reveal their plans for and orientation toward future
capital undertakings. Additional data on the final revised
Kaduna LG development plan for the three-year period
1977-80[38] provide an historical benchmark against which the
Fourth Plan proposals can be assessed.

Total Cost Figures

Figures showing the total estimated cost of all
capital development projects proposed by the Kaduna and
Bauchi LGs for inclusion in the Fourth National Development
Plan, along with estimates for the approved projects
contained in the Kaduna Local Government's revised 1977-80
development plan, are presented in Table 6.1. These
figures must be treated cautiously for comparative
purposes since the Kaduna 1977-80 data refer to approved
rather than proposed projects, exclude all local education
projects, and cover a shorter time period relative to the
Fourth Plan exercise. The cost estimates initially
submitted for the Bauchi proposals, which later would be
substantially reduced by the local government, also
encompass a longer time frame.[39]

TABLE 6.1
Total Estimated Cost of All Proposed Capital
Development Projects; By LG and Plan Period

LG + Period (No. of Projects)	All Projects: Total Estimated Cost (₦'000)	Mean Cost Per Project (₦'000)
Kaduna 1977-80 (N=24)	2,920.0	121.6
Kaduna 1981-85 (N=66)	12,086.2	183.1
Bauchi 1980-85 (N=64)	34,355.5	536.8

Comparative analysis of the data found in Table 6.1 is
still useful, however, after these qualifications are taken
into account. We observe, for instance, that both the
total estimated cost and the mean cost per project for the
Kaduna Local Government's Fourth Plan proposals show only
modest increases over its revised 1977-80 development plan
when consideration is given to the inclusion of education
projects, the likelihood that some projects would later be
disapproved or reduced in scale and expenditure at higher
levels in the planning process, and inflationary trends.
The most striking observation drawn from Table 6.1 relates
to the especially high total and mean cost of the Bauchi
LG's preliminary proposals. Discounting the additional
year covered by the Bauchi cost estimates does not alter
the finding that its first submission required more than
twice the total estimated expenditure proposed for all of
the Kaduna LG's capital development projects. This result
is exactly the opposite of what one would expect based on
population considerations, since the Kaduna Local
Government Area is estimated to encompass at least twice as
many people as the Bauchi LGA incorporates.

Moreover, the Bauchi Local Government's proposals
evidenced a greater predilection toward high-cost projects
in comparison with those selected by the Kaduna Local
Government. This difference possibly resulted in part from
Bauchi's relatively recent designation as a capital city

and administrative perceptions that this status required large-scale infrastructural investment. In any event, both LGs generally favored large-scale projects over small-scale undertakings in their Fourth Plan proposals. The Table 6.1 figures, moreover, only constitute estimates of capital outlays. The escalating recurrent expenditures which accompany most new development projects are likely to equal or even exceed total capital expenditures over the plan period (see Caiden and Wildavsky, 1980:241). The recurrent plan estimates reported separately by the Bauchi Local Government exceeded ₦12 million for its proposed education projects alone. They are broken down as follows: personal emoluments (₦9,588,800); equipment (₦1,783,200); other materials (₦1,052,000).[40]

In sum, the data compiled in Table 6.1 suggest that the Kaduna Local Government adopted a relatively conservative and cautious approach to development planning for the first half of the new decade. The Bauchi LG's initial plan proposals, in contrast, evidenced lack of realism on the part of local planners. To cite one of several common projects that further illustrate this point, the Bauchi Local Government initially proposed to purchase eight ambulances, while the Kaduna LG's plan only called for the acquisition of two ambulances.

After the Federal Government notified state authorities that it had placed a ₦400 million ceiling on all Bauchi State projects (inclusive of local government proposals) for the entire plan period, the Ministry for Local Government returned the Bauchi LG's plan with instructions to revamp it so that combined capital and recurrent expenditures did not exceed ₦15 million. The total capital cost estimate contained in the final plan for the Bauchi LGA which officials at the state level approved and forwarded to Lagos amounted to ₦10,946,000. Planners only eliminated three projects entirely. However, one of the schemes dropped, construction of a new central market in Bauchi town, allowed a ₦10 million reduction in overall capital expenditures for the revised plan period (Bauchi Local Government, 1980). Table 6.2 shows that local planners cut a further ₦11 million by eliminating the new motor park project and scaling back three other undertakings (electrification schemes and the construction of classrooms and teachers' houses). The Bauchi Local Government managed to reduce its initial plan proposal by more than ₦23 million. Nevertheless, the cost of its final five-year development plan rivaled the total sum proposed for the much more populous Kaduna LGA.

Table 6.2
Cost Estimate and Project Differences in the Bauchi LG'S Initial and
Final Fourth Plan Proposals

	Project	-Estimated Costs-	
		Preliminary 1980-85 (₦'000)	Final 1981-85 (₦'000)[b]
(1)	**Projects with Reductions from Preliminary Proposal**		
	Vehicle purchases	128	118
	Land compensation	800	500
	Furnishings for vet clinics	30	18
	Purchase (15;10) refuse collection vans	289	190
	Purchase (8;5) ambulances	705[a]	51.5
	Purchase (10;5) vehicles	150	100
	Purchase agric vehicles & equipment	68.5	51.8
	Electrification schemes	3,500	500
	Inter-village roads	1,000	500
	Construction eight staff quarters	80	50
	Construction one mighty incinerator	30	--
	Construction (899;500) classrooms	4,790.5	2,750
	Construction (1450;500) teachers' houses	5,150	2,200
	Construction (2580;1000) toilets	1,485	742
	Construction new market	10,000	--
	Construction new motor park	3,000	--
	Subtotal Estimated Expenditures	31,206	7,771.3
(2)	**Projects with Additions to Preliminary Proposal**		
	Purchase office equipment	19.1	30.9
	Purchase one Bedford Kit car	--	9
	Refence orchard at Bauchi	--	4
	Subtotal Estimated Expenditures	19.1	43.9
(3)	Total Estimated Costs (1 + 2)	31,225.1	7,815.2
(4)	Net Reduction From Initial to Final Proposal		23,409.9

[a]This figure first appeared as ₦705,000 instead of ₦75,000 -- presumably
due to typographical error.
[b]SOURCE: Bauchi Local Government (1980).

Principal Sector

     Sectoral, or functional, analysis is a widely utilized
and potentially valuable method of evaluating the projected
impact of a development plan (see Conyers and Hills, 1984:
198). The Federal Government's Guidelines for the Fourth
Plan specified certain sectoral priorities which serve as a
useful benchmark against which to measure local government
capital development proposals. The Guidelines assigned
first priority to agricultural production and processing,
followed by expansion of the economic infrastructure sector
(particularly electric power, water supply, and telecommunications). The other sectors which should receive
priority attention according to this document are transportation (secondary and tertiary roads; public transport
within urban areas), education, health, and housing.
Education, health, agriculture, and transportation again
received specific mention in the Guidelines, along with
community development, as the sectors in which local
governments are expected to concentrate their development
projects during the 1981-1985 plan period.
     The data set forth in Table 6.3 show that the four
categories relied upon by state and local governments for
sectoral analysis of LG plans in Bauchi State[41] are too

Table 6.3
Estimated Cost of Bauchi LG's Initial and
Final Proposed Capital Development Projects;
By Sectors Identified by State and Local Governments
for Planning Purposes in 1980

Sector	-Initial Proposals-		-Final Proposals-	
	Est. Cost (₦'000)	% Total Est. Cost	Est. Cost (₦'000)	% Total Est. Cost
Administrative	2,186.6	6.5%	1,897.2	17.3%
Economic	5,285.9	15.8	1,785.9	16.3
Social	12,694.5	37.8	6,901.0	63.0
Regional Development	13,373.0	39.9	362.0	3.3

Source: Figures compiled by the Bauchi State Ministry for Local
        Government, April, 1980; Bauchi Local Government (1980).

broad to permit meaningful assessment of the extent to which development proposals are consistent with the national and local priorities specified in the Guidelines.[42] We are only able to discover that the heaviest cuts in the Bauchi Local Government's initial proposals occurred within the regional development (₦13 million in reductions) and social (₦6 million) categories. The proportion of total planned expenditures devoted to social-sector projects increased from 38 per cent to 63 per cent, while the share of the total expenditure pie allotted to administrative-sector projects more than doubled.

Distributing project proposals and accompanying cost estimates into ten more narrowly differentiated sectoral categories (Table 6.4) facilitates analysis of the relative emphasis placed on different sectors in a local government's proposed development plan and comparisons with national priorities and expectations.[43] The cost figures found in Table 6.4 indicate that the proposed Kaduna and Bauchi development plans assigned the majority share of anticipated capital investment resources to three of the priority sectors identified in the Federal Government's Guidelines: infrastructure (including housing), education, and transportation. However, three other recommended priority sectors received relatively little financial support under the Kaduna and Bauchi proposals (a total of 16% and 5%, respectively). These are agricultural production (given top priority in the national Guidelines), health, and community development.[44] The failure to emphasize agricultural projects is particularly disturbing in the Bauchi case, given the predominantly rural character of two of the three districts in that LGA, and less serious with respect to Kaduna Local Government, where most agricultural projects support the activities of urban-based absentee farmers.[45]

The data presented in Table 6.4 also are revealing in terms of the relationship between the number of planned projects within each sector and their total estimated cost. In the proposed Kaduna plan for 1981-85, there are only slight variations between the proportion of all projects planned for and the proportion of total estimated costs assigned to most sectors. The chief exceptions are (1) the education sector, which accounted for 14 per cent of all project proposals but would receive 24 per cent of total estimated expenditures over the four year period, and (2) the culture, information, sports, and recreation sector, to which planners assigned the most projects (10) but the least financial resources (2% of total estimated costs).

The Bauchi Local Government plan, in contrast, is marked by considerable variation in nearly every sector between the proportion of projects and the percentage of total estimated costs. The education, infrastructure, and economic sectors enjoy a higher percentage of anticipated total financial resources than one would expect based on the proportion of projects devoted to those areas, while the converse applies to the administrative, health, agricultural, and cultural/recreational sectors. One virtue of this type of analysis, then, is that it reveals the plan sectors where high-cost projects are concentrated.

Finally, several suggestive insights emerge through historical and comparative analysis of the figures assembled in Table 6.4. The Kaduna Local Government's 1981-1985 proposals indicate a move away from spending on

TABLE 6.4
Comparative Analysis of Local Government Development Plan Proposals (Projects and Estimated Cost); By Principal Sector

	- LG + Years -			
	Kaduna 1977-80			
Principal Sector	No. Projs.	% All Projs.	Est. Cost (₦'000)	% Total Est. Cost
Education	-	-	-	-
Bureaucracy, administration	4	16.7%	867	29.7%
Infrastructure & amenities[a]	3	12.5	248	8.5
Transport[b]	6	25.0	926	31.7
Economic[c]	2	8.3	190	6.5
Health	5	20.8	309	10.6
Sanitation	3	12.5	280	9.6
Agric. Production	0	0.0	0	0.0
Culture, Inform-ation & Recreation[d]	0	0.0	0	0.0
Community Development	1	4.2	100	3.4

(continued)

TABLE 6.4 (continued)
Comparative Analysis of Local Government Development Plan
Proposals (Projects and Estimated Cost); By Principal
Sector

Principal Sector	No. Projs.	% All Projs.	Est. Cost (₦'000)	% Total Est. Cost
		Kaduna 1981-85		
Education	9	13.6%	2854.9	23.6%
Bureaucracy, admin.	8	12.1	1893.0	15.7
Infrastructure & amenities[a]	6	9.1	1743.0	14.4
Transport[b]	9	13.6	1745.0	14.4
Economic[c]	7	10.6	890.2	7.4
Health	7	10.6	874.0	7.2
Sanitation	4	6.1	758.5	6.3
Agric. Production	4	4.5	810.0	6.7
Culture, Inform- ation & Recreation[d]	10	15.2	217.1	1.8
Community Development	3	4.5	300.5	2.5
		Bauchi 1980-85		
Education	1	1.5%	4790.5	13.9%
Bureaucracy, admin.	7	10.8	157.9	.5
Infrastructure & amenities[a]	11	16.9	10719.0	31.2
Transport[b]	7	10.8	4350.0	12.7
Economic[c]	12	18.5	10420.1	30.3
Health	7	10.8	1298.0	3.8
Sanitation	7	10.8	1952.0	5.7
Agric. Production	9	13.8	425.0	1.2
Culture, Inform- ation & Recreation[d]	4	6.2	243.0	.7
Community Development	0	0.0	0.0	0.0

[a]Includes housing, water supply, electricity.
[b]Includes roads, street lights, motor parks.
[c]Includes markets, slaughter slabs, forest & grazing
 reserves, employment skills.
[d]Includes sports.

projects in the administrative and transport sectors in
comparison with its approved 1977-80 development plan, and
moderate increases in proposed funding for infrastructural
and agricultural sector projects. In comparison with the
Kaduna 1981-85 plan, the Bauchi LG proposed to allocate
smaller proportions of its capital-development funds to
administrative, education, and agricultural-production
projects, and higher proportions to the economic and
infrastructural sectors.

Principal Objective

To aid in proposal evaluation, each project also is
reported in this study according to its primary objective.
This method of analysis focuses on the principal objective
or purpose of plan schemes and services without regard to
sector. For instance, the construction of teachers
quarters constitutes an education sector project, but the
primary objective behind the activity is administrative
(attracting and retaining teaching staff). Similarly,
projects falling in the economic sector are intended by a
local government primarily as revenue-earning undertakings
(e.g., markets) or as non-revenue earning social services
(e.g., home economics training). Transport sector projects
include road construction and revenue-raising (e.g., motor
parks) activities. The eight-fold classification scheme
utilized in preparing Table 6.5 allows for these and other
distinctions to be made according to the principal
objective or purpose of each proposed project.
The data found in Table 6.5 reveal the utility of
analyzing projects by objective in addition to sector. For
instance, a comparison of Tables 6.4 and 6.5 shows immediately that many projects which are basically administrative in purpose are not uncovered through sectoral
analysis. Using the primary objective of a project as
one's point of departure indicates that 17 projects com-
manding 19 per cent of total estimated plan expenditures
are oriented toward administrative ends in Bauchi (Table
6.5). Yet, only seven projects accounting for less than
one per cent of total estimated costs fall strictly within
the bureaucratic sector (Table 6.4). Similar increases in
administrative projects and costs are uncovered with
respect to the Kaduna LG's 1981-1985 proposals when project

analysis according to purpose is juxtaposed against the
sectoral classification scheme.

TABLE 6.5
Comparative Analysis of Local Government Development Plan
Proposals (Projects and Estimated Cost); By Principal
Objective

		- LG + Years -	
		Kaduna 1977-80	
Principal Objective	No. Projs.	% All Projs.	Est. Cost (₦'000)	% Total Est. Cost
Administration	5	20.8%	1057	36.2%
Social service provision	11	45.8	747	25.6
Revenue generation	2	8.3	190	6.5
Roads	6	25.0	926	31.7
Education	-	-	-	-
Production economic goods	0	0.0	0	0.0
Control; maintain order	0	0.0	0	0.0
Aesthetic	0	0.0	0	0.0
		Kaduna 1981-85	
Administration	19	28.8%	4172.0	34.5%
Social service provision	19	28.8	2223.9	18.4
Revenue generation	5	7.6	910.7	7.5
Roads	7	10.6	2121.0	17.5
Education	8	12.1	2285.5	18.9
Production economic goods	1	1.5	200.0	1.7
Control; maintain order	6	9.1	83.1	.7
Aesthetic	1	1.5	90.0	.7

(continued)

TABLE 6.5 (continued)
Comparative Analysis of Local Government Development Plan
Proposals (Projects and Estimated Cost); By Principal
Objective

| | | - LG + Years - | | |
| | | Bauchi 1980-85 | | |
Principal Objective	No. Projs.	% All Projs.	Est. Cost (₦'000)	% Total Est. Cost
Administration	17	26.2%	6434.9	18.7%
Social service provision	23	35.4	8311.0	24.2
Revenue generation	14	21.5	13428.1	39.1
Roads	2	3.1	1020.0	3.0
Education	2	3.1	4810.5	14.0
Production economic goods	5	7.7	342.0	1.0
Control; maintain order	1	1.5	6.0	a
Aesthetic	1	1.5	3.0	a

a = less than .1%

In addition, the classification method used in Table
6.5 facilitates evaluation of local government plan
proposals according to issues of major concern to students
of development administration. Government bodies are
usually held to retard development to the extent that they
emphasize control-oriented functions, maintenance of the
status quo, administrative overhead, and (in some quarters)
urban road construction.[46] Development-oriented government
structures, in contrast, would devote greatest attention
and the largest share of their scarce resources to the
provision of social services (including education),[47] and
to projects aimed at promoting productive economic activities directly or generating public revenues that will
exceed investment and maintenance expenses over the short
or long term and, thereby, help to defray the high costs of
providing social services. The Central Planning Office's

Fourth Plan guidelines stated that the "project selection
exercise must ... attempt to strike a reasonable balance
between the need to satisfy the demand for various welfare-
raising or amenity type of projects and those aimed at
raising productivity and incomes more directly ..."
(Nigeria, C.P.O., 1979:38).

By applying these criteria for distinguishing between
static and dynamic government structures to the data at
hand in Table 6.5, one observes that the Bauchi Local
Government's overall proposals placed far more emphasis
than those selected by the Kaduna LG on developmental
objectives. Thus, nearly 40 per cent of the Bauchi LG's
total proposed expenditures are devoted to revenue-
generating projects, compared with 7.5 per cent of the
Kaduna Local Government's total estimated plan costs.[48]
According to Ahmad Abubakar (1979:141), the most common
complaints registered by people living in the Bauchi LGA
"are about shortage of water and roads." Water needs and
other frequently mentioned complaints about public
services, with the exception of health-related matters and
rural road construction, are addressed in terms of the
relative distribution of capital expenditures proposed in
the Bauchi 1980-85 development plan.

Local planners allocated more than half of the Kaduna
LG's proposed 1981-85 expenditures to administration and
road construction/maintenance projects. The same pattern
is observed with respect to the Kaduna Local Government's
final 1977-80 plan expenditures. In contrast, the Bauchi
LG's Fourth Plan proposal only calls for 22 per cent of
total expenditures to be spent on these two objectives.
Overall, the Kaduna Local Government is more vulnerable
than the Bauchi Local Government is to criticism that it
failed to operationalize the CPO's recommended balance
among welfare and income-generating projects.

Given that education has consumed the lion's share of
the capital expenditure budgets of many local governments
(Aliyu and Koehn, 1982:67), it is surprising to discover
from Table 6.5 that both the Kaduna and the Bauchi LGs
proposed to undertake a relatively small number of
education projects during the plan period and to devote
less than 20 per cent of their total plan expenditures to
the pursuit of educational goals. This outcome is
consistent with the Federal Government's Fourth Plan policy
guideline (p. 70) that "while the importance of education
will continue to be recognized, steps will be taken to
ensure that it does not take up a disproportionate amount
of available resources."

Principal Project Type and Method

Local governments should also find it useful for
policy-making purposes to analyze all capital projects and
expenditures proposed for inclusion in the development plan
by the predominant type of activity and method of executing
the work involved. In Table 6.6, the projects and costs
proposed in the Kaduna and Bauchi development plans are
distributed among six distinct categories based on the
principal type of activity that would be carried out.
The results show that a majority of the proposed
projects are of the construction type in both local
governments, with the Bauchi development plan containing a
slightly higher proportion of construction projects and
anticipated expenditures on new construction activity in

TABLE 6.6
Comparative Analysis of Local Government Development Plan
Proposals (Projects and Estimated Costs); By Principal Type
of Expenditure

| | | - LG + Years- Kaduna 1977-80 | | |
Principal Type of Expenditure	No. Projs.	% All Projs.	Est. Cost (₦'000)	% Total Est. Cost
Construction	16	66.7%	2402	82.3%
Equipment	3	12.5	380	13.0
Maintenance, repairs[a]	1	4.2	30	1.0
Vehicle Purchases	4	16.7	108	3.7
Land Purchases	0	0.0	0	0.0
Other[b]	0	0.0	0	0.0
		Kaduna 1981-85		
Construction	38	57.6%	9348.1	77.3%
Equipment	13	19.7	1425.1	11.8
Maintenance, repairs[a]	5	7.6	668.5	5.5
Vehicle Purchases	6	9.1	474.0	3.9
Land Purchases	1	1.5	100.0	.8
Other[b]	3	4.5	70.5	.6

(continued)

TABLE 6.6 (continued)
Comparative Analysis of Local Government Development Plan
Proposals (Projects and Estimated Costs); By Principal Type
of Expenditure

		- LG + Years-		
		Bauchi 1980-85		
Principal Type of Expenditure	No. Projs.	% All Projs.	Est. Cost (₦'000)	% Total Est. Cost
Construction	41	63.1%	30653.0	89.2%
Equipment	5	7.7	120.9	.4
Maintenance, repairs[a]	1	1.5	1000.0	2.9
Vehicle Purchases	9	13.8	1503.0	4.4
Land Purchases	9	13.8	1079.6	3.1
Other[b]	0	0.0	0.0	0.0

[a]Includes replacement parts and materials
[b]Includes travel, ceremonial expenses, matching funds

comparison to the Kaduna plan. Both LGs also proposed to
allocate only a small proportion of total capital investments to maintenance activities, repairs, and the purchase
of replacement parts and materials (particularly in relation to expenditures on construction and equipment),
although the proportion devoted to maintenance increased
slightly in the Kaduna 1981-85 proposal over its 1977-80
approved plan (from 1% to 5.5%). In any event, expenditures on maintenance items (as well as the 'other'
category) fit more appropriately in a local government's
recurrent estimates rather than under new capital investments on development activities. Finally, the Kaduna LG
intended to expend a much higher proportion of total
plan funds on equipment than the Bauchi Local Government
did (12 per cent versus less than 1 per cent), while the
Bauchi proposals placed greater emphasis on land purchases.
In general, both plans devoted disproportionate attention
to new construction work relative to other essential types
of development activity. A. Y. Aliyu (1980a:12-13) points
out that "the heavy emphasis on building projects would

seem to arise from a basic misconception among federal,
state and local government officials alike that social or
economic development is best indicated by the number of
physical structures constructed."

In assessing the likely impact of a local government's
development plan, it is useful to devote attention to
certain "side effects" of public investment which frequently are as important for the local (and national)
economy as the specific project undertaken. Two factors
which merit special consideration in this regard are
employment and self reliance.[49] The national Guidelines
referred to both factors in its section (p. 19) on
objectives for the Fourth Plan period. Included among the
five newly introduced objectives for the 1981-1985 National
Development Plan were "reduction in the level of under
employment" and "greater self reliance--that is, increased
dependence on our own resources in seeking to achieve the
various objectives of society." These objectives are
clearly relevant at the local level. By fostering additional employment opportunities, LGs directly raise the
income levels of residents and promote their material
welfare, which in turn stimulates growth in the local
economy (see, for instance, Curry, 1987:263). At the same
time, reliance on indigenous resources and products reduces
dependence on foreign or externally based producers and
conserves scarce capital and foreign exchange.

Although not a perfect measure, both objectives are to
a large extent promoted by labor-intensive projects and
retarded by capital-intensive undertakings. Labor-
intensive activities clearly provide wide employment opportunities, tend to utilize local resources (including
human resources), and rely on indigenous and sustainable
techniques or, at most, intermediate forms of technology
which need not be imported (see Bryant and White, 1982:149;
Hellinger, et al., 1983:45).[50] Therefore, the capital-
versus labor-intensive distinction is employed as another
indicator of the nature and likely impact of the Kaduna and
Bauchi development plans. Projects classified as capital-
intensive include all purchases (vehicles, equipment, and
land) and construction projects where the percentage expenditure on machinery, equipment, and materials (often imported) is likely to exceed the percentage spent on salaries
and wages. Examples taken from the Kaduna Local Government's proposals for 1981-85 are the purchase of ambulances
and tractors (Project Nos. 15 and 25), construction of an
office block, veterinary clinic, and stores (Project No.
26),[51] and construction of TV viewing centres (Project No.

31). Projects considered to be labor-intensive include
those designed to promote community self-help activities by
providing simple tools such as shovels, diggers, wheel
barrows, etc. (No. 9), and construction projects utilizing
"the traditional system" (Nos. 20-22) or relying primarily
upon manpower (e.g., the construction and maintenance of
feeder roads in rural areas).[52]

The distinction between capital- and labor-intensive
methods of executing development projects has been used in
constructing Table 6.7. The results indicate that at least
three-fourths of the projects and expenditures proposed or
approved in the 3 plans under review require predominantly
capital-intensive methods of implementation. Indeed, the
Bauchi Local Government's 1980-85 proposed plan called for
96 per cent of total costs to be expended on capital-
intensive projects. A slight shift on the part of the
Kaduna Local Government away from expenditure on capital-
intensive projects and toward the funding of labor-
intensive activities is observed by comparing its 1981-85
proposals with the 1977-80 approved revised plan.

TABLE 6.7
Comparative Analysis of Local Government Development Plan Proposals
(Projects and Estimated Costs); By Principal Method

Principal Method	No. Projects	% All Projects	Estimated Cost (₦'000)	% Total Est. Cost
Kaduna 1977-80				
Capital-intensive	18	75.0%	2393.0	82.0%
Labor-intensive	6	25.0	527.0	18.0
Kaduna 1981-85				
Capital-intensive	50	75.8	8971.5	74.2
Labor-intensive	16	24.2	3114.7	25.8
Bauchi 1980-85				
Capital-intensive	52	80.0	33024.4	96.1
Labor-intensive	13	20.0	1331.1	3.9

Nevertheless, the overriding conclusion to be drawn from
the data found in Table 6.7 is that capital-intensive
methods of operation far outweighed labor-intensive
techniques in the development planning undertaken by the
Kaduna and Bauchi LGs for the Fourth Plan period.

Principal Beneficiaries

A compelling argument can be made that the most
important factor in assessing a development proposal is the
likely beneficiary. The reduction of inequality lies at
the heart of development (see Conyers and Hills, 1984:29).
This goal can be conceptualized in three distinct ways for
the purpose of development plan analysis: (1) principal
immediate project beneficiaries (intra-versus extra-
organizational); (2) principal socio-economic background
of intended project beneficiaries; and (3) principal location (urban versus rural) of project beneficiaries.[53] The
first step in a development-planning strategy directed
toward reducing poverty and inequality is for those
involved in the project-selection process to raise "who
benefits?" questions. The objective is better informed and
more sensitive judgment (Chambers, 1978:216-217;
Montgomery, 1979:59).

Public Versus Staff. Few would disagree that development planning should be geared primarily toward meeting the
needs of citizens rather than the bureaucracy. Nevertheless, a sizeable share of public sector capital investment
may in practice be devoted to meeting the immediate needs
of administrators for such facilities as office buildings
and equipment, vehicles, and staff housing (see Aliyu,
1980a:12; Aliyu and Koehn, 1982:25-26; Graf, 1986:107;
Koehn, 1983a:18-19; Osoba, 1979:73). While public administrators can play an important role in promoting development
and a certain level of facilities is essential for the
effective and dedicated performance of this function,
important policy questions arise concerning the wisdom of
allocating a major share of scarce capital resources to
internal bureaucratic needs rather than to direct public
service programs. By analyzing a proposed local government
development plan in terms of each project's principal immediate beneficiary, policy makers and planners are able to
assess the extent to which it concentrates on bureaucratic
or public needs. In addition, this method of analysis
allows policy makers to plan and enforce the desired ratio

of new capital project investments among the two categories of immediate beneficiaries.

A breakdown of the Kaduna and Bauchi proposals according to principal immediate project beneficiary is given in Table 6.8. Construction projects have been classified as internally or staff-oriented whenever a major part of the work involves the building of offices. Local government administrators also are held to be the foremost beneficiary of projects that call for vehicle or equipment purchases, or the construction of staff housing. The public is deemed to benefit from all other projects.

The results shown in Table 6.8 indicate that nearly twice as high a proportion of both projects and proposed capital investments served staff interests in the Kaduna Local Government's 1981-1985 plan in comparison with figures for the Bauchi LG's initial plan. Indeed, the proportion of total expenditures allocated to staff-oriented projects in the Kaduna 1981-1985 proposed plan showed a slight increase over its approved 1977-1980 plan.

TABLE 6.8
Breakdown of Local Government Development Plan Proposals
(Projects and Estimated Costs); By Principal Immediate Beneficiaries

Principal Immediate Beneficiaries	No. Projects	% All Projects	Estimated Cost (₦'000)	% Total Est. Cost
		Kaduna 1977-80		
Public	19	79.2%	1863	63.8%
Staff	5	20.8	1057	36.2
		Kaduna 1981-85		
Public	41	62.1	7498.7	62.0
Staff	25	37.9	4587.5	38.0
		Bauchi 1980-85		
Public	49	75.4	27990.6	81.5
Staff	16	24.6	6364.9	18.5

The 4-1 ratio of capital investments on projects directly benefitting citizens versus staff achieved in the Bauchi Local Government's 1980-85 proposals clearly represents a more "people-oriented" development approach than the 3-2 ratio encountered in the Kaduna case. Nevertheless, even higher capital-expenditure ratios than 4-1 certainly can be justified as targets for planners to strive to attain in dynamic local governments.

Elite Versus Mass. An official objective of Nigerian development planning is attaining "more even distribution of income among individuals and socio-economic groups" (Nigeria, FMNP, 1979:19). It is crucial for policy makers to assess the extent to which progress in this direction is likely to be achieved through a local government's proposed plan. The main challenge is to design projects that benefit the poorest residents (Bridger and Winpenny, 1983:106). This aspect of development planning is measured in part by the classification scheme described above, whereby the principal immediate beneficiaries of proposed projects (the administrative elite versus all others) form the basis of analysis. A more complete picture of the likely impact of a development plan can be secured by examining projects according to the socio-economic background of their principal intended beneficiaries.

Two broad categories are used in this analysis for the purpose of providing a rough indication of the likely distributional and redistributional impact of proposed LG capital-development programs. Projects are defined as elite-oriented when they mainly promote the interests or meet the needs of the relatively wealthy classes of local society. A good example is the construction of district head residences (Project No. 2001). In addition to projects aimed at serving administrative needs, other examples of elite-directed undertakings are the purchase of machinery used to empty septic tanks (Project No. 1002), the provision of or improvements to parks, gardens, and open spaces (No. 1049), and the installation of street lighting (No. 1065). Projects designed to benefit relatively poorer rural and urban residents, such as education, feeder roads, health and sanitation schemes, are categorized as mass-oriented in terms of principal intended beneficiaries.[54]

When the Kaduna and Bauchi development plans are analyzed along these lines, the findings (Table 6.9) prove interesting in a couple of respects. First, in comparison comparison with the 1977-80 plan, the Kaduna Local

Table 6.9

Breakdown of Local Government Development Plan Proposals
(Projects and Estimated Costs); By Principal Socio-economic
Background of Intended Beneficiaries

Principal Intended Beneficiaries	No. Projects	% All Projects	Estimated Cost (₦'000)	% Total Est. Cost
		Kaduna 1977-80		
Elite	6	25.0%	1107.0	37.9%
Mass	18	75.0	1813.0	62.1
		Kaduna 1981-85		
Elite	31	47.0	6022.1	49.8
Mass	35	53.0	6064.1	50.2
		Bauchi 1980-85		
Elite	24	36.9	7300.5	21.2
Mass	41	63.1	27055.5	78.8

Government's 1981-85 proposals revealed a decided shift
away from mass-oriented and toward elite-oriented
activities both in terms of the proportion of projects
falling within each category and in terms of capital
investment allocations. Planners devoted about half of the
Kaduna 1981-85 total estimated capital expenditures to
elite-based projects. In sharp contrast, the Bauchi
1980-85 proposed plan allocated roughly 80 per cent of
total estimated capital costs to projects principally
intended to benefit the masses.

The utility of multidimensional impact analysis is
particularly striking at this point. Most local government
planners and policy makers would be hard pressed to justify
and defend allocating half of all new capital investments
to elite-oriented projects. Impact analysis clearly
reveals the distorted sense of development plan priorities
which guided the Kaduna LG's Fourth Plan proposals.
Indeed, responsible policy makers are even likely to find

the 4 to 1 cost ratio which prevailed in the initial Bauchi
proposals to be indefensibly generous to the elite minority
residing in the local community.

Rural Versus Urban. Another important dimension of
development planning concerns the location of intended
project beneficiaries. The distribution of proposed
projects and expenditures among the rural and urban areas
falling within a local government's jurisdiction constitutes a prime locational consideration for policy makers
and planners.[55] By concentrating capital-development
projects in rural areas, most LGs in Africa would be
assigning priority to the residential location of the vast
majority of their populace. Concomitantly, a rural-
oriented development plan might result in some reduction in
migration to urban centres, an explicit objective of
Nigeria's Fourth National Development Plan (Nigeria, FMNP,
1979:19).
        In most cases, it is relatively easy for planners to
determine whether the impact of a given plan proposal is
principally likely to benefit rural or urban inhabitants.
Rural-oriented projects include agricultural schemes and
equipment, inter-village roads, forest reserves, and all
other programs specifically designated to take place or be
located in a village or other rural setting. Urban
projects are those located at the local government
headquarters, most sewage evacuation and piped water
schemes, and all other programs scheduled to take place or
be located in a city or town. Another group of projects,
however, defies classification as predominantly rural-or
urban-oriented. These are projects which may be located
both in rural and urban settings, or involve the purchase
of equipment which will be used in rural as well as urban
environments. In this analysis, such rural/urban projects
are treated under the category "both."[56]
        In Table 6.10, each project and its estimated cost is
classified and reported on the basis of the residential
location of the principal beneficiaries. As one would
expect, given the predominantly urban character of the
Kaduna LGA, its proposed 1981-85 projects are more likely
to benefit city dwellers than is the case with the Bauchi
Local Government's proposed plan, which covers two
primarily rural districts in addition to Bauchi town. In
comparison with its 1977-80 approved development plan,
moreover, the Kaduna Local Government's 1981-85 submission

Table 6.10
Breakdown of Local Government Development Plan Proposals
(Projects and Estimated Costs); By Principal Location

Principal Location	No. Projects	% All Projects	Estimated Cost (₦ 000)	% Total Est. Cost
		Kaduna 1977-80		
Rural	2	8.3%	58	2.0%
Urban	7	29.2	1332	45.6
Both	15	62.5	1530	52.4
		Kaduna 1981-85		
Rural	7	10.6	1554.7	12.9
Urban	50	75.8	8317.0	68.8
Both	9	13.6	2214.5	18.3
		Bauchi 1980-85		
Rural	17	26.2	1659.1	4.8
Urban	28	43.1	15490.9	45.1
Both	20	30.8	17205.5	50.1

aimed to undertake a smaller proportion of projects that
serve all residents and, in their place, to increase the
proportion of activities and total capital investment
devoted to specifically rural- and urban-oriented
projects.[57]
    The most noteworthy results that emerge from treatment
by the principal location of intended beneficiaries concern
the Bauchi plan proposals. Roughly one out of every four
proposals initially selected by the Bauchi Local Government
for inclusion in the Fourth Plan would exclusively benefit
rural residents, and planners allotted less than five per
cent of total estimated costs to these projects. Given the
rural character of much of the LGA, the Bauchi proposals
reveal a surprising inclination to select urban and joint
development projects over strictly rural ones. It is

difficult to justify an urban bias of such magnitude in
this case (also see Chambers, 1978:209-210, 217).

The most expensive projects which the Bauchi LG
proposed to execute during the plan period are those
intended to serve both rural and urban dwellers. The mean
cost per project in the "both" category is ₦860,275, while
it is ₦553,246 for all urban programs and only ₦97,594 for
all rural projects. Policy makers are advised to probe
deeper into the sources of such cost discrepancies. It is
possible that they are justified in part by the nature of
urban versus rural projects (e.g., piped water supply
versus boreholes) or by the scale of the operations
involved. On the other hand, it is normally more expensive
to serve a geographically dispersed population than a
densely settled one. The appropriateness of large-scale
urban projects must be taken into consideration by local
planning authorities.

Distribution by District

The issue of balanced development among geographical
regions is important for local governments, as well as at
the state and national levels, for economic and political
reasons (see Nigeria, FMNP, 1979:19). Development planning
can and should be used as an administrative opportunity to
bring about a more equitable distribution of services among
and within the territorial subdivisions comprising a LGA.
This endeavor requires, at the initial and subsequent
stages of the planning process, that attention be devoted
to the location and status of existing services and
amenities, as well as to assuring reasonable geographical
balance in the provision of new projects during the plan
period.[58]

For illustrative purposes, it suffices to analyze a
local government's proposed development plan on a district-
by-district basis. With a longer list of projects before
them and greater specificity in submitted program descrip-
tions regarding the precise location of projects, however,
policy makers are advised to shift the level of comparative
analysis one step closer to the grass roots; i.e., to the
village level.[59]

A complete breakdown by district of the Kaduna and
Bauchi proposals is presented in Table 6.11. In
constructing this table, the estimated costs of projects
intended to serve all districts are divided equally among
them, while proposed expenditures on projects located
unevenly in two or more districts are distributed in
proportion to the extent of the activity to be undertaken
in each. For example, the costs of purchasing tools for
self-help efforts at each district headquarters in Kaduna
(Project No. 1009) are spread evenly over the five
districts, while proposed expenditures on dispensary and
clinic equipment (Project No. 1001) are allocated only to
the three districts in which clinics are to be equipped
during the 1981-85 period (with Tudun Wada receiving half
of the estimated amount since 2 of the 4 clinics involved
are located in that district).

Several suggestive findings can be drawn from the
comparative figures reported in Table 6.11. First, the
approved 1977-80 development plan for the Kaduna Local
Government shows great disparity in the costs of projects
located in different districts. Out of the total estimate
for all projects affecting the various districts, nearly
half (43 per cent) is concentrated in the central city
district of Doka, with one-fourth allocated to Tudun Wada,
and the other three districts sharing what remains. The
proposed Kaduna plan for 1981-1985 exhibits greater balance
in the distribution of estimated project costs among the
five districts. Moreover, the local government made a
deliberate effort this time to locate various development
projects in areas where similar facilities were not already
available.[60] As a result, Doka and Gabasawa (G.R.A.)
districts received the smallest share of projected plan
expenditures for 1981-85. Once again, however, the Kaduna
LG assigned a disproportionate amount of capital-
development funds to proposed projects in Tudun Wada
district.

The Bauchi results reveal the gross disparity which
exists in the estimated costs of proposed projects located
in Bauchi town in comparison with the two outlying
districts of that LGA. Although one subsequently eliminated project (construction of a new ₦10 million central
market) inflated the initial figures, the Bauchi plan still
calls for more investment on projects sited in the town
than the combined totals for schemes located in the two
other districts when the central market project is excluded
from consideration. This planning bias prevailed in spite
of the fact that the LG had neglected residents of Galambi

Table 6.11
Breakdown of Local Government Plan Proposals
(Projects and Estimated Costs); By District

Kaduna 1977-80

District (1979 est. pop. 'ooo's)[a]	No. Projs.	Est. Cost (₦'000)	% Total Est. Cost
Kawo (102)	9	242.9	11.6%
Gabasawa (45)	8	183.0	8.8
Makera (60)	9	215.9	10.3
Tudun Wada (105)	10	553.7	26.5
Doka (131)	8	890.5	42.7
TOTAL	-	2086.0	99.9

Kaduna 1981-85

Kawo (102)	31	1592.4	21.1%
Gabasawa (45)	27	1222.8	16.2
Makera (60)	26	1264.4	16.8
Tundun Wada (105)	32	2231.7	29.6
Doka (131)	24	1222.4	16.2
TOTAL	-	7533.7	99.9

Bauchi 1980-85

Bauchi Town (87)	20	14708.1	79.5%
Zungur (69)	14	1900.3	10.3
Galambi (40)	12	1882.7	10.2
TOTAL	-	18491.1	100.0

[a]Kaduna State, Ministry of Economic Development
(1977:10a); Bauchi Local Government.

in the community development activities it undertook from
1976-1979 (see Aliyu and Koehn, 1982:26-28).

The figures in Table 6.12 are provided primarily in
order to illustrate how data on estimated project costs by
district can be correlated with other important
characteristics of plan proposals. For this purpose, the
principal method and objective of project proposals are
broken down on a cost basis for each local government
district. The first set of results shows that Kaduna Local
Government planners balanced projected expenditures on
capital- and labor-intensive projects by district. In
contrast, the Bauchi LG concentrated funding for
labor-intensive projects in the two outlying districts.
The virtual absence of labor-intensive development
proposals for Bauchi town is quite puzzling. Mobilizing
community labor is an important development strategy in
urban as well as rural areas.

In terms of objectives, the major findings with regard
to the Kaduna 1981-85 proposals are (1) that projected
expenditures on administrative projects are much higher in
Tudun Wada and Doka districts than in the others; (2) that
the proportion of proposed expenditure on the provision of
social services is highest in Kawo (27%) and Makera (26%)
districts, while for roads it is highest in Kawo (35%) and
Doka (33%) and for education it is highest in Gabasawa
(40%) and Makera (34%); and (3) that proposed funds for
revenue-generating projects are concentrated in Kawo,
Makera, and Tudun Wada districts. In fact, the 1981-85
development plan slated no expenditures at all on
revenue-generating projects in Doka. The Bauchi results
indicate that expenditures on revenue-generating projects
would be concentrated in Bauchi town, while planners
devoted much higher proportions of total estimated project
costs to social service provision and roads in Zungur and
Galambi.

Implementing Department

The development plan proposals submitted by LGs in
Kaduna and Bauchi states lend themselves readily to a final
grouping that focuses on the department within the local
government organizational structure which would be assigned
responsibility for implementing or executing the proposed
capital-development project.[61] Analysis along these lines
might prove helpful, among other purposes, in identifying
the most dynamic department or departments within each

Table 6.12
Breakdown of Proposed Local Government Plan Expenditures; By District, Method, and Objective

Districts - Kaduna 1981-85

	Kawo			Gabasawa			Makera		
	₦'000	% Total Dist. Expend.	% Total Item Expend.	₦'000	% Total Dist. Expend.	% Total Item Expend.	₦'000	% Total Dist. Expend.	% Total Item Expend.
Prin. Method									
Cap-intens.	1290.5	81.0%	21.6%	908.5	74.3%	15.2%	993.0	78.5%	16.6%
Labor-intens.	301.9	19.0%	19.2%	314.3	25.7	20.0	271.4	21.5	17.3
TOTALS	1592.4	100.0%	21.2%	1222.8	100.0%	16.2%	1264.4	100.0%	16.8%
Prin. Objective									
Admin.	116.0	7.3%	9.8%	111.0	9.1%	9.4%	111.0	8.8%	9.4%
Soc. Serv.	432.6	27.2	24.8	2261.6	21.4	15.0	330.1	26.1	18.9
Revenue	98.9	6.2	31.8	55.2	4.5	17.8	78.3	6.2	25.2
Roads	560.3	35.2	26.4	310.3	25.4	14.6	303.3	24.0	14.3
Education	384.7	24.2	17.8	484.7	39.6	22.5	429.7	34.0	19.9
Other	0	0.0	0.0	0	0.0	0.0	12.0	.9	100.0
TOTALS	1592.4	100.1%	21.2%	1222.8	100.0%	16.2%	1264.4	100.0%	16.8%

	Tudun Wada			Doka		
	₦'000	% Total Dist. Expend.	% Total Item Expend.	₦'000	% Total Dist. Expend.	% Total Item Expend.
Prin. Method						
Cap-intens.	1786.3	80.0%	30.0%	985.3	80.1%	16.5%
Labor-intens.	445.4	20.0	28.4	237.1	19.4	15.1
TOTALS	2231.7	100.0%	29.6%	1222.4	100.0%	16.2%
Prin. Objective						
Admin.	636.0	28.5%	53.6%	213.0	17.4%	17.9%
Soc. Serv.	488.9	21.9	28.0	231.4	18.9	13.3
Revenue	78.3	3.5	25.2	0	0.0	0.0
Roads	543.8	24.4	25.6	403.3	33.0	19.0
Education	484.7	21.7	22.5	374.7	30.7	17.4
Other	0	0.0	0.0	0	0.0	0.0
TOTALS	2231.7	100.0%	29.6%	1222.4	100.0%	16.2%

Districts - Bauchi 1980-85

	Bauchi Town			Zungur			Galambi		
	N'000	% Total Dist. Expend.	% Total Item Expend.	N'000	% Total Dist. Expend.	% Total Item Expend.	N'000	% Total Dist. Expend.	% Total Item Expend.
Prin. Method									
Cap-intens.	14647.7	99.6%	84.8%	1319.7	69.4%	7.6%	1306.7	69.4	7.6%
Labor-intens.	60.4	.4	5.0	580.6	30.6	47.7	576.0	30.6	47.3
TOTALS	14708.1	100.0%	79.5%	1900.3	100.0%	10.3%	1882.7	100.0%	10.2%
Prin. Objective									
Admin.	40.0	.3%	21.6%	65.0	3.4%	35.1%	80.0	4.2%	43.2%
Soc. Serv.	1551.7	10.5	39.7	1186.7	62.4	30.4	1166.7	62.0	29.9
Revenue	13094.0	89.0	98.8	83.0	4.4	.6	71.0	3.8	.5
Roads	20.0	.1	2.0	500.0	26.3	49.0	500.0	26.6	49.0
Education	0	0.0	0.0	10.0	.5	50.0	10.0	.5	50.0
Other	2.4	a	2.1	55.6	2.9	49.2	55.0	2.9	48.7
TOTALS	14708.1	99.9%	79.5%	1900.3	99.9%	10.3%	1882.7	100.0%	10.2%

a = less than .1%

local government and in order to avoid overloading a single
unit with too many development responsibilities.

When the data gathered for this study are classified
by department (Table 6.13), we observe that planners in

TABLE 6.13
Breakdown of Local Government Development Plan Proposals;
By Implementing Department

Implementing Department	No. Projects	% All Projects	Estimated Cost (₦'000)	% Total Est. Cost
		Kaduna 1981-85		
Agriculture and Natural Resources	4	6.1%	900.0	7.4%
Central Administration[a]	24	36.4	2918.3	24.1
Education	11	16.7	3441.9	28.4
Health and Social Welfare	11	16.9	1762.0	14.6
Works	16	24.2	3084.0	25.5
		Bauchi 1980-85		
Agriculture and Natural Resources	19	29.2	638.1	1.9
Central Administration[a]	19	29.2	11992.1	34.9
Education	3	4.6	11425.5	33.3
Health and Social Welfare	13	20.0	2120.0	6.2
Works	11	16.9	8179.8	23.8

[a]Includes Treasury, Information, and Community Development.

both LGs concentrated expenditures on development projects
assigned to the Education, Central Administration, and
Works departments. Under the 1981-85 Kaduna plan, the
Central Administration Department emerged as the most
active executing department in terms of number of proposed

capital-development projects. The 1980-85 Bauchi development plan granted the Agriculture and Natural Resources Department numerous project-implementing responsibilities. Curiously, however, the Bauchi LG allotted less than 2 per cent of total capital investment to this department. The mean proposed expenditure on projects to be implemented by the Bauchi Local Government's Department of Agriculture and Natural Resources amounts to ₦33,500, while, by comparison, the mean for Education Department projects comes to ₦3,808,500 -- about 100 times as much. In general, Bauchi planners spread project executing responsibilities more evenly among the 5 departments in numerical terms, while the Kaduna plan formulators did a better job of balancing expenditure responsibility.

CONCLUSIONS AND LESSONS

The Fourth Plan (1981-85) marked the first time that local governments in Nigeria participated in national development planning. In general, LG officials in Kaduna State exhibited greater seriousness of purpose and played a more active role in the process than their counterparts did in Bauchi state. In both states, however, those involved at the local level failed to follow basic planning prin-ciples, such as conducting cost-benefit or cost-effectiveness analysis, carrying out feasibility studies and basic research, securing public participation at the design stage,[62] considering alternative approaches to achieving desired objectives, using decision matrices, establishing meaningful criteria for project appraisal, and providing detailed project estimates of recurrent costs and anticipated revenues. In view of these deficiencies and the current lack of professionally trained planners at the local government level, this chapter presented a readily comprehensible and easily executed scheme for evaluating projects and illustrated its use and utility through impact analysis of the initial proposals for the Fourth Plan period submitted by two Nigerian LGs. The lessons which emerge from this analysis are likely to assume even greater importance in the future given the AFRC's recent moves to introduce popularly elected LG councils led by a powerful chief executive, to release local governments from state ministry control, and to allocate federal revenues directly to the LG level.

Nigeria's experience suggests that local-level planners and policy makers in Africa could strengthen their role in development planning by adopting a clearly defined, multidimensional perspective when evaluating project proposals. The key is understanding local priorities, asking the right questions at an early stage in the project selection process (Chambers, 1978:216), and critically assessing proposals in terms of likely impact. In light of its emphasis upon applying appropriate evaluative criteria from the start of the development-planning process, elements of the approach illustrated here also should prove useful at the state and national levels (see, for instance, Okafor, 1984:250, 254-256).

Local governments can play a vital role in identifying the pressing needs of residents, incorporating local priorities into the national development planning framework, and promoting and defending citizen interests at higher levels of policy review and budgetary allocation. The effective performance of these roles requires a conducive process and the application of appropriate planning methodologies. A simple and meaningful proposal-evaluation scheme offers a powerful tool for enhancing the involvement of elected policy makers in the plan-formulation process.

In the case studies reviewed here, career officials exercised the central policy-setting roles. A people-centered approach to development planning also requires that local councillors and administrators promote wide popular participation in the project identification and selection stage as well as at other critical decision-making junctures (Adeniyi, 1980:169; Bryant and White, 1982:111, 246; Conyers and Hills, 1984:50; Hellinger, et al., 1983:31, 44-45; Gran, 1983:20-22).[63] Based upon a review of African and Latin American rural experience, George Honadle (1982:175) affirms that "the practice of development administration could be greatly improved by incorporating village-held knowledge into project designs and evaluations." In particular, people living in the local government area should be involved in "social assessment, identification of projects, and discussion of their feasibility" (Bryant and White, 1982:246).[64] Studies consistently show that "if the people who will be affected by a project or a program are involved in its planning, they are more likely to accept it when it is introduced and sometimes even to make a contribution to its establishment or maintenance through some sort of self-help effort because they will identify with it" (Conyers and Hills, 1984:222; also see Caiden and Wildavsky, 1980:25).

The types of planning choices uncovered through analysis of the Kaduna and Bauchi Fourth Plan proposals reflect the top-down nature of the plan-formulation process itself. An entirely different orientation is likely to result when there is effective public participation in local development planning and planners and policy makers develop and apply appropriate multidimensional expenditure-impact criteria in the submission, analysis, and selection of project proposals.

NOTES

1. A grant awarded to the author by the Joint Committee on African Studies of the Social Science Research Council and the American Council of Learned Societies assisted the research reported here.

2. Dudley (1982:232-236) refers to the Second National Plan as a "technocratic description of the bureaucrat's conception of the economically rational."

3. The above discussion is based on Yeye (1980:9-10); D.A. Ajala, Principal Planning Officer (Sectoral), Kaduna State Ministry of Economic Planning, interview held on 16th May, 1980; United Nations Economic Adviser, Kaduna State Ministry of Economic Planning, interview held on 16th May, 1980. According to Ajala, Ministry advisers emphasized the importance of agricultural projects, water supply, rural roads, primary education, health schemes, and maternity centers.

4. This discussion is based on interviews conducted by the author on 27th March 1980 with the Secretary to the Kaduna Local Government and on 16th May 1980 with the Senior Assistant Secretary who coordinated the planning exercise in Kaduna.

5. Including certain development projects which had been identified as priorities by the local government early in 1979.

6. The FGPC took this action approximately two weeks after its first meeting. Village and district heads again attended the second FGPC meeting.

7. Interview with the U.N. Economic Adviser, Kaduna State Ministry of Economic Development, 16th May 1980; interview with the Acting Secretary, Kaduna Local Government, 27th March 1980.

8. Ministry of Economic Development staff consulted with officials in the Ministries of Education regarding education projects, Works on transportation sector proposals, and Local Government on projects proposed by central administration departments. During these consultations, officers examined the LG's proposals for consistency with the state government's plans and policies, as well as for standardized cost estimation.

9. Both local government and Ministry estimates are provided on the summary sheets.

The officials involved expected that only projects assigned "high priority" by the state would be executed during the plan period. In the unlikely event that unanticipated additional government revenues would be forthcoming, certain projects in the "medium priority" category would be selected for funding over "low priority" projects.

10. Members of the State Planning Committee (constituted in April, 1980) included government representatives, farmers, businessmen, and Ahmadu Bello University staff. There were no commissioners to serve on this committee in Kaduna State. For the list of SPC members in Kaduna, see New Nigerian, 24 May 1980, p. 6.

Although the federal government placed a ceiling on the total cost of a state's plan proposals, Kaduna State Ministry for Economic Development officials based their actions on their understanding that the ceiling figure excluded expenditures associated with local government development plans. Interview conducted with the Permanent Secretary, Ministry of Economic Development, Kaduna State, 16th April 1980.

11. The official deadline for state submissions, originally set for 31st March 1980, had to be extended several times.

12. On the crucial role played by field agents from the Ministries of Agriculture and Natural Resources, Works and Transportation, Education, and the state Health Management Board in the plan-preparation process, see Aliyu and Koehn (1982:57-58). The Ministry for Rural and Community Development prepared its own separate list of community-development proposals for inclusion in the Fourth Plan. In formulating its plan proposals, Ministry officials supposedly consulted with local government officers in order to avoid program duplication. M. M. Abdullahi, Head of the Ministry's Community Development Division, maintained (interview, 9th April 1980) that most of the Ministry's projects were located in remote rural areas

where local governments did not propose development projects of their own.

13. Bala Muhammed Yalwa, Supervisor for Community Development, Bauchi Local Government, interviews held on 21st February 1980 and 9th April 1980.

14. Letter to the Ministry for Local Government from the Bauchi Local Government Treasurer (for the Secretary), 31st January, 1980. Ref. No. BLGA/S/ FIN/2/316.

15. Hence, one can only assume that no local government in Bauchi State carried out feasibility studies, engaged in cost-benefit or cost-effectiveness analysis, evaluated objectives in terms of the relative merits of alternative projects and approaches, or prepared implementation and resource-mobilization plans. These shortcomings abound in spite of contrary advice given to Bauchi LG officials by Central Planning Office representatives at the 1978 Seminar (see Nigeria, Central Planning Office, 1979:40-41).

16. Shortage of trained planning staff is a problem at all levels of government in Nigeria (Adeniyi, 1980:174; Adeniji, 1984:26-27); also see Chapter 7 of this book. Planning tasks are typically assigned to administrative officers who are exposed to ad hoc advice prior to plan preparation and later redeployed without playing any role in plan implementation.

17. Interview conducted with officials at the Ministry for Local Government, Bauchi State, 9th April 1980.

18. For further details, see Koto (1980:9,13,14,21, 22, and Appendices D,E,F).

19. Interviews with officials in the Ministries for Local Government and Finance and Economic Development, April 1980.

20. The Committee consisted of the state commissioners along with the Governor. Interview with the Senior Assistant Secretary, Bauchi State Ministry of Finance and Economic Development, 9th April 1980.

21. Local officials defended these acknowledged deficiencies in development planning on the basis that they had to operate under extreme time pressures due to the narrow deadline (one month from initiation of the process to submission of proposals) insisted upon by the Ministry of Economic Development. Interview with the Senior Assistant Secretary, Kaduna Local Government, 16th May 1980.

22. For details regarding these planning devices, see Conyers and Hills (1984:80-84, 135-139); Kent and

McAllister (1985:40-57, 72); and Bridger and Winpenny
(1983:6-12). Kent and McAllister (1985:26-31, 79-83) also
offer useful ideas concerning what a project justification
should contain. Conyers and Hills (1984:51), among others,
stress that planning should be approached as a learning
process.

23. The Political Bureau pointed out in 1987 (p. 205)
that the Nigerian public has "never been involved in the
conception and planning of ... projects." On the
importance of public participation, see Bryant and White
(1982:111).

24. Regional development planning in Nigeria has been
criticized on similar grounds by Adeniyi (1980:163-170) as
well as for lack of clear definition of objectives, the
adoption of unrealistic programs, inadequate feasibility
studies, and "the absence of appropriate ways and means of
involving and identifying the people with the development
plans and the development process...." G. O. Orewa and
J. B. Adewumi (1983:216-222) rue the absence of a reliable
system of matching state government grants, the lack of
borrowing authority and an adequate loan fund, and the
failure to link project proposals with specific revenue
sources and development plans with capital budgets; also
see Adebayo (1981:169-170); Oyovbaire (1985:174-175,
226-227, 85); A. Phillips (1985:255). The arbitrary and
unreliable nature of Bendel State grants for community
development projects is documented in Okafor (1984:254-
256). He shows the need for state governments to establish
criteria for funding self-help activities and to analyze
"the development impacts of the projects."

25. I am indebted to Richard A. Hay, Jr. for these
suggestions.

26. Chambers (1978:212-15) advocates "selection
procedures in which the assumptions are always clear and
which ... can be understood by a noneconomist layman
decision-maker."

27. In the words of Nigeria's Political Bureau (1987:
116), "in a society with endemic shortage of skill and
technical know-how, the simpler the operation for achieving
result, the better." For related arguments, see Bryant and
White (1982:66); and Rondinelli (1982:50, 69).

28. Furthermore, there are instances where the Kaduna
Local Government's proposals call for the same project
to be carried out by different departments. For example,
both the Central Administration and the Works Department
proposed to construct staff quarters (Project Nos. 46 and
62) and to improve market stalls (Project Nos. 34 and 48).

29. Likewise, they will be of little benefit in
evaluating plan implementation. Adewumi (1977:11) points
out that, historically, expenditures on development schemes
in Nigeria tend to be lumped under the "miscellaneous
sector." One result is that development plans "have not
registered concrete achievements in people's minds."

30. Another useful list of important project
characteristics that could serve as an easy-to-implement
set of criteria for proposal evaluation during the
planning process can be found in Cohen and Uphoff (1977:
120-121). From Caiden and Wildavsky (1980:309-310), as
well as Chambers (1978:210-211), comes the idea of project
size as an important criterion of evaluation. Small-scale
schemes are held to possess clear advantages over large-
scale ones -- particularly at the local government level.

31. Roemer and Stern (1975:2-3) suggest that planners
assign weights to evaluative criteria and use comparable
units of measurement as an aid for choosing among
alternative proposals. They recommend that the planning
staff then rank each project according to its total score
on "the various and sometimes conflicting development
objectives" and include those with the highest scores in
the development plan or propose them for support.

32. Other important considerations include the
distribution of existing projects and services, the degree
of local need, and the reduction of poverty. According to
one commentator on local government in Bauchi, for
instance, "some areas are looking for primary schools and
are even prepared to have them through self-help, [while]
others seem to have too many of them" (Abubakar, 1979:
141). Planners also must recognize that the redistribution
of land and power is often a precondition for effective
development planning (see, for example, Chambers,
1978:209).

33. A complete and numbered list of project titles,
cost estimates, and coding decisions for both of the local
government plan proposals analyzed here can be found in the
Appendix to Koehn (1980:51-57). Those with more intimate
knowledge of the nature of local government projects may
justifiably disagree with some of the categorization
decisions reached by the author.

34. The population of the Kaduna Capital Territory is
estimated to be growing at an average annual rate of
approximately 10 per cent, while Bauchi's rate of population increase is estimated at about 7 per cent (see
Seymour, 1979:4; Dar al-handasah Consultants, 1978:III.1,
3).

35. The estimated population size of the Bauchi LGA
in 1979 was 182,957; for Kaduna in 1979, it was 442,235.
Both estimates are projected from 1963 census figures, with
Bauchi results calculated at a 2.5 per cent rate of annual
increase and Kaduna's annual population increase estimated
at 7.0 per cent (figures provided by the Bauchi State
Ministry for Local Government and the Kaduna State Ministry
of Economic Development).

36. The principal sources are "Bauchi Local
Government Development Plan, 1980-85" (Ref. No. BLGA/S/FIN/
2/316); Kaduna Local Government Forms A, Annex I submitted
for Kaduna State Fourth Development Plan, 1980-85. I am
indebted to Richard A. Hay, Jr. for useful suggestions
regarding the method of project analysis and for assistance
in the computerized data processing at Ahmadu Bello
University which made this study possible.

37. Unfortunately, due to incomplete data on the
completed forms, it proved impossible to analyze proposals
by the relative priority (high, medium, low) assigned to
each by the local government.

38. These data are drawn from Kaduna State, Kaduna
State Local Governments Revised Development Plan 1975-80
(Revised 1977-80) (Kaduna: n.p., n. d.), pp. 29-30.
Education projects, formerly the responsibility of Local
Education Authorities, are not included in the figures for
the revised 1977-80 Kaduna development plan.

39. The Bauchi LG based its preliminary proposals on
a five-year plan period (1980-85), subsequently reduced to
four years when the Federal Government decided to extend
the Third Plan through the end of 1980 (M. D. Abdu, Chief
Inspector, Bauchi State Ministry for Local Government,
interview held on 21st February 1980). The Ministry later
required the Bauchi LG to bring its plan proposals in line
with the ₦15 million ceiling on capital and recurrent costs
imposed on all local governments in the state.

40. Further analysis of the estimated recurrent costs
attached to proposed LG development projects is not
possible here due to gaps in available documentation.
However, recurrent expenditure estimates can and should be
evaluated by local government planners and policy makers in
the same ways that capital costs are analyzed and assessed
(see Conyers and Hills, 1984:195).

41. The Kaduna state and local governments also used
broad sectoral categories for local planning purposes.

42. Mabogunje (1972:5) further criticizes the type of
economic sector planning relied upon in Nigeria's first two

national development plans as "far removed from the understanding of the great majority of Nigerians."

43. In Table 6.4 and all subsequent tables presenting data extracted from the local government plan proposals, readers can check the way in which the author classified each proposed project by consulting the full coding scheme included as an appendix in Koehn (1980:51-57).

44. With regard to the latter, the Kaduna and Bauchi LGs appear to have ignored the Federal Government's explicitly expressed expectation that local governments not neglect "traditional areas of emphasis such as the encouragement of self-help efforts among local communities, which have served as an effective means of mobilizing the people for development ...." The relationship between aided self-help activities and local government community development programs is discussed in Onokerhoraye (1984: 172-173, 176-177).

45. Interviews with the Secretary and Senior Assistant Secretary, Kaduna Local Government.

46. Bicycle paths, pedestrian walkways, and subsidized mass transportation systems are likely to be more appropriate options if planners seek to address the needs of the majority of Nigeria's urban dwellers (see Frishman, 1986:63).

47. Onokerhoraye (1984:6-7, 10) uses the concept "social services" in much the same way as it is applied here, emphasizing community or individual welfare. My definition excludes recreation and tourism, however. The importance of social services (which include the education, infrastructure, health, sanitation, and community development categories of Table 6.4) is stressed in the Federal Government's Guidelines (p. 57) with respect to urban development programs, which are at stake in both of the LGAs investigated here. Furthermore, the "communique" issued at the conclusion of the National Conference on Local Government and Social Services Administration in Nigeria held at the University of Ife from 18-21 February 1980 recommended that "the provision and maintenance of basic social services should be seen as the primary responsibility of local governments" and that LGs should play a more central role in "the definition, planning, and execution of basic social services."

48. By emphasizing revenue-generating projects, the Bauchi LG's initial development plan embodied the Ministry for Local Government's higest priority objective for local governments in Bauchi State. M. D. Abdu, Chief Inspector, interview, 21 February 1980.

49. It might also prove fruitful to evaluate development plans with respect to the extent of their reliance upon private contractors versus direct project execution by public employees. This method of analysis could be used by policy makers to control the proportion of projects and expenditures assigned to private contractors, thereby promoting competitive bidding and performance on available projects, enhancing the local government's ability to enforce quality control and cost-accountability measures, and reducing unnecessary expenditures and corruption. The available data do not permit analysis along these lines in this work, but it would be easy to require in future planning exercises that a project proposal justify why it cannot be undertaken "in house" before a contract award would be authorized.

50. In addition, labor-intensive projects are likely to benefit more low-income residents than are capital-intensive undertakings. Thus, this criterion reinforces the project beneficiary standard introduced below. Planners should endeavor to select programs which maximize such linkages.

51. It proved difficult to categorize certain types of construction projects from afar due to lack of information on the ratio of expenditures on equipment and materials versus labor costs. For instance, the erection of prefabricated housing units is likely to be capital- (and import-) intensive while the construction of cement block structures using local materials is likely to be predominantly labor-intensive.

52. One means of determining the principal method likely to be employed on a given construction project is through evaluation of the type of work (capital- versus labor-intensive) typically performed by the category of contractor that is likely to bid on the project.

53. One also might fruitfully analyze development plan proposals in terms of those who would be adversely affected or "underdeveloped" by proposed projects (see Kent and McAllister, 1985:15; Conyers and Hills, 1984:33).

54. Chambers (1978:216-220) provides a five-fold scheme for classifying potential beneficiaries and suggests that (1) groups be ranked according to degree to which they benefit and (2) planners provide an "index of unit costs" by dividing total estimated project expenditures by the number of people who will benefit directly from it.

55. Chambers (1978:217) suggests more refined analysis according to the following categories:

extra-rural, cattle posts, small villages, large villages, and urban centers.

56. In several cases, the principal location of project results could not be ascertained by the author given the scant information provided by the local government in its plan proposals (for example, Bauchi Project Nos. 45, 52). Such projects have been coded "both" in this paper, although local officials should be able to classify many of them as predominantly rural or urban given their greater familiarity with the proposals.

57. The relative mean cost per project also changed dramatically, as the following comparative figures for the Kaduna LGA reveal:

- Mean Est. Cost (₦) -

Project Type	1977-80	1981-85
Rural	29,000	222,100
Urban	190,286	166,340
Both	102,000	246,056

58. Ursula Hicks (1965:21, 32-33) warns against "spreading investment too thin in order to please local interests...." Yet, she argues that small, widely scattered health treatment centers possess important advantages over institutions that are larger in size and offer more skilled attention, but are fewer in number. Francis Okafor (1984:257) makes a different point. He maintains that the principal of spatial equity in development demands that financially depressed communities receive favored attention and assistance.

59. This would be consistent with the Political Bureau's (1987:121) recommendation that the primary settlement area be the principal "target of government development policies"(also see Hellinger, et al., 1983:43).

60. This is apparent from a number of the project justifications submitted by the Kaduna Local Government.

61. It is safe to assume that in many cases the implementing department also originally initiated these project proposals.

62. In Nigeria, government officials are prone to inform rather than to involve the public (Nigeria, Political Bureau, 1987:205-206).

63. For specific suggestions along these lines that focus on community opinion leaders and the "natural community," see Adamolekun (1979a:11-12).

64. David Korten (1980:485-494, 497-500) provides a useful discussion of the learning and adapting processes utilized in successful community development programs.

RECOMMENDED READING

There are a number of outstanding works on national development planning. Two classics are Albert Waterston, Development Planning; Lessons of Experience (Baltimore: Johns Hopkins Press, 1965); and Naomi Caiden and Aaron Wildavsky, Planning and Budgeting in Poor Countries (New Brunswick: Transaction Books, 1980). Highly recommended recent studies are Diana Conyers and Peter Hills, An Introduction to Development Planning in the Third World (Chichester: John Wiley & Sons, 1984); and Tom Kent and Ian McAllister, Management for Development; Planning and Practice from African and Canadian Experience (Washington, D.C.: University Press of America, 1985). A comprehensive study of national planning in Nigeria, which includes an insider's critique, is Pius N. C. Okigbo's National Development Planning in Nigeria, 1900-1992 (London: James Currey, 1989).

Local government involvement in development planning has been largely ignored. The best published source for ideas appropriate at this level is Robert Chambers, "Project Selection for Poverty-Focused Rural Development: Simple is Optimal," World Development 6, No. 2 (February 1978):209-219. A frequently cited study of evolving community involvement in development is David C. Korten's "Community Organization and Rural Development: A Learning Process Approach," Public Administration Review 40, No. 5 (October 1980):480-511.

7

Local Government
Organization and Staffing

For most students of public policy and administration
in the United States, local government in Africa remains a
mystery. Neither the structure nor the performance of
public institutions at the grass-roots level have received
much attention in the published literature. The basic
"picture" of local administration is particularly fuzzy.

The goal of this chapter is to provide a detailed
"snapshot" of one LG organization in anglophone Africa.
The focus here is on structure and staffing.[1] The selected
subject is the Bauchi Local Government. Nigerian local
governments possess relatively standardized structural
characteristics and tend to experience similar personnel
problems. For comparative perspective, readers should bear
in mind that the Bauchi Local Government Area encompasses
the state capital and consists of two rural districts and
one urban district of modest population size.

Local governments are supported and constrained by
external and internal factors. State and national
government interventions constitute one of the most
decisive exogenous influences. In Africa, these higher
levels generally have retained control over LG personnel
matters and inhibited autonomous development administration
at the grass roots (see Smith, 1985:189; Hodder-Williams,
1984:177; Koehn, 1988b). Crucial internal considerations
include the organization of local activities and the
utilization of available skills (Vengroff, 1983:6).
Organizational structure and personnel administration at
the LG level have largely escaped critical analysis in
spite of their importance for local policy making and
performance. This chapter aims to rectify that omission.

We begin by examining structural impediments to effective
development administration.

ORGANIZATIONAL STRUCTURE

The Bauchi LG emerged from Nigeria's 1976 local
government reform with only six departments. Several
previously existing departments under the Bauchi Local
Government Authority remained intact, but as units of
diminished status administered by section heads who had to
report to the head of department. We shall also observe
that considerable change in organizational structure and
staffing occurred as a result of incorporation of the local
education authority into the Bauchi LG in 1978.

The four supervisory councillors had to divide
responsibilities for six departments.[2] They assigned
Central Administration to the chairman and Works and Survey
to a second councillor. The two remaining supervisory
councillors dealt with two departments each: Treasury (or
Finance Department) and Medical/Health in one case, and
Education and Natural Resources in the other.

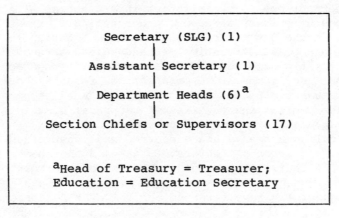

Figure 7.1 Hierarchical Structure of the
Bauchi Local Government

Figure 7. 1 shows the top levels of the formal administrative hierarchy of the Bauchi Local Government in the
initial post-reform period. The specific sections
reporting to each department head are set forth in Figure

7.2. It is clear from this figure that the Bauchi LG had adopted the standard organizational structure found throughout Nigeria. This arrangement is heavily influenced by British colonial practice.

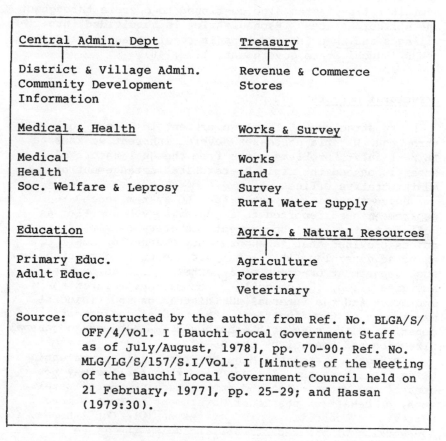

Central Admin. Dept

District & Village Admin.
Community Development
Information

Medical & Health

Medical
Health
Soc. Welfare & Leprosy

Education

Primary Educ.
Adult Educ.

Treasury

Revenue & Commerce
Stores

Works & Survey

Works
Land
Survey
Rural Water Supply

Agric. & Natural Resources

Agriculture
Forestry
Veterinary

Source: Constructed by the author from Ref. No. BLGA/S/ OFF/4/Vol. I [Bauchi Local Government Staff as of July/August, 1978], pp. 70-90; Ref. No. MLG/LG/S/157/S.I/Vol. I [Minutes of the Meeting of the Bauchi Local Government Council held on 21 February, 1977], pp. 25-29; and Hassan (1979:30).

Figure 7.2   Arrangement of Departments and Sections in the Bauchi LG

Rural Orientation

The strong rural orientation along with the emphasis on maintenance functions reflected in Figure 7.2 can be attributed in large measure to the fact that most of the departments and sections found in the Bauchi Local Government in 1979 trace their administrative origins to

the former Bauchi Local Government Authority. Out of the
six units carrying departmental status, only one (Medical/
Health) could be categorized in 1977 as predominantly
oriented toward providing social services for the rapidly
expanding urban population of Bauchi town. The district
administration system also continued to operate throughout
the northern states. Each district is subdivided into
villages and then into hamlets in rural areas or wards in
towns (Bauchi State Government, 1979:140).

Structural Defects

The structure of local government in Bauchi and
throughout Nigeria possesses several inherent weaknesses.
In particular, problems arise from the dual nature of
supervision and the rigid hierarchical arrangement of
administrative offices.

Under the initial post-reform LG system, each
department head reported to a supervisory councillor as
well as to the local government secretary (SLG). State
edicts provided that "a supervisory councillor shall ...
exercise general political but not executive direction over
such department or group of departments ... as may be
assigned ... by the council."[3] On matters related to
employees and the internal administration and financial
management of the department, the head of department
constituted the responsible authority -- subject to general
direction and control by the SLG.[4]

In Bauchi State, the Ministry for Local Government's
Guidelines state that "a Supervisory Councillor may not
control any votes, sign and authorize payments or receive
money on behalf of the Local Government." Furthermore,
supervisory councillors are not responsible for "the day to
day working of the department, including the issue of Local
Purchase Orders (L.P.O.'s) or Jobbing Orders, etc." Indeed,
the Bauchi Guidelines continue, "a Supervisory Councillor
may not authorize LPOs or Jobbing Orders or instruct that
an L.P.O. or Jobbing Order be issued to a particular
contractor." Instead, "all Local Purchase Orders and
Jobbing Orders must be counter-signed by the Secretary to
the Local Government [along with the head of department]."
Finally, the Guidelines authorize supervisory councillors
to "give orders to executive heads of local government
departments on policy issues only, but not on the internal
management of the departments."

While these provisions made it clear that the supervisory councillor possessed far less authority than the former portfolio councillor who exercised executive responsibility,[5] the problem of dual supervision remained. In the first place, there is rarely a sharp distinction between political decisions and administrative matters and direction. In addition, one can readily envision situations where policy directives would require actions that contradict executive instructions. Although there are no guidelines for department heads to follow in such cases of conflicting orders from different supervisors, LG staff proved only weakly accountable to councillors in the post-reform period since the latter played no role in their appointment or promotion and could not discipline, transfer, or dismiss them. Nevertheless, department heads did find themselves in awkward situations where they faced conflicting orders or encountered a supervisory councillor who acted in the familiar style of the portfolio councillor.[6] These situations proved particularly tense and demanding when a head of department received a conflicting order while on tour with the supervisory councillor. Alex Gboyega (1979b:38-39) correctly foresaw that it would not be "realistic to expect councillors with political responsibility for a group of departments to limit their activities to political direction only."

Another structural defect is that local government employees in Nigeria must operate under a strict administrative hierarchy headed by the SLG. Critical theorists have attacked public administrative systems built on rigid adherence to hierarchy of authority on grounds that they retard the internal and external flow of communication and lead to lackluster performance (Denhardt, 1984:172; Hummel, 1987:53, 251; Peters, 1989:136-137; also see Bryant and White, 1982:185).[7] In Nigeria, deference to hierarchy, coupled with reluctance on the part of top officials to delegate authority, results in work overload at the top and in perpetual backlogs (Cohen, 1980:75-76; also see Hyden, 1983:146-147).

A. D. Yahaya (1980:126-127) forcefully argues that job satisfaction is likely to be a more decisive factor in the recruitment and retention of qualified local government staff than conditions of service. Yahaya sees employee job satisfaction as dependent upon opportunities for autonomy, discretion, initiative, and creativity.[8] These are behavioral qualities which hierarchy typically stifles (see Peters, 1989:136). Coralie Bryant and Louise White

(1982:198) argue further that, as change agents, development administrators "need to know that they have some discretion and can modify their tasks and respond to the preferences of the public." Thus, public managers are advised to implement consensus-building processes for goal setting and decision making (Harmon, 1987:54-56). Consensus building has the further advantage of being a familiar procedure for dispute resolution in Africa.

Functional Responsibilities

In Nigeria, the number of departments at the local government level is controlled by state government and rarely exceeds six (Yahaya, 1980:133). The Bauchi LG assigned functional responsibilities among its departments in standard and predictable fashion (see Figure 7.3). After 1978, the Education Department handled primary as well as adult education. The Agriculture and Natural Resources Department assumed responsibility for agricultural, forestry, and veterinary functions. The Medical and Health Department dealt with health care and education as well as social welfare and rehabilitation.[9] Treasury is primarily involved with budget preparation and the collection of local fees and charges. The Works and Survey Department is responsible for rural water supply, intervillage roads, and other (primarily rural) public construction projects. In addition to general management responsibilities, community development, public information, and village and district administration functions all fall under the Central Administration Department.

FIGURE 7.3
Principal Functions Assigned to Each Department in the
Post-Reform Bauchi Local Government

Central Administration

Overall management and direction of the LG
Community development
Assist in assessment and collection of community tax,
property tax and other rates
Dissemination of information to the public

(continued)

FIGURE 7.3 (continued)
Principal Functions Assigned to Each Department in the
Post-Reform Bauchi Local Government

Central Administration (continued)

Regulation and licensing of bicycles, hand carts, and
other self-propelled vehicles
Provision of public reading rooms
Control and collection of revenue from private forest
estates
Naming of roads and numbering of plots and buildings
Registration of births, deaths, and marriages
Market and market stall construction
Construction of cemeteries and burial grounds
Resolution of land-allocation disputes

Treasury

Assessment and collection of taxes and fees
Budget preparation
Disbursement of payments
Stores management
Commercial undertakings

Medical and Health

Operation of dispensaries, clinics, maternity and
leprosy centers, and workshops for the blind
Control of sanitary conditions in markets and motor
parks
Conduct sanitary inspections
Regulation of slaughter houses and slabs
Control vermin
Maintenance of public conveniences
Collection and disposal of refuse and night soil
Regulation of cemeteries and burial grounds
Provision of community and recreation centers
Public health and sanitation education
Provision of ambulance service
Regulation of liquor sales

(continued)

FIGURE 7.3 (continued)
Principal Functions Assigned to Each Department in the
Post-Reform Bauchi Local Government

Works and Survey

Rural and semi-urban water supply (boreholes, wells)
Regulation and control of buildings in non-urban areas
Construction and maintenance of inter-village roads
Construction of public conveniences
Electrification projects
Construction of motor parks, taxi parks, and market
   stalls
Conduct land surveys and participate in land-use
   planning

Education

Operate primary schools
Provision of adult education and literacy training

Agriculture and Natural Resources

Assist state government in distribution of fertilizer
Provision of veterinary clinics
Provision of parks, gardens, and open spaces
Establishment and regulation of grazing grounds
Control and keeping of animals
Establishment and operation of fuel plantations, forest
   reserves, and orchards
Promotion of poultry farming
Construction of slaughter slabs and houses

Source: Hassan (1978:32-34); interview with Adamu Aliyu,
Permanent Secretary, Bauchi State Ministry for Local
Government, 29 December 1978; Aliyu and Koehn (1982);
"Bauchi Local Government Development Plan, 1980-85" (Ref.
No. BLGA/S/FIN/2/316); Bauchi State Government
(1979:140).

Structural Alternatives

P. C. Morelos (1979:81, 83, 78, 75) presents an
intriguing sketch of an alternative organizational
structure that merits serious consideration in Nigeria and
elsewhere in Africa. The principal structural alterations
proposed by Morelos are a reduction in the number of
departments from six to four and the creation and staffing
of a planning unit directly attached to the LG Secretary.
Structural consolidation would join Medical & Health and
Education into a Department of Social Development Services
and merge Works & Survey and Agriculture & Natural
Resources into a new Economic Development Services
Department. The new Planning Unit would formulate
development plans, strategies and policies, town plans, and
accompanying budget estimates; provide technical assistance
to departments and LG committees; undertake project
studies, research, data collection, and statistical
analysis; prepare proposals for loans, investment, and
external assistance; and monitor plan and budget implementation. The utility of such a unit at the local government level in Nigeria is particularly striking given the
additional planning responsibilities which have been
assigned to the third tier (see Chapter 6).
Finally, Morelos (1979:79) proposes the establishment
of a Planning and Budgeting Committee to be chaired by the
SLG and to include all heads of departments and the
planning staff director as members. Unfortunately, Morelos
makes no provision for councillor involvement, even though
it would be appropriate for elected council members to
comprise a majority on such an important committee. The
Planning and Budgeting Committee could propose and assess
policies and strategies for development of the LGA; review
and evaluate proposed plans, programs, budgets, and
projects; and periodically study and assess plan impacts
and results.
Local government organizations also must be sensitive
to political considerations. Foremost among these,
according to the 1987 report of Nigeria's Political Bureau
(1987:186-188), are the needs to mobilize people, promote
popular participation in local government, and decentralize
LG functions. With these goals in mind, the Political
Bureau proposed an elaborate scheme for the creation of
subordinate development area councils (maximum of 5 per

LG), village or neighborhood committees, and functional
subcommittees and associations. Its report identifies the
"village, clan, autonomous community or urban neighborhood"
as "the primary unit of local government" (also see
Adamolekun, 1979:11-12). Village or neighborhood
committees would possess important functional
responsibilities -- including rural land allocation,
community development, and monitoring the activities of
public officials. Local residents would select committee
members who, in turn, elect a chairperson. The chairpeople
of the village and neighborhood committees comprise the
development area council (Nigeria, Political Bureau,
1987:117-119).[10] The Babangida administration accepted
most of these proposals in 1987 (Nigeria, Federal
Government, 1987:39-41).

PERSONNEL ADMINISTRATION

Personnel matters rank among the most interesting and
revealing issues in the study of public administration.
With capable employees, local governments can participate
responsibly in a wide range of development functions. In
this section, we examine the structure of personnel
decision-making authority, staffing patterns, recruitment
and training issues, and certain important staff characteristics. The following section analyzes the link between
personnel costs and recurrent expenditures. These issues
again are illustrated by reference to the Bauchi Local
Government.

Appointment and Promotion

Local government staff appointments and promotions are
handled in accordance with schemes of service (and staff
regulations) established by the Ministry for Local
Government and implemented by the state's Local Government
Service Commission.[11] In addition, the Guidelines for
Local Government Reform charge the Ministry with the nearly
impossible duty of ensuring that LGs possess staff in
sufficient quantity and quality to discharge their assigned
functions and to provide the services for which they are
responsible.

In Bauchi, two subcommittees on establishment functioned in the post-reform period. The Secretary chaired the first subcommittee, with all six department heads serving as members. This subcommittee on establishment dealt with applications and candidate interviews for positions at grade level 06 and above. The other subcommittee came under the Assistant Secretary. All of the department heads again constituted its membership. The junior staff establishment subcommittee considered personnel matters affecting employees at grade level 05 and below.[12] This group submitted its recommendations to the Secretary, who possessed final authority to make junior staff personnel decisions (Ibrahim, et al., 1979/80: 15-16).[13] The SLG could and did alter recommendations made by the junior staff establishment subcommittee.

Resolutions of the senior staff establishment subcommittee had to be approved first by the LG council and, finally, by the Local Government Service Commission (Ibrahim, et al., 1979/80:16). Thus, senior local government officers ultimately owed their initial appointments and future assignments/promotions to decisions made at the state level (primarily by the Local Government Service Commission). Lack of any significant disciplinary powers left councils in the position of merely forwarding adverse judgments on to the LGSC.

An official objective behind assigning authority for LG personnel decisions to a state government body is the desire to insulate such matters from political consider-ations (Yahaya, 1980:127, 131).[14] Recognition of the political roles performed by higher administrators at the local (as well as state and federal) level can be useful, however (Yahaya, 1980:127, 131). In actuality, then, increased job security for LG staff may be the most impor-tant underlying reason behind the centralization of personnel administration (Yahaya, 1980:131).[15] Whatever the rationale, this arrangement provoked council chairmen from Bauchi and four other northern states to complain at a 1978 meeting that senior local government staff "give their loyalty to the Local Government Service Boards and not to their Councillors."[16] The council chairmen strongly recommended that such state boards and commissions be "scrapped" and that LGs be delegated final authority over "recruitment and discipline of all categories of their staff."[17]

Staff Size and Distribution

We now turn to an analysis of the size and distribution of the Bauchi Local Government's staff in the immediate post-reform period. The number of staff employed by each department in the first and second years following the 1976 reform is presented in Table 7.1. One receives a clear picture from the figures reported in this table of the dramatic consequences associated with incorporation of

TABLE 7.1
Bauchi LG Staff Situation; By Department (1977, 1978)

| | | Year | | | | |
| | | Sept. - 1977[a] - | | | Aug. - 1978[b] - | |
Department	Perm.	Daily Paid	Total	% of Total	Perm.	% of Total
Central Admin.	105	10	115	26.7%	93	8.4%
Treasury	8	17	25	5.8	41	3.7
Medical and Health	40	43	83	19.3	49	4.4
Works and Survey	26	98	124	28.8	47	4.2
Agric. & Nat. Resources	40	31	71	16.5	61	5.5
Education	6	6	12	2.8	817	73.7
TOTAL NO. LG EMPLOYEES	225	205	430	99.9%	1108	99.9%

[a]Source: Hassan (1979:31); Ref. No. BLGA/Est/27/60.
[b]Source: compiled by the author from Ref. No. BLGA/S/OFF/4, Vol. I, pp. 70-90, 38-42.

the Bauchi Local Education Authority into the structure of local government. In 1977, Education constituted the smallest department within the Bauchi Local Government; its twelve employees devoted their attention exclusively to adult education activities. One year later, the infusion of primary school teachers and other LEA staff had resulted in nearly a 140-fold increase in the number of permanent employees working in this department. Overnight,

Education had moved from a marginal department to one which housed nearly three-fourths (74 per cent) of all employees of the Bauchi LG and consumed the lion's share of its recurrent personnel expenditures. This reflects a nation-wide trend. A manpower survey based on LG estimates for 1978-79 revealed that two-thirds of all established positions at the local government level had been dedicated to school teachers, headmasters, and educational superintendents (Orewa, 1978:1-4).

The number of permanent staff assigned to the Treasury also expanded appreciably, from 8 to 41, within the space of a year. The conversion of daily workers (mainly security guards) to permanent staff status accounts for a large part of this expansion. All other departments, with the exception of Central Administration, experienced modest increases in the size of their permanent work force. The largest decline in total employment came in the Works and Survey Department.

Table 7.2 depicts the distribution of the permanent staff of the Bauchi Local Government in August 1978 across three grade-level categories that correspond to the Bauchi State Local Government Service Commission's designation of

TABLE 7.2
Grade Level Held by Bauchi Local Government Staff in 1978; By Department

DEPARTMENT	13-17		06-12		01-05	
	No.	%Total	No.	%Total	No.	%Total
Central Administration	1	1.1%	3	3.2%	89	95.7%
Treasury	0	0.0	2	4.9	39	95.1
Medical and Health	0	0.0	1	2.0	48	98.0
Works and Survey	0	0.0	2	4.3	45	95.7
Agric. and Nat. Resources	0	0.0	1	1.6	60	98.4
Education	0	0.0	132[a]	16.2	685	83.8
Entire LG	1	.1	141	12.7	966	87.2

[a]Includes 54 who had been recommended for approval at this level in 1978.
Source: Compiled from File No. BLGA/S/OFF/4, Vol. I, pp. 70-90, 38-42.

higher, middle, and junior-grade officers. One immediately
observes a preponderance of junior-grade officers serving
in 5 out of the 6 departments. When Education is excluded
from consideration, only 9 middle-grade officers are found
in the Bauchi LG, and they constitute less than 5 per cent
of the combined personnel strength of these five departments. The Secretary to the Local Government held the only
higher grade post (GL 13) in the Bauchi administrative
structure.[18] In 1978, then, the Bauchi Local Government
certainly could not be characterized as a top-heavy
bureaucracy.

Chief Administrative Officer

By 1980, four different individuals had served as
Secretary to the Bauchi Local Government -- an average of
one per year. This is not particularly unusual. Roughly
half of the local governments in the northern states had
already experienced 4 or more different post-reform
secretaries by 1979. Three or more individuals had been
appointed to the post of secretary in the capital city LG
of all 10 northern states. The widespread practice of
secondment undoubtedly promoted rapid turnover at the top
levels of local government bureaucracy. In general,
though, lateral mobility is the norm throughout the upper
reaches of all levels of government service in Nigeria and
"sticking with a program or job to see it through is not
part of Nigerian bureaucratic culture" (Cohen, 1980:77).
The third person to serve as the chief local
government administrative officer, Alhaji Baba Ma'aji
Abubakar, had completed middle school and had undertaken a
management training course in Kaduna. An indigene of
Bauchi town, he had been born into a prominent local family
some forty years prior to assuming the post of Secretary to
the Bauchi LG. His previous administrative experience
included service with the Ministry for Local Government at
the outset of the post-reform period (Ibrahim, et al.,
1979/80:21). The fourth individual to serve as Secretary
to the Bauchi Local Government, Alhaji Muhammadu Dan,
worked his way up through the administrative ranks of local
government service. Prior to assuming the post in February
1980, he had served as senior assistant secretary in the
Bauchi LG.[19]
The first three secretaries to the Bauchi Local
Government all served on secondment from state government
service and subsequently tranferred back into prestigious

posts at that level. Alhaji Baba Ma'aji Abubakar secured
appointment to the State Electoral Commission (New
Nigerian, 16 February 1980, p. 11). By late 1979, his two
predecessors as SLG had been appointed to the state posts
of Permanent Secretary, Ministry of Trade and Industry and
(Acting) Accountant General (Ibrahim, et al., 1979/80:21).

Recruitment and Training

The effectiveness of a local government organization
is dependent, in large measure, on its ability to attract
and retain competent staff in adequate numbers. The Review
Commission on the Local Government Service concluded, with
reference to the pre-reform situation, that inability to
recruit sufficient technical, professional, and
administrative staff constituted "'the greatest single
factor that persistently inhibits the effectiveness of
local authorities ...'" throughout Nigeria (cited in
Oyediran, 1979:51).[20] The members of the Commission
believed that the existing salary and benefit discrepancies
between the same category of public employees at the local
level vis-a-vis the state and federal services had the most
adverse effect on LG recruitment efforts.[21] Therefore,
the Commission recommended, and the FMG accepted, that
salaries and conditions of service of all categories of
local government staff be "harmonzied" with those of the
civil service (see Rowland, 1979:92; Oyediran, 1979:
51-52).[22]
At the state level, committees then set about
reviewing LG positions and harmonizing them with equivalent
posts in the state civil service (Kaduna State Government,
1979:199).[23] On its part, the Federal Government
advertised in the national media that:

> local governments are now in a position to offer
> and guarantee general conditions of service and
> security of tenure no less attractive than those
> offered by the Federal and State Governments. The
> career expectations and retiring benefits of local
> government staff have now been brought at par with
> those of the Federal and State Officials. Further-
> more, the new status conferred on Local Government as
> the third tier of government puts local government
> staff on the same social pedestal as those [sic] of
> federal and state officials.[24]

In the wake of the major recruiting efforts launched by
Local Government Service Boards, a number of applicants
sought LG posts and some state and federal civil servants
applied for transfer to the local level -- particularly in
their area of origin (Kaduna State Government, 1979:199;
Adamolekun and Rowland, 1979:298).[25]

Nevertheless, the overall shortage of qualified
personnel remained a problem at the third tier. From 1
October 1979 to 1 October 1980, for instance, the LGSB only
managed to recruit 8 post-graduate, 28 graduate, and 18
Higher National Diploma (H.N.D.) officers into the unified
service that staffs all 14 LGs in Kaduna State (Kaduna
State, Office of the Governor, 1980:25). The most serious
needs are in the technical and professional ranks. Few LGs
have succeeded in attracting and retaining experts in
adequate numbers to staff their principal line departments
-- Health, Works, Agriculture, and Education (Adamolekun
and Rowland, 1979:298; Nwankwo, 1984:74; Stock, 1985:
477).[26] Most urgently required are persons with
appropriate qualifications and experience to fill positions
at the local level as service delivery program managers,
engineers, planners, health staff, technicians of various
types, and valuation officers. Such individuals are in
short supply throughout Nigeria and are eagerly and
aggressively recruited by the federal and state
governments, parastatals, and private enterprises (Rowland,
1979:92, 99; Adamolekun, 1983:83). Aside from funding
constraints, the two major inherent factors that retard
further progress in LG staff recruitment are (1) the
widespread and often accurate perception that career
prospects are not as attractive as those existing at higher
levels of government service and (2) lack of official
recognition that "the Local Government Service is
essentially a rural service and, as such, requires some
special incentives and attractions if qualified and
experienced persons are to be recruited and retained."[27]

Like most other LGs, the Bauchi Local Government found
it difficult to recruit qualified staff. Its inability to
attract trained manpower in the required numbers produced
critical staff shortages and necessitated the hiring or
promotion of individuals into jobs they were not fully
and/or formally qualified to perform. All six department
heads complained about personnel shortages in 1979. This
situation primarily resulted from a dearth of qualified
candidates rather than from limited establishments (see
Hassan, 1979:32). A quick review of their establishments
indicates that most, if not all, departments of the Bauchi

LG encountered difficulties filling approved positions of a highly skilled or supervisory nature.[28] The presence of only one medical and one agricultural specialist among its staff is of particular concern. The recruitment of semi-skilled workers also presented problems in certain areas. For instance, the Bauchi Local Government had filled only 9 of the 14 dispensary assistant (with 5 of the 9 on training course), 11 of the 16 dispensary attendant, and 3 of the 14 female dresser positions found on the Medical and Health Department's 1980 establishment.[29]

Local government staff throughout Bauchi State also experienced a shortage of educational and training opportunities (see Bauchi State Government, 1979:141).[30] As one result, a serious qualifications gap existed at higher levels of the LG bureaucracy. The requirement that department and section heads hold a relevant first (bachelors) degree had to be waived in every case. A diploma or certificate in a related training course became

TABLE 7.3
Number of Bauchi LG Staff Who Had Attended Established Training Courses; By Type (as of August 1978)

Type of Training Course	No. Staff Attended
Diploma in LG	5[a]
Intermediate Studies in LG	4[b]
Livestock Superintendent	1
Agricultural Field Overseer	6
Veterinary Assistant	2
Forester	2
Arabic Teachers College (Kano & Gombe)	30[c]
Clerical	12[d]
Typist	1
TOTAL	63

[a]Includes 3 in Treasury and 2 in Central Administration
[b]Includes 2 in Treasury
[c]All in Education Department
[d]Includes 6 in Central Administration and 4 in Treasury
Source: File No. BLGA/S/OFF/4, Vol. I, pp. 38-42.

250

the de facto paper qualification for such posts.[31] Even
after this loosening of standards had been applied, further
substitutions often had to be allowed. For instance, both
the treasurer and the revenue officer employed by the
Bauchi LG in 1980 possessed a diploma in local government
rather than in finance.[32]

The training gap affected all levels of administration
within the Bauchi Local Government. The scarcity of
efficient typists provides one example (Bauchi State
Government, 1979:142). The extent of the problem is
revealed by evidence that only 63 persons out of a total
staff that numbered in excess of 1000 permanent employees
had undergone some form of specialized training by 1978
(Table 7.3).[33] This number shrinks to 20 when teacher and
clerical training programs are removed from consideration.
Treasury turned out to be the most fortunate department; 9
of its 41 permanent employees (22 per cent) had undergone
some training by late 1978.[34]

The detailed staff profile presented in Table 7.4
gives specific definition to many of the observations made
thus far in this chapter regarding the practices and
problems of personnel administration in the Bauchi Local
Government. In the case depicted (the main office of the
Central Administration Department), we find a paucity of
higher-level officers, an unfilled establishment, a
qualifications gap, and can identify specific training
needs.

TABLE 7.4
Selected Staffing Profile: Central Administration Department
Office (1980)

Position (Establishment)	Required Qualifics	Qualifics Person in Post	Special Training Attended
SLG (1)	B.A. in soc. sci. or humanities; higher[advanced] dip. in LG	Higher dip. in LG	HDLG
Sr. Asst. Sec. (0)	"	No establishment	–
Asst. Sec. (2)	"	Vacant posts	–
Higher Exec. Officer (1)	National dip. in LG	Vacant post	–

(continued)

TABLE 7.4 (continued)
Selected Staffing Profile: Central Administration Department
Office (1980)

Position (Establishment)	Required Qualifics	Qualifics Person in Post	Special Training Attended
Exec. Officer (2)	National dip. in LG or promotion based on performance in clerical cadre	Experience only; vacant post	None
Asst. Exec. Officer (0)	"	No estab- lishment	-
Sr. Cler. Officer (1)	Sec. school leaver's certif.	Vacant post	-
Cler. Officer (2)	"	Sec. School leaver's certif. Vacant post	Clerical training course
Typist (2)	Primary 7 complete	Primary 7 complete	None
Messenger (9)	Primary 7 complete	Experience only	None
TOTAL = 20 (6 vacant)			

Source: Information provided by staff of the Bauchi State Local
Government Service Commission, 30 June 1980.

Staff Characteristics

Table 7.5 unveils several interesting dimensions of
the Bauchi Local Government's staffing situation in the
immediate post-reform period. First, the data confirm that
the work force consisted of a particularly large proportion
(65 per cent) of former Bauchi Local Government Authority
(L.G.A.) and Native Authority (N.A.) employees. Indeed,
every one of the department and section heads in the four
departments for which such information is available had
worked under the previous system of local administration.
This is important because such officers are likely to
possess strong ties to traditional authorities. With
respect to junior-grade staff, Treasury housed the highest
proportion among these four departments of former L.G.A.
and/or N.A. personnel (74 per cent). In only one

department (Medical and Health) did the number of freshly recruited employees approach half of the total staff size in 1978.

The figures found in Table 7.5 also shed light on the length of service and age structure of the Bauchi Local Government's staff in 1978. Over half of the employees at GL 05 and above and roughly 30 per cent of those holding lower grade positions had secured their first appointment (to the Bauchi Native Authority) prior to Nigeria's independence. Officials appointed during colonial rule still held half or more of all supervisory positions in the new local government in all of the departments investigated with the exception of the Treasury.[35] Individuals appointed to the permanent staff of the Bauchi LG after 1970 constituted less than 8 per cent of all reporting middle- and higher-grade officers and only 56 per cent of the total work force assigned to these four departments in 1978.

TABLE 7.5
Employment History of Bauchi LG Permanent Staff, Four Departments (1978)

Dept./Section/GL (# persons in post)	% First Employed By N.A./ L.G.A.	% First Employed By Post- Reform LG	% First Employed Under Colonial Rule	% 10 or More Yrs. Local Service By 1980
AGRIC./NATURAL RESOURCES				
Dept. Office				
GL 08 (1)	100.0%	0.0%	0.0%	100.0%
GL 01-04 (9)	55.6	44.4	22.2	33.3
Agric. Section				
GL 05 & above (0)	–	–	–	–
GL 01-04 (21)	52.4	47.6	14.3	14.3
Forestry Section				
GL 05 (1)	100.0	0.0	100.0	100.0
GL 01-04 (18)	83.3	16.7	50.0	61.1

(continued)

TABLE 7.5 (continued)
Employment History of Bauchi LG Permanent Staff, Four
Departments (1978)

Dept./Section/GL (# persons in post)	% First Employed By N.A./ L.G.A.	% First Employed By Post- Reform LG	% First Employed Under Colonial Rule	% 10 or More Yrs. Local Service By 1980
AGRIC./NATURAL RESOURCES				
Veterinary Section				
GL 05 (1)	100.0	0.0	100.0	100.0
GL 01-04 (10)	80.0	20.0	30.0	40.0
Dept. Total				
GL 05 & above (3)	100.0	0.0	66.7	100.0
GL 01-04 (58)	67.2	32.8	29.3	36.2
Subtotal all				
GLs (61)	68.9	31.1	31.1	39.3
MEDICAL & HEALTH				
Dept. Office				
GL 08 (1)	100.0	0.0	100.0	100.0
GL 01-04 (10)	50.0	50.0	20.0	50.0
Medical Section				
GL 05 (1)	100.0	0.0	100.0	100.0
GL 02-4 (11)	36.4	63.6	9.1	18.2
Health Section				
GL 05 (1)	100.0	0.0	100.0	100.0
GL 02-04 (19)	63.2	36.8	31.6	36.8
Soc. Welfare & Leprosy Section				
GL 05 & above (0)	-	-	-	-
GL 02-03 (6)	33.3	66.7	16.7	16.7
Dept. Total				
GL 05 & above (3)	100.0	0.0	100.0	100.0
GL 01-04 (46)	50.0	50.0	21.7	32.6
Subtotal all				
GLs (49)	53.1	46.9	26.5	36.7

(continued)

TABLE 7.5 (continued)
Employment History of Bauchi LG Permanent Staff, Four
Departments (1978)

Dept./Section/GL (# persons in post)	% First Employed By N.A./ L.G.A.	% First Employed By Post-Reform LG	% First Employed Under Colonial Rule	% 10 or More Yrs. Local Service By 1980
TREASURY				
Dept. Total				
GL 05 & above (3)	100.0	0.0	0.0	66.7
GL 01-04 (23)[a]	73.9	26.1	26.1	47.8
Subtotal all				
GLs (26)	76.9	23.1	23.1	50.0
WORKS & SURVEY				
Dept. Total				
GL 05 & above (4)	100.0	0.0	50.0	100.0
GL 01-04 (43)	60.5	39.5	37.2	51.2
Subtotal all				
GLs (47)	63.8	36.2	38.3	55.3
TOTAL FOR ALL				
4 DEPTS.				
GL 05 & above (13)	100.0	0.0	53.8	92.3
GL 01-04 (170)	61.8	38.2	28.8	40.6
Grand Total all				
GLs (183)	64.5	35.5	30.6	44.3

[a]Excludes the 15 market security guards, an unknown number
of whom had worked for the Bauchi Native and/or Local
Government Authority on a daily contract basis, who were
first designated as permanent staff in 1978.
Source: Ref. No. BLGA/S/OFF/4, Vol. I, pp. 70-90.

    The high proportion of older, former L.G.A. personnel
employed in 1978 undoubtedly reflects both reliance on a

narrow geographical base in previous recruitment practices
and a preference on the part of former officials for
continued service in the state capital local government
area where they resided over the other four outlying LGs
that had been carved out of the former Bauchi Local
Government Authority (Ibrahim, et al., 1979/80:ii). In any
event, the outcome placed a considerable burden on the
post-reform Bauchi Local Government in terms of over-
enrollment in the lower grades. This situation concomitantly limited the LG's capacity to recruit younger and more
qualified candidates from outside the traditional structure
of local administration into the demanding new roles
envisioned at all levels under the reformed system. Along
with the former L.G.A. staff, the Bauchi LG inherited old
work habits, archaic customs and procedures, entrenched
informal relationships, rural orientations, etc. (see
Hassan, 1979:31; Adewumi, 1979:110, Aquaisua, 1980:100).
In short, the absorption of most former L.G.A. staff members by the newly created Bauchi Local Government greatly
inhibited prospects for change in its initial years of
operation.

PERSONNEL COSTS AND RECURRENT EXPENDITURES

    Budget-allocation practices often provide one of the
most revealing indicators of the nature and orientation of
governmental operations. Thus, from Table 7.6, which
focuses on recurrent expenditures, we gain a clearer fix on
certain crucial operational consequences resulting from the
staffing apparatus sustained by the Bauchi LG and its
predecessor, the Bauchi Local Government Authority.
Several interesting findings emerge from analysis of the
data found in this table. First, staff salaries and
certain emoluments (principally transport and housing
allowances) consumed approximately 80 per cent of all
recurrent expenditures budgeted by the Bauchi LGA during
the 1975-77 financial years.[36] This proportion fell to 55
per cent in the first full fiscal year of operation for the
post-reform local government, but climbed back to 78.4 per
cent for 1978-79 following incorporation of the Local
Education Authority.
    With the exception of Central Administration, the
Bauchi Local Government devoted smaller shares of its
recurrent spending to salaries and emoluments for all

TABLE 7.6
Staff Salaries and Emoluments (S&E) as a Proportion of Bauchi LGA/LG's
Total Recurrent Expenditure: 1975/76-1978/79

| | --Year and Expenditure Category--
L.G.A./L.G. | | | | | |
| | ---------1975-76--------- | | | --------1976-77--------- | | |
DEPARTMENT	S&E ('000s₦)	Total Recurr. Expend ('000s₦)	S&E % of Total	S&E ('000s₦)	Total Recurr. Expend ('000s₦)	S&E % of Total
Central Admin.	253.4[a]	333.0[a]	76.1%	275.4	386.1	71.3%
Finance (Treasury)	24.9	42.3	58.9	26.6	47.3	56.1
Agric. & Nat. Res.	120.5	145.0	83.1	123.5	159.8	77.2
Medical & Health	162.8	213.7	76.2	181.9	279.7	65.0
Works & Survey	120.5	156.8	76.8	124.1	162.8	76.1
Education	121.2	129.0	94.0	111.5	122.4	91.0
Other	252.0[b]	252.0[c]	100.0	214.0[b]	214.0[c]	100.0
TOTAL	1,055.3	1,271.8	82.9	1,056.9	1,372.1	77.0
	----------1977-78--------			---------1978-79--------		
Central Admin.	149.4	185.4	80.5%	138.4	162.0	85.4
Finance (Treasury)	36.1	164.4	21.9	43.0	179.0	24.0
Agric. & Nat. Res.	108.2	158.6	68.2	75.8	111.1	68.2
Medical & Health	91.2	154.0	59.2	83.2	137.4	60.5
Works & Survey	85.8	208.2	41.1	89.3	142.7	62.5
Education	18.6	24.1	77.2	1,062.5	1,107.0	90.8
Other	-	-	-	-	-	-
TOTAL	489.2	894.6	54.6	1,492.2	1,902.2	78.4

[a]Includes "special services."
[b]Pensions and gratuities paid to retired employees (revised estimate
 used for 1975-76).
[c]Excludes contributions to other authorities and subordinate councils,
 "government share of taxes."

Source: Bauchi Local Authority Treasury, Annual Estimates/Accounts for
1975/76, 1976/77, "Summary of Recurrent Expenditures," Form L.A.T.62; Bauchi
Local Government Treasury, Annual Estimates/Accounts for 1977/78, 1978/79,
"Summary of Recurrent Expenditure," Form L.G.T.62.

departments in 1977-78 than had been the practice in the
Bauchi L.G.A. This pattern continued in 1978-79 with
respect to the Finance, Agriculture and Natural Resources,
Medical and Health, and Works and Survey Department -- even
though a sizeable number of daily workers had been converted to permanent staff during this period. Moreover,
10 (77 per cent) of the 13 supervisory staff at GL 05 and
above and 36 of the permanent employees at GL 01-04 working
in these departments had been upgraded to a higher salary
level in 1977. In 1978, a further 27 junior-grade
employees received an increase in grade level, bringing the
total number of staff at GL 01-04 who had been upgraded
within the first two years by the newly established local
government to 63 (or 40 per cent of those holding
appointments at GL 01-04 in 1976).[37] Much of the impetus
for the upgrading of staff came from new Ministry for Local
Government directives requiring uniform (generally higher)
entry grade levels for qualified staff occupying similar
positions.[38]

It is the assumption of responsibility for primary
education, then, that most dramatically altered the
distribution of recurrent expenditures by the Bauchi Local
Government. Allocations to the Education Department alone
constituted 61.5 per cent of all recurrent costs in
1978-79, whereas they had amounted to only 2.7 per cent
of the total one year earlier. As a direct result of the
incorporation of new responsibilities in the area of
primary education, the Bauchi LG's recurrent budget for
1978-79 greatly exceeded the total amount incurred by the
Bauchi Local Government Authority in 1976-77. In
particular, the merger required the new local government to
assume the massive, escalating personnel expenses
associated with running primary schools. In 1978-79,
salaries and emoluments for Education's personnel
comprised 91 per cent of that department's recurrent
budget, 71 per cent of all S&E costs, and over half (56 per
cent) of the entire Bauchi Local Government's recurrent
expenditure proposal.[39]

Heavy personnel costs triggered a series of budgetary
crises for the new Bauchi Local Government. Thus, the
Bauchi State Ministry for Local Government rejected its
initial 1978-79 budget proposals without even calling a
meeting of the Ministerial Local Government Estimates
Committee to consider them. The Ministry took this highly
unusual action because the LG's expenditure estimates did

not include a single major capital-development project due
to the consuming burden of staff costs. It based its
outright rejection of these estimates on the contention
that failure to provide for important capital-development
expenditures would thwart one of the primary purposes of
the local government reform. In its second budget
submission to the Ministry, the Bauchi LG incorporated
major capital-development projects. It succeeded in this

TABLE 7.7
Recurrent Costs as a Proportion of Total Approved
Expenditures for 1979-80; All Bauchi State Local
Governments

Local Government	Recurrent Expends.(₦)	Total (Recurrent & Capital) Expends.(₦)	Recurrent as % of Total Expends.
BAUCHI	2,782,075	3,245,075	85.7%
Katagum	3,073,277	3,895,867	78.9
Akko	3,838,206	5,198,592	73.8
Gombe	2,240,303	2,968,811	75.5
Gamawa	2,180,085	3,885,785	56.1
Dass	804,247	1,006,817	79.9
Jama'are	892,633	1,080,696	82.6
Tafawa Balewa	2,291,793	2,834,813	80.8
Darazo	1,836,066	2,882,566	63.7
Toro	1,857,033	3,081,133	60.3
Alkaleri	2,046,804	2,631,504	77.8
Dukku	2,317,729	3,556,334	65.2
Tangale/Waja	3,895,448	4,239,648	91.9
Shira	2,494,906	3,937,868	63.4
Misau	2,602,441	3,649,241	71.3
Ningi	1,588,226	2,853,471	55.7
All LGs	36,741,272	50,948,221	72.1%

Source: Bauchi State, Ministry for Local Government
(1980:1-2).

endeavor only by implementing drastic reductions in recurrent expenditures through the involuntary retirement of many of the older staff members who had transferred to the LG from the Bauchi Local Government Authority.[40]

In spite of the mass retirement exercise undertaken as a result of the Ministry's negative reaction to its initial budget submission for 1978-79, recurrent costs still consumed the preponderant share (86 per cent) of the Bauchi Local Government's approved expenditure estimates for 1979-80.[41] The 86 per cent share allocated by the LG to recurrent expenditures in 1979-80 is considerably higher than the 72 per cent average figure recorded for all 16 local governments in the state (see Table 7.7).[42] In general, however, Nigerian local governments typically devote such a high proportion of total expenditures to recurrent costs that little remains for new capital and development initiatives.[43]

CONCLUSIONS AND LESSONS

In this chapter, we examined problems of organizational structure and staffing at the local government level in Nigeria. Reference to the Bauchi LG facilitated understanding of these aspects of public administration by providing specific micro-level details about a concrete situation. The importance of local control over staffing and other personnel decisions emerged from this discussion.

The most important immediate conclusion reached from the expenditure analysis presented here is that the principal source of the burden placed upon the Bauchi LG's financial resources had not been removed by the 1978-79 retrenchment exercise. By far the largest share of its budget remained tied down on staff costs associated with the operation of primary schools. The vast sums which must be devoted to personnel costs in the education area have drained away needed funds from school construction and maintenance and from the supply of basic textbooks, materials, and equipment (Adebayo, 1981:163).

African local governments will continue to possess severely limited budgetary flexibility as long as they must bear full responsibility for paying the salaries of primary school teachers, headmasters, and headmistresses. Education certainly is crucially important and highly desired. However, commitments to primary school teachers

quickly become totally consuming of LG resources. The
troubling, but inescapable lesson to be drawn from the
findings presented in this chapter is that local
governments in Africa must withdraw from the educational
staffing function if they are to accomplish any other
important development objectives. In Nigeria, the
Babangida administration recently accepted the Political
Bureau's recommendations (1987:86-87, 117) that "the
federal government should bear full responsibility for
paying salaries of all teachers in the primary schools
throughout the country." Local governments would retain
responsibility for constructing, equipping, and maintaining
primary schools (Nigeria, Federal Government, 1987:28).
This is a promising development in that implementation
would enhance the ability of LGs to allocate resources to a
diverse set of local development needs.

   If the Nigerian experience provides an accurate gauge,
exogenous and endogenous pressures are interacting in a
complex and shifting way at the LG level in Africa.
Efforts by local governments to enhance their overall
performance capacity have been constrained by a weak and
diminishing local revenue base and by the misallocation of
available budget resources. In particular, too many local
institutions have fallen into the trap of acquiring an
immense and potentially insurmountable burden of recurring
personnel expenditures. There is a mounting sense of
urgency over the need to resolve this intractable problem
-- especially in the popular area of public education.

   At the same time, conflicting signals are coming from
national governments. On the one hand, there is renewed
interest in genuine decentralization and, at least in
Nigeria, a commitment to allocating nationally collected
revenues directly to the grass roots.[44] Moves in these
directions have been offset, on the other hand, by state
and federal monopolization of the most lucrative and
fastest growing revenue-collection sources (Frishman, 1980;
Abdullah, 1980:243) and by the recent blanket application
on the part of central government authorities of structural adjustment measures such as mandatory reductions in
LG expenditures and wage cuts or freezes (You and
Mazurelle, 1987:25; Nsingo, 1988:82; Egwurube, 1989:39).
Yet, the critical micro-level analysis conducted in
Chapters 6 and 7 of this book exposes the fallacious nature
of key structural adjustment assumptions regarding the performance capacity of public institutions in Africa. Powerful, well-organized, capably managed, development-oriented

government structures currently are not available to carry out new policies -- particularly at the local level.

Throughout urban and rural Africa, indigenous development requires capacity building at the grass roots. Local organizations must be equipped to plan and execute new programs in response to changing demand, to maintain existing infrastructure and services (Rondinelli, 1981:597), and to promote sustained growth of the local economy. Such efforts require considerable resources. The need to mobilize additional public resources to meet the challenge of creating effective local institutions has been ignored by the proponents of structural adjustment (You and Mazurelle, 1987:25). Cultivating reliable local sources of revenue collection (including property taxes) and developing effective means of attracting qualified personnel who are eager to serve in Africa's rural and semi-rural LGs are major components of this challenge.

NOTES

1. For an analysis of post-reform LG functional performance in the northern states of Nigeria, see Aliyu, Koehn, and Hay (1980). In May 1989, the AFRC entrenched about 150 new local government areas, along with the existing 301 LGs, in Part I of the First Schedule to the New Constitution (see West Africa, 15-21 May 1989, p. 784; 22-28 May 1989, p. 847).

2. In 1987, the FMG decreed that the reconstituted LGs would have a minimum of 5 supervisory councillors, one of whom would serve as deputy chairman of the council (Nigeria, Federal Government, 1987:31). Application of the 1988 Civil Service Reform to the LG level has changed the name of the Central Administration Department to "Personnel Management" and of the Treasury to "Finance, Supplies, Planning, Research, and Statistics." LG staff also are now required to specialized in a chosen field of administration (Egwurube, 1989:16).

3. In some cases, local governments also established functional committees, chaired by the supervisory councillor, to serve in an advisory capacity. See the discussion of the Badagry LG's Health and Social Welfare Committee in Olomolehin and Ndanusa (1980:193).

4. The 1987 revisions in the LG system concentrated executive as well as political powers in the council chairman (Nigeria, Federal Government, 1987:31). See

the arguments in favor of this approach advanced earlier
by Bala J. Takaya (1980:64, 66-67).

5. As background to the 1976 LG reform, the Public
Service Review Commission recommended that the portfolio
councillor system be abolished and that the multiple
committee structure be replaced by a single all-purpose
executive committee. The Commission based both recommendations on the finding that the portfolio councillorship and
the multiple committee system frequently result in inter-
ference by council members in day-to-day administrative
functions (reported in Oyediran, 1979:50; and in Gboyega,
1979b:37).

6. For instance, the Perm Sec, Ministry of Local
Government in Niger State, felt compelled to issue a 1977
circular which stated in part that "some councillors do not
fully appreciate their roles in the running of their Local
Government councils.... A supervisory councillor should
not undertake any 'executive duties'.... He can, however,
demand to know (through the head of department) any
information about the department and he has the right to
check and see that the head of department is properly and
efficiently carrying out his responsibilities" (also see
Ibrahim, 1980:165).

7. Ebong Ikoiwak's research among civil servants
working in the equally hierarchical federal service of
Nigeria lends support to the latter conclusion. He found
organizational constraints, including defects in
supervision/staff control, associated with impaired
effectiveness (Ikoiwak, 1979:227, 247-248; also see Sani,
1980:107-108). The colonial legacy of highly legalistic
and formalistic organizational structures and managerial
attitudes is especially pronounced in francophone Africa
(see, for instance, Nellis, 1986:37).

8. Other factors likely to promote job satisfaction,
according to Yahaya (1980:126-127), are the allocation of
strategic functions to local governments and the
establishment of a relationship with state government that
is "based on mutual interdependence and guidance instead of
superiority and dominance."

9. On these three departments, see Bauchi State,
MLGCD (1977:123). In the Badagry LG, the Public Health
Department also handles immunizations and school health
(Olomolehin and Ndanusa, 1980:193).

10. Many of these recommendations are based on
suggestions advanced earlier by A.D. Yahaya (1982:62-64).
These measures also closely parallel recommendations set
forth in Hyden (1983:95-96).

11. The Head of Service issues the Civil Service
Rules and prepares the scheme of service at higher levels
of government (Akpan, 1982:123).

12. Interview conducted with the Council Chairman and
the Secretary, Bauchi Local Government, on 28 December
1978; Ibrahim, et al. (1979/80:15-16, 22).

13. The Secretary exercised greater authority over
junior staff personnel matters in Bauchi than in the other
9 northern states (see Aliyu and Koehn, 1982:54).

14. Habibu Sani (1980:117-18) contends that political
victimization is a likely outcome of local control over
personnel matters. On the other hand, N. A. Aquaisua
(1980:96) reports that in Cross River State "there exist
instances where secretaries of Local Governments have
successfully resisted all attempts by local politicians and
interests to interfere and influence appointments and
promotions ...."

15. However, see Aquaisua (1980:100) on the
importance of family ties and the feelings of insecurity
associated with posting to another LGA; also see Sani
(1980:105, 113).

16. The chairmen also complained that the LGSC has
"unnecessarily delayed" the appointment of local government
staff at GL 06 and above (minutes of the Conference of
Chairmen of Local Government Councils in Zone D - Bauchi,
Benue, Bornu, Gongola and Plateau States held in Bauchi on
13-14 April, 1978; p. 8 of main report).

17. For a persuasive scholarly defense of this
position, see Elaigwu (1982:141). He adds that "the secretary to the LG should be appointed by the LGC and should be
responsible to its employees." However, such complaints
and suggestions did not lead to changes in the personnel
procedures covering senior local government officers in
Bauchi State by 1980. Council pressures for greater
decentralization in personnel decision making also ended up
being frustrated in Benue State (see Mohammed, 1983:3-9).

18. These findings are consistent with the LG staffing
pattern for all of Cross River State. In that state,
senior staff (GL 13 and above) constituted less than 1 per
cent of all LG employees. Middle management (GL 08-12)
amounted to 4.6 per cent, and low-level employees (GL
01-05) provided 87.7 per cent of total LG employment
(Aquaisua, 1980:96; also see Sani, 1980:124).

19. New Nigerian, 16 February 1980, p. 11; interview
conducted with the Chief Inspector, Bauchi State Ministry
for Local Government and Cultural Affairs, on 21 February
1980.

20. The problem has been most serious in the northern states.

21. For a number of reasons, regional and state governments had succeeded in enticing qualified LGA staff to join their ranks (see Adewumi, 1979:106).

22. There are two sides to the issue of harmonization. A.D. Yahaya (1980:138) maintains that, in comparison to state officials, local government officers have "responsibility over a smaller territorial area [and] are involved in fairly simple and less complex activities ...."

23. They generally equated the SLG with the position of ministry division head; i.e., GL 14 or 15 (Yahaya, 1980:133).

24. Nigerian Herald of 27 June 1977, as reported in Rowland (1979:92). In Cross River, LGSB officials took this message around the state via "enlightenment tours" (Aquaisua, 1980:100).

Nevertheless, the AFRC only approved a comprehensive scheme of LG service in 1988. The approved LG scheme, which is based on the work of the Oyeyipo Commission, establishes parity with state and federal civil services across all staff levels (see Egwurube, 1989:14-17).

25. The presumed advantages of employing indigenes are the higher degree of personal commitment and local citizen acceptance which they bring to LG service.

26. A common response to this situation throughout Africa has been "temporary" reliance upon civil servants seconded from higher levels of government. See, for example, Rondinelli (1981:603) on Sudan.

27. "Conclusions and Recommendations of the National Conference on Local Government, September 19-23, 1977" in Adamolekun and Rowland (1979:283); also see Adamolekun (1983:83); Aquaisua (1980:99). On the perceived status of LG employment, see Adeogun (1982:180-181); Sani (1980:105, 112).

Throughout Africa, governments encounter great difficulty attracting competent individuals for rural public service (see, for instance, Rothchild and Olorunsola, 1983:8). Following the 1974 revolution, Ethiopia's military rulers succeeded in posting a new cadre of educated and progressive administrators to key field positions within the rural local government system (Cohen and Koehn, 1980:278). Later on, however, these development administrators "sided with student activists in the agitation that swept the countryside ... and had to be removed" (Markakis, 1987:297n).

28. In terms of supervisory posts, the main unfilled vacancies existing in 1978 were section heads for agriculture and social welfare. Ref. No. BLGA/S/ OFF/4, Vol. I, pp. 70-90.

29. Figures provided by the Bauchi State Local Government Service Commission, 30 June 1980. One means of dealing with this staffing problem is by recruiting and training "primary health workers." The primary health worker is selected from and by the local community to engage in health education, treat minor ailments, and ensure safe water supply and proper sanitation practices (Olomolehin and Ndanusa, 1980:198).

30. Adeogun (1982:179-80) reports that LG staff in five other northern states also found it difficult to secure a place in the available training programs.

31. The supervisors for community development and information should possess a certificate in the related course of study. They are appointed at GL 06. Department heads holding a diploma who have attained 10 years of (satisfactory) post-qualification experience in their field are placed on GL 07, while those with 15 or more years of satisfactory service are given substantive appointments on GL 08. Interviews conducted with staff of the Bauchi State Local Government Service Commission, 30 June 1980. For a list of certificates/diplomas held by Bauchi LG department heads in late 1979, see Ibrahim, et al. (1979/80: 22).

32. Z.A. Abdullah (1980:247-248, 258-259) could not find a single professional accountant among the persons serving as LG treasurer in the northern states in the late 1970s.

33. It also is relevant that "many of the existing staff received their basic training more than a decade ago and have since had no further training of any type" (Bauchi State Government, 1979:142).

34. In 1980, the person holding the position of chief accountant in the Treasury took leave to attend the Diploma in Finance course (interviews, Bauchi State, LGSC, 30 June 1980). In addition to courses offered at the Institute of Administration, local government staff in Bauchi State participated in training programs operated by the Kaduna Polytechnic, the Staff Training Center in Potiskum, the Institutes of Health and Agricultural Research at A.B.U., the State School of Agriculture and Animal Husbandry, the Ministry of Health's School of Nursing and Midwifery, and the Azare Staff Training Center (Bauchi State Government, 1979:141-142). On training programs for LG employees in Cross River State, see Aquaisua (1980:97).

35. The widest disparity occurred in the Medical and Health Department, where 100 per cent (3 of 3) of those at GL 05 or above traced their service back to the days of colonial rule while only 22 per cent (10 of 46) of those at GL 01-04 had initially been appointed during that period.

In addition, both the secretary to the local government and the executive officer serving in the main office of the Central Administration Department in 1980 had first been appointed to local service during colonial rule and had accumulated 24 and 30 years of experience, respectively. The clerical officer had served for 10 years, the typists for an average of 3 years, and the messengers from 3-30 years (interviews conducted with staff of the Bauchi State Local Government Service Commission, 30 June 1980).

36. Although the 80 per cent rate would leave few funds for recurring needs of a non-personnel nature (e.g., supplies, operating and maintenance expenses), it actually underestimates staff-related overhead costs. This occurs because the budgeting category "other charges" (that is "other than" salaries and emoluments) in fact has included such manifestly personnel-related costs as pensions paid to retired employees (after 1976/77), final gratuities, medical and health expenses, and car, motorcycle, and bicycle loans. By 1980, the Ministry for Local Government had assumed complete responsibility for the payment of pensions to retired local government staff. To finance the pension scheme, the Ministry made reductions (at source) in each LG's federal and state grant allocations. According to the then Commissioner for Local Government, "the deduction is made on the basis of 5% of total teachers' salaries and 15% of the total of other personal emoluments expenditure of each of the local governments" (Mohammed, 1980:7).

37. Ref. No. BLGA/S/OFF/4, Vol. I, pp. 70-90; also see minutes of the Establishments Committee meeting held on August 31, 1977 (Ref. No. MLG/LG/S/157/S. I/Vol. I, pp. 87-91).

38. See the directive on "Local Government Estimates 1978/79" issued by the Permanent Secretary, Bauchi State Ministry for Local Government and Community Development, on 22 August 1977 (Ref. No. MLG/LG/EST/1/121, p. 122).

39. Assuming that the share of recurrent estimates devoted to Education Department personnel remained the same, then the costs of education staff encumbered at least half of the Bauchi LG's entire (capital and recurrent) budget for 1979-80. In contrast, rural community councils

in the Kaolack and Fatick regions of Senegal devoted only 14 per cent of their annual budget to education (Vengroff and Johnson, 1987:287).

40. Interview conducted with D. D. Motomboni, Zone Inspector for Bauchi Zone, Inspectorate Division, Bauchi State Ministry for Local Government, 22 February 1980.

41. Approved supplementary estimates later increased the Bauchi LG's capital-expenditure budget for 1979-80 from ₦463,00 to ₦619,500. Motomboni, interview.

42. The sum assigned by the Bauchi Local Government to recurrent expenditures for 1979-80 (₦2.8 million) constituted a 146 per cent increase over the amount contained in its rejected budget submission for the previous financial year, and a 263 per cent increase over the approved figure for 1978-79. Bauchi State, Ministry for Local Government (1980:1).

43. Also see the figures for selected LGs in Anambra, Benue, and Imo states reported in Nwankwo (1984:74-75).

44. In 1987, moreover, the AFRC accepted the Political Bureau's recommendation (1987:120, 122) that the amount of revenue allotted to LGs from the federation account be raised from 10 per cent to "not less than" 20 per cent (Nigeria, Federal Government, 1987:59).

RECOMMENDED READING

Students interested in local government structure are directed to Donald C. Rowat (ed.), International Handbook on Local Government Reorganization (Westport: Greenwood Press, 1980). Although the dated nature of comparative treatments of local government is a persistent problem, there are useful chapters in this book on francophone and anglophone systems and on Ghana, among others. Another valuable study, which relates local organizational structure to the challenges of development, is Milton J. Esman and Norman T. Uphoff, Local Organizations; Intermediaries in Rural Development (Ithaca: Cornell University Press, 1984). For a detailed analysis of urban and rural local government structure and staffing, consult John M. Cohen and Peter H. Koehn, Ethiopian Provincial and Municipal Government; Imperial Patterns and Postrevolu-tionary Changes, Monograph No. 9 (East Lansing: Michigan State University, 1980). A detailed francophone study is Richard Vengroff, Development Administration at the Local Level: The Case of Zaire, Monograph No. 40 (Syracuse: Maxwell School of Citizenship and Public Affairs, 1983).

Readers interested in the critical theorist's approach to the analysis of organizational structure and action might begin with Robert B. Denhardt's Theories of Public Organization (Monterey: Brooks/Cole Publishing Company, 1984) -- especially chapter 7. Denhardt believes that "democratic outcomes require democratic processes" -- in the workplace as well as in elections (p. viii). Another excellent and comprehensive source that treats traditional bureaucratic structure critically is Ralph P. Hummel, The Bureaucratic Experience, 3rd edition (New York: St. Martin's Press, 1987).

There are a few helpful sources, though they are rapidly becoming dated, that discuss post-reform local government organization and staffing in Nigeria. Readers should consult 'Ladipo Adamolekun and L. Rowland (eds.), The New System of Local Government in Nigeria; Problems and Prospects for Implementation (Ibadan: Heinemann Educational Books, 1979); and Suleiman Kumo and A. Y. Aliyu (eds.), Local Government Reform in Nigeria (Zaria: Institute of Administration, A.B.U., 1980).

8

Public Administrators, Policy Management, and Public Sector Reform in Africa

By focusing on administrative processes and policy outcomes, Chapters 1-7 compiled considerable evidence concerning the ways in which public administrators consistently manage to play central roles in the formulation and execution of public policy at all levels of government. They are likely to continue to act in these capacities in the future. Our understanding of this policy-management role must be enlarged through inquiry concerning the objectives pursued by administrative decision makers and through considered assessment of the impact of their involvement. The conscience of the bureaucrat cannot be salved by claiming that career administrators merely implement decisions reached by other actors in the political system (Dvorin and Simmons, 1972:36, 39).

Toward what ends have public administrators applied their extensive and persistent influence? What results can be attributed to their intervention? How can we characterize their actions?

The student of public policy and administration in Africa encounters three divergent schools of thought that address the issues raised by these questions. The traditional, uncritical commentator views the public service as the glue that prevents the political system from coming unstuck. From this perspective, administrative officials toil with insufficient remuneration in the service of the public's interests.[1] The radical critic sees career bureaucrats responding to the beat of a different drummer. According to this view, multinational corporate interests prevail and dependency in the periphery is sustained in large measure due to complicit behavior on the part of the indigenous bureaucracy. In short, local public administrators assume the role of compradore agents of global

capitalism. The third position delivers an equally
sweeping indictment of the public service. However, those
holding this perspective, which some label the
"contractocracy school,"[2] direct the brunt of their criticism at the individual bureaucrat rather than at world
capitalist economic relations. Over the course of its successful effort to accumulate power and consolidate control
over policy making and implementation, "the personal goals
of the executive stratum slowly displace those mandated by
external political elites" (Fischer and Sirianni, 1984:13).
Unchecked acquisitiveness lies at the heart of the problem
with public administration from this point of view.

THE NIGERIAN CASE

Unfortunately, important questions about the ways in
which bureaucratic power are used tend not to be seriously
addressed in textbooks on Nigerian public administration
(see Wilks, 1985:273). The first aim of this concluding
chapter is to apply the three competing explanations of
bureaucratic behavior to the Nigerian context and to draw
reasoned conclusions about the nature of administrative
performance at all levels of government to date. The deep
involvement of career officials in decision making throughout Nigeria's post-independence political history means
that policy analysis will provide an important basis for
investigating this issue. In the nearly three decades
following Nigeria's independence, public administrators
have constituted the only political actors to exercise
uninterrupted influence over the fashioning of public
policy. Thus, the nature of state policy, as well as
implementation outcomes, are matters largely shaped by the
actions of career bureaucrats and by the decision-making
processes and procedures which they have instituted.

Development Administrators

The first possibility to consider is that public
administrators in Nigeria have utilized their power to
promote national and local development. Such officials
should be found pursuing public interests in a selfless
manner. Although not always (or even often) successful in
achieving the demanding objectives which they have adopted,
bureaucrats of this ilk merit the title "development
administrator" based upon the dynamic and change-oriented

nature of their activities. Genuine development administrators advance policies that enhance national autonomy
and self reliance and promote economic growth along with
greater social equity.

The case for classifying public officers in Nigeria as
development administrators is brief. There are two main
reasons for this assessment. First, bureaucratic rules and
procedures typically have blocked or restrained change,
rather than promoted it. Second, the types of "development" programs and projects selected and pursued by public
administrators generally have neither enhanced national
autonomy nor primarily benefitted the poorer segments of
the population. Nwosu (1977:65) maintains, for instance,
that under the influence of higher civil servants, federal
and state agencies generally pursue goals and projects that
are "remote from the basic needs of Nigerians, the majority
of whom live in rural areas." Failed projects and wasted
development opportunities characterized the oil-boom period
(Nigeria, Political Bureau, 1987:35). At the grass roots,
Bala Takaya (1980:64) contends that post-reform local
government "is now physically nearer to the people, [but]
it is not people oriented."

At critical junctures in Nigeria's political history,
however, federal civil servants have demonstrated their
commitment to the concept of a strong and effective central
government (Williams and Turner, 1978:149, 154; Aliyu,
1977:266-267; Rothchild, 1987:121). Key permanent secretaries, in particular, have been credited with intervening
decisively to preserve and extend the authority of the
federal government vis-a-vis the regions/states (see
Luckham, 1971:312; Tukur, 1970b:21; Williams, 1976a:44;
Ciroma, 1980:7; Phillips, 1989:440). M. J. Balogun
(1983:88) suggests that "the bureaucratic class kept the
machinery of government going" when the civil war broke
out. Public servants also played highly visible policy-
shaping roles that are acknowledged to have encouraged successive military regimes to create additional states out of
the subnational units that had comprised the federation
(Yahaya, 1978:206; Williams and Turner, 1978:154; Campbell,
1978:76, 97; Oyovbaire, 1980:273).[3] Phillip Asiodu
(1979:89, 94) credits administrators with advancing
detailed programs for return to civilian rule in the early
1970s and with promoting government control of the
economy -- including the country's petroleum resources.[4]
These actions are frequently cited in characterizations of
the career bureaucracy as a cohesive and effective institution dedicated to promoting political stability and

national unity (Ayida, 1979:220, 222, 229; Adamolekun, 1978a:40; Akinyele, 1979:237; Garba, 1979:1; Yahaya, 1978:212).

Certainly, there have been individual cases of dedicated, development-oriented administrative leadership at all levels of government. Such behavior is generally regarded to be the exception rather than the rule in the Nigerian context, however. Ebong Ikoiwak's study (1979:247), for instance, found an "absence of commitment to the public interest" among federal civil servants at all levels of employment (also see Nigeria, Political Bureau, 1987:111). Takaya (1980:64, 67-68) suggests that civil servants generally resist change and that a number of local government staff, many of whom formerly served traditional authorities and/or the pre-reform LGAs, still "see themselves as rulers" rather than as public servants. These assessments of prevailing orientations at national and local levels carry serious implications. In the absence of a supportive and reinforcing environment, even development-minded administrators become frustrated and discouraged.

It is incorrect, therefore, to argue that a shortage of experienced, skilled manpower, or inadequate executive capacity, constitutes the principal constraint on development (Forrest and Odama, 1978/79:129n; Ayida, 1979:222-223; Anosike, 1977:30-46; Ibodje, 1980:282; Hyden, 1983:148). The future of development policy and administration in Nigeria depends more upon bureaucratic orientations and the uses of power than it does upon possession of technical skills.

Agents of Underdevelopment

The harshest critics contend that the Nigerian public bureaucracy is a "parasitic class" (Anise, 1980:23) that exploits the wealth of the nation and the labor of its dispossessed masses in alliance with and for the principal benefit of foreign capitalist interests (see Osoba, 1979: 64-67, 70, 74; Dudley, 1982:116; Joseph, 1977:11; Collins, Turner, and Williams, 1976:191-192; Ake, 1985b:32). According to this formulation, public administrators act primarily as agents of national underdevelopment.

Private firms are heavily dependent upon government contracts in economies dominated by public sector capital investments.[5] In addition, foreign corporations require official cooperation in order to operate profitably in the periphery. On its part, to ensure a continued supply of

the foreign capital, technology, and expertise which under-
pin its dominant position in the domestic legal and social
order, the bureaucratic segment of the ruling class "seeks
further integration, in its own interest, with its protec-
tor, the international bourgeoisie" (Ojo, 1985:166). In
order to strengthen their control over the policy-making
process, public administrators in dependent states forge
alliances that link the top levels of government organizations with private foreign enterprises (see Gould, 1980:31,
34-35). The same individuals even fill multiple roles.
Thus, Michael Watts and Paul Lubeck (1983:112-13) refer to
the "rising, highly educated professional and technical
groups who mediate between state employment and servicing
international firms, while at the same time maintaining
their own private firms and investments."

The principal area of disagreement among the critics
concerns the beneficiaries of bureaucratic activity. The
central issue is whether the economic policies and pursuits
of state officials promote the interests of multinational
capitalist firms or the development of an autonomous indigenous bourgeoisie. What we want to know is: "in whose
interest does the Nigerian state operate?" (Beckman,
1982b:38).

In exchange for the award of a lucrative contract,
market access, the allocation of public housing or use
rights over desirable plots of rural and urban land,
fraudulent mobilization fees, non-enforcement of a law or
regulation, or some other form of economic advantage
dispensed by the state, public decision makers at all
levels of government have been rewarded handsomely by
representatives of foreign and local firms[6] (see Williams
and Turner, 1978:156; Dudley, 1982:116; Joseph, 1978:235;
Collins, 1977:135-136; Ake, 1985a:195; Bangura, 1986a:
45-46; Andrae and Beckman, 1986:218-219; Othman, 1984:442,
450-452; Nigeria, Political Bureau, 1987:58; Adeogun,
1980:117-121, 130; Adebayo, 1988:66; Agbese, 1988a:279-280;
Marenin, 1988:223; Forrest, 1986:21; 1987:336; Aina,
1982:74; Phillips, 1989:441). Corruption of the public
bureaucracy "is an essential factor in the reproduction of
the links of dependency and exploitation" in Africa (Gould,
1980:8). It enables global capitalist institutions to
secure cooperation from administrative officers in such
vital operational areas as oil sales, import permits,
foreign investment, profit repatriation, monetary policy,
and credit repayment. In return for maintaining political
and economic conditions that are conducive for capital
accumulation by western corporations, including low wages,

suppression of opposition, and even actions that hold back indigenous capitalist development,[7] the bureaucratic element of the ruling class is invited to share in the surplus that accrues to the beneficiaries of administrative decisions (Ake, 1985a:198; Gould, 1980:35; Beckman, 1979:11; Andrae and Beckman, 1986:217, 219; Berry, 1983:255).

Clearly, many public servants in Nigeria have effectively employed the power of the state to establish a strong position for themselves in the private sector and, thereby, to consolidate their central strategic location within the domestic bourgeoisie. The salaries and fringe benefits issued to public servants pale in significance by comparison with the more indirect opportunities for individual enrichment that government service offers given the nature of the Nigerian political economy. Thus, for many, the principal attraction attached to holding a strategic position in the public bureaucracy is "the opportunity which it may provide for income, experience, and contacts, which can be used contemporaneously and subsequently in independent economic activity" (Williams, 1976a:37; Williams and Turner, 1978:164; Berry, 1983:255-258, 265; Heinecke, 1979:2-3). Tom Forrest (1987:334) notes, for instance, that many "senior civil servants retire from the public service at a relatively young age to take executive positions in the private sector."

In short, public administrators have not hesitated to employ their strategic position for the purpose of personal capital accumulation. In this sense, the state has functioned as an effective "instrument for the privatization of societal resources ... " (Agbese, 1988a:273; Forrest, 1987:336). The promulgation and implementation of government decrees promoting the indigenization of enterprises operating in Nigeria provided one of the most expedient avenues for accomplishing this result. Analysis of the outcome of the indigenization exercise is particularly revealing with respect to the chief beneficiaries of bureaucratic involvement in economic policy making in Nigeria.

Indigenization Outcomes. The Nigerian Enterprises Promotion ('Indigenisation') Decree (NEPD) of 1972 mandated that "by March 1974, a range of economic activities, then dominated to varying degrees by non-nationals, should be partly or wholly owned by indigenes" (Collins, 1977:128). NEPD essentially excluded foreigners from the ownership of small-scale domestic business operations and required 40 per cent Nigerian participation in the 36 types of large-scale manufacturing, service, processing, and commercial industries listed in its second schedule. Decree No. 3 of 1977 expanded the scope of Schedule II and increased the indigenous equity participation requirement to 60 per cent effective 31 December 1978 (Collins, 1977:145; Biersteker, 1987:159-244).

In furtherance of NEPD objectives, the federal government acquired a 40 per cent share in foreign-owned commercial banks and mandated that at least 40 per cent of their total loans be allotted to indigenous borrowers (Collins, 1977:128-130). High-level civil servants ranked among the main beneficiaries of subsequent increases in the volume of bank lending to Nigerians (see Dudley, 1982:118-119). Recipients used their newly acquired capital to purchase shares in Schedule II firms at highly favorable prices. Thus, through privileged access to information as well as credit, public officials secured a place at the "core of the Nigerian emergent share-owning class" (Collins, 1977:141, 131-133, 137; Williams, 1976a:32-33; Joseph, 1978:229). Billy Dudley (1982:117-118) cites the example of the Chief of Banking Operations at the Central Bank of Nigeria who gained ownership of a foreign electrical and plumbing firm under the indigenization program. This company subsequently received nearly 6 million naira in contracts and subcontracts awarded by the Central Bank and its major contractor. By 1979, moreover, more than half of the high-level (GL 14-16) Kaduna State civil servants interviewed by Patrick Heinecke (1979:8) reported that they were company shareholders, and about 20 per cent of those at GL 08-11 owned shares in private firms. The principal beneficiaries of Nigeria's indigenization exercises have been "the state and a few indigenous businessmen ...; the vast majority of the country's population has been excluded entirely from the process, from start to finish" (Biersteker, 1987:299, 278).

It is important to recognize that the deeper involvement of public officials and indigenous capitalists in multinational companies as passive shareholders or nominal "directors" which NEPD fostered also had the effect of providing foreign-dominated firms "greater security in their operations without the concomitant danger of lessened control ... " (Joseph, 1978:230; 1977:9-11). Based on extensive research, Thomas Biersteker (1983:193-202) shows that "virtually every transnational corporation involved in Nigeria has found ways to neutralize the indigenization requirements."[7] In his incisive analysis of the impact of Nigeria's indigenization measures, Paul Collins (1977:143,

131-137) concluded that "the overall picture which thus
emerges from these various patterns of co-operation and
clientage is one of a tightening nexus between government
and foreign capital .... The state must now protect even
more the interests of foreign capital in which the local
bourgeoisie has a stake" (also see Nigeria, Political
Bureau, 1987:58). Bjorn Beckman (1982b:48-49) adds that
"the scope for attacking capital on an anti-foreign plat-
form was reduced ...."

The Price of Bureaucratic Overhead. In a different
vein, a compelling argument also can be made that the costs
associated with bureaucracy foreclose the pursuit of other
development options that promise greater benefits for the
majority of citizens who do not hold government jobs. Much
of Nigeria's oil wealth has been siphoned into bureaucratic
expansion, entrenchment, and enrichment. Federal, state,
and local governments have incurred escalating recurrent
budgetary commitments for staff salaries and emoluments as
a result of relentless growth in the size of all public
service ranks. The Udoji awards alone added nearly one
billion dollars in annual recurrent expenses to the federal
government's estimates (also see Adamolekun, 1982:32).
Wages and salaries for members of the armed forces and
police consume a particularly large proportion of the
immense federal and state government allocation on defense
and security (Freund, 1979:93; Uche, 1977:177-178). At the
local government level, teachers' salaries have absorbed an
inordinate share of expenditures following the delegation
of responsibility for operating the Universal Primary
Education (UPE) scheme to this level (see Chapter 7; Aliyu
and Koehn, 1982:67; Kwara State, 1980:13-15). Some of
these payments have been diverted into internally unproductive investments in overseas real estate, foreign bank
accounts, and medical treatment abroad (see West Africa, 5
March 1984 on the latter issue). Legally and illegally
acquired wealth also allows many public administrators to
"buy out" of the inefficient and inadequate service systems
they operate, thereby reducing their personal incentive to
improve performance (Stren, 1988b:222). In sum, excessive
bureaucratic overhead costs and corruption perpetuate
underdevelopment.

Self-aggrandizing Functionaries

In Chapter 2, we read that public servants in Nigeria
have advanced and supported major public policy initia-

tives, particularly during periods of military rule. The
policies shaped by administrative intervention include
decrees, edicts, legislation, and budget allocations
affecting the ownership of business enterprises, the allocation of rural and urban land-use rights (D. Okpala,
1979:38), the construction of public housing (Onibokun,
1976:22), house rents (Dudley, 1982:120), wage and salary
levels, and agricultural-development strategies. In each
case, independent analysts evaluating the impact of these
measures have discovered that government bureaucrats, particularly higher administrative and professional officers,
consistently rank among the principal beneficiaries of
policy execution and program implementation.[9] For
instance, project staff have emerged as one of the few
domestic beneficiaries of the capital-intensive, large-
scale irrigation schemes pursued by successive federal
administrations for the avowed, and mostly unrealized, purpose of increasing domestic food production (Nigeria,
Political Bureau, 1987:36; Palmer-Jones, 1980:2, 7, 10-11;
Wallace, 1979a:10-18, 31; 1978/79:67-72; 1980b:8; Koehn,
1982a:265-267). The mechanized Bakalori scheme discussed
in Chapter 3 provides a case in point. River Basin authorities, international firms and consultants, and a leading
businessman and member of the Sokoto aristocracy with ties
to Impresit Bakalori and the Fiat tractor assembly plant in
Kano managed to profit from this expensive and unproductive
undertaking (Andrae and Beckman, 1985:112; also see Bernal,
1988:96, 103, on Sudan).

Furthermore, major proportions of the new capital
investment expenditures authorized in recent years at all
levels of government have been devoted to the construction
of staff housing, office buildings, and rest houses, to the
purchase of imported, labor-saving machinery and equipment
used exclusively for administrative convenience, and to
unproductive overseas purchases of military equipment
(Freund, 1979:93-94; Aliyu, 1980a:87, 89; Koehn,
1980:24-38, 51-57; Graf, 1986:107; Agbese, 1988a:283). The
establishment of twelve states out of the four former
regions in 1967, the development of seven additional state
capitals in 1976 and two others in 1987, and the creation
of numerous new local government units under the 1976
reform edicts and following the return to civilian rule all
required massive outlays of public funds for the employment
of locally based personnel, the construction of new headquarters and staff residences, and the provision of basic
infrastructural facilities and elite services (see Osoba,
1979:73; Williams, 1976a:36; Aliyu and Koehn, 1982:25-26;

Collins, 1980b:11-12; Oyovbaire, 1985:47; Graf,
1986:118).[10] It is reasonable to conclude, then, that
higher public servants have acted primarily out of vested
interest in enhancing bureaucratic authority, wealth, and
privileges at the centre of the political system and/or
expanding employment and advancement opportunities in their
home areas (Yahaya, 1978:206; Williams and Turner,
1978:154; Campbell, 1978:76; McHenry, 1984:8, 10; Ikoiwak,
1979:93; D. Okpala, 1979:38)[11] rather than out of personal
devotion to national development and the interests of the
masses.

Bureaucratic self-aggrandizement comes in many different shapes and forms. To begin with, there is the
matter of compensation. Higher public servants receive
salaries far in excess of national per capita income. In
1980, following the Udoji awards and the establishment of a
new minimum wage, even the lowest level civil servant
(GL 01) earned roughly 4 times more than the nation's estimated per capita income. All public servants benefit from
generous pension and gratuity packages (Balogun, 1983:196).
Moreover, government employees have secured for themselves
a measure of real income protection through a host of
untaxed subsidies (Williams and Turner, 1978:164; Otobo,
1986:118, 121; Balogun, 1983:195-196; Joseph, 1978:235).
The subsidies have included home-leave allowances, free use
of agency vehicles, transportation grants, low-interest car
loans, overseas educational leaves at full pay (plus transportation, tuition expenses, and generous maintenance and
shipping allowances), free health care and medical
supplies, and the provision of low-rent, furnished government housing. Most of these costly benefits can be traced
to contact with the British colonial service (Nigeria,
Political Bureau, 1987:109; Anise, 1980:33; Williams,
1976a:32; Barnes, 1982:6-7).

The higher echelons of government service have been
favored consistently by wage commissions and have benefitted the most from official salary and fringe-benefit
policies (Otobo, 1986:114-117). Ladun Anise (1980:29-33)
estimates that, added together, government subsidies result
in "real shadow wages" for the average high-level public
servant that are "twice as high as the direct, taxable
nominal wages quoted on the official payrolls" (also see
Collins, 1980a:320-321). The allocation of subsidized
government housing alone can easily result in an annual
savings of U.S. $10,000 over the market rental price.[12]
Nominal rents (about $600 per year), coupled with (now
discontinued) low-interest automobile loans in excess of

$10,000, have constituted the ranking civil servant's main hedges against the rampant inflation which has plagued Nigeria's economy in recent years.[13] The availability of additional allowances also has enabled the top-level bureaucrat to withstand in relatively painless fashion the protracted delays in salary payment which have occurred in several states (Otobo, 1986:121).

Extended family demands, the needs of one's home village, and the desire to accumulate material goods fitting an elite lifestyle create financial pressures that the administrator cannot satisfy even with increased salary, lucrative benefits, and subsidized housing (Cohen, 1980:77-78; Otobo, 1986:126). Thus, public servants have utilized bureaucratic position and resources to promote personal or family economic gain less directly as well. Widespread corruption has been documented at all levels of government (see Diamond 1984:6-10, 47-48; Aquaisua, 1980:100; Sani, 1980:104; Adamolekun, 1983:83-84; Toyo, 1986:231-232; Nwabueze, 1985:302; New Nigerian, 9 October 1984, p. 1; Aina, 1982:70; Forrest, 1986:21; Nigeria, Political Bureau, 1987:213).[14] Many public administrators in Nigeria, moreover, aspire to become wealthy traders or contractors (Joseph, 1978:34; Dent, 1978:122; Berry, 1983:258). If their business prospers, they may eventually resign from public service. Until that occurs, public servants at all levels of government utilize privileged access to information and decision makers to further private business interests (see Dent, 1978:123; Adeogun, 1980:132n; McHenry, 1984:8; Agbese, 1988a:277; Berry, 1983:265; also see Lancaster, 1988:34). Strategic "inside" posts in government are particularly valuable when privatization is under serious consideration.

On yet another front, civil servants have manipulated state land-allocation policies and procedures in ways that have further entrenched their elite economic standing and expanded profit-making opportunities. Via special access to information, credit, and fellow decision makers, public servants have acquired valuable rural and urban real-estate holdings. One manifestation of bureaucratic power in this area has been the dedication of expropriated land on the periphery of rapidly growing cities and in rural areas to "insider" commercial agricultural and real estate speculation ventures (see Kaduna State Government, 1981:19-21, 30-31, 39-40; Wallace, 1980b:2-7, 11; Palmer-Jones, 1980:10-11; Dudley, 1982:119; Lubeck, 1979:40; Ega, 1979:288, 296-297; Francis, 1984:14).

Throughout Africa, high-level bureaucrats have shown
particular interest and success in securing the most
desirable residential and commercial/industrial plots in
major urban centers (Hodder-Williams, 1984:171). Personal
favoritism, inside information, the effective application
of pressure, and the ability to satisfy requirements that
allocated land be "developed" (i.e., built upon) within a
short period of time have been effectively utilized by
public servants to accumulate urban land-use rights for
residential, rental, and business purposes (see Koehn,
1984; 1987; Ojo, 1977:269, 271-274; D. Okpala, 1979:32;
Dudley, 1982:272; Agbese, 1988a:278; Osoba, 1979:75;
Lubeck, 1979:39; Joseph, 1978:228; Frishman, 1977:295-398).
It is interesting in this connection that top-level administrators primarily invested their backdated Udoji wage
awards in urban real estate (Collins, 1977:141). Moreover,
higher civil servants are in a superior position to meet
prevailing land-allocation requirements since they qualify
for government housing loans and are preferred commercial
bank loan customers. Between 1976 and July 1980, for
instance, the Bauchi State Ministry of Finance's staff
housing loan scheme approved 559 out of 920 loan applications, awarding a total of 6,795,000 naira (in amounts up
to 5 times the applicant's yearly salary repayable in
15-year installments) to indigenes employed by state
government ministries, corporations, and boards at GL 06
and above.[15] Finally, the overwhelming emphasis on
moderate- and high-income residences evident in the units
constructed by public housing boards during this period
constituted another dimension of the explicit government
policy of promoting owner occupation by civil servants (D.
Okpala, 1979:35-36).

Following the 1975 coup, the press published lists
revealing extensive acquisition of land in the Lagos area
by top civil servants and other politically influential
individuals (Collins, 1977:141; also see I. Okpala,
1979:17-20). In spite of such revelations and promulgation
of the Land Use Decree in 1978, public servants remained
one of the main beneficiaries of state government land-
allocation policies and practices during the
Murtala/Obasanjo regime.[16] The main change involved the
incorporation of more intermediate-level public administrators into the web of post-Decree beneficiaries (see
Chapter 5).

LESSONS FROM NIGERIA

Although burned by their intimate association and
identification with official levers of political power
under the Gowon regime, shaken by the mass purges of 1975
and 1984, and threatened by the loss of formal authority
under provisions of the 1979 Constitution, civil servants
managed to preserve their prominent status position in
Nigerian society (Ayida, 1979:221; Beckett and O'Connell,
1977:8; Anise, 1980:23-35). Higher public servants in particular, remain politically powerful and economically privileged. Military and civilian rulers alike find that the
bureaucrat's skills, expertise, insights, and advice are
indispensable in policy making. Senior public administrators undeniably "overreached" their conventional roles on
various occasions, especially under General Yakubu Gowon.
Nevertheless, extensive bureaucratic involvement in determining public policy generally is viewed as inevitable and
even informally sanctioned. This situation presents a formidable barrier to change in public administration and
public policy. The main problem is that the object
requiring reform (the public service) constitutes the subject which is typically relied upon to undertake the reform
(Takaya, 1980:68-69).
The most important and controversial issue with
respect to the public bureaucracy concerns its role commitments. Will civil servants utilize their elite social
status and powerful policy-shaping position in the political system primarily to engage in further self-aggrandizement, to pursue narrow class and group interests, to aid
and abet multinational capitalist interests? Or, will they
promote the design and execution of effective public
projects that enhance indigenous, sustained economic
development and bring about substantial improvements in the
living conditions encountered by the vast majority of the
country's rural and urban population?

Commitment to Ruling Class Interests

Relative to many other countries, Nigeria is fortunate
in that its petroleum reserves offer an opportunity to
attain modest national development objectives and address
the basic needs of the citizenry. However, the public

sector's performance record to date offers little ground
for future optimism. The student of public administration
and public policy making in Nigeria cannot ignore evidence
that bureaucratic power has been utilized primarily to
promote self interest and to facilitate external penetration of the economy (also see Nigeria, Political Bureau,
1987:111). Usually, both objectives can be pursued
simultaneously. As Beckman (1982b:45) maintains, "the
primary role of the Nigerian state is to establish, maintain, protect and expand the conditions of capitalist
accumulation in general [i.e., both multinational and
domestic]." In his view (p. 50), the Nigerian state can no
longer be seen as acting in a compradore capacity because
"the relations of domination originating from outside have
been built into the fabric of domestic class relations."
Solidarity within the ruling class effectively greases
bureaucratic wheels at the top levels of national, state,
and local government.

　　Nevertheless, the critical perspective applied in this
book reveals that public administrators have most enthusi-
astically and effectively employed their social status and
political influence to shape public policies in directions
that promote their own personal and collective interests.[17]
Occasionally, moreover, fractures are apparent within the
ruling class (Ake, 1985b:31; Turner, 1986:61).[18] The
shifting divisions that arise provide an opening for certain forms of economic competition which are allowed on the
political agenda. For instance, Williams and Turner
(1978:159-160) contend that conflicts between administrators and technocrats regarding "whether authority over
state corporations should rest with officials of the
ministries or with officials of the [public] corporations"
are not simply the result of intra-bureaucratic rivalries,
but also involve a struggle over the lucrative personal
rewards to be derived from occupying the most strategic
positions governing the state's relations with foreign
firms (also see Stren, 1985:65; Schumacher, 1975:227). In
the infrequent struggles which arise over the distribution
of wealth and privilege, high-level administrative and
professional civil servants and parastatal officials often
manage to prevail over their domestic and foreign competitors in the private sector.[19] The latter depend upon
the career bureaucrat's inside knowledge and cooperation to
secure a desired policy or favor (Otobo, 1986:110-111;
Watts and Lubeck, 1983:113). When contractocracy interests
prevail, the state often undermines long-term strategies of
capitalist development (Andrae and Beckman, 1986:222;

Biersteker, 1983:202-203, 205; Falola and Ihonvbere,
1985:xi; Forrest, 1986:5).[20] Divisions within the ruling
class and among actors in the international system are
responsible for the limited degree of autonomy and the
unpredictability one encounters in peripheral societies
such as Nigeria that are organized along state-capitalist
lines and possess natural resources which are in demand at
the center of the world economy (also see Beckman,
1982b:42; Callaghy, 1987:88).

Absence of Restraint

We also have observed that at critical threshold
points in Nigeria's political history, backlashes have
occurred against the scope and depth of bureaucratic power.
The chief catalytic factor behind these episodic upheavals
is not the arrogation of power alone, but the pervasive and
blatant manner in which career officials have discredited
themselves through virtual single-minded devotion to their
own vested interests (see Williams and Turner, 1978:163;
Adedeji, 1981:804; Nwosu, 1977:81; Otobo, 1986:126). The
ruling class has shown remarkably little capacity or inclination to restrain itself. Preoccupation with accumulating
personal wealth and status has led civil servants to engage
in role behavior that at best neglects and at worst promotes exploitation of the vast majority of Nigeria's rural
and urban populace. Moreover, the steps taken to transform
bureaucratic orientations and behavior in Nigeria have
failed to produce noticeable improvements (Chapter 1). The
resulting negative reaction to administrative performance
and erosion of public confidence in government underlies
reductions-in-force (RIFs) and calls for privatization
(Wilks, 1985:268; West Africa, 3 August 1987, p. 1483;
Nigeria, Political Bureau, 1987:62-63, 106, 110; Esman and
Uphoff, 1984:280)[21] as well as low staff morale and motivation. These are all glaring symptoms of the trouble with
public administration throughout the continent.

PUBLIC ADMINISTRATION IN AFRICA: A CRITICAL ASSESSMENT

An emerging consensus among those who study public
administration in Africa holds that structural performance
is deteriorating and that bureaucratic policy makers frequently act in ways that are inimical to the needs of the
rural and urban poor (Stren, 1988b:243; Hyden, 1983:xii,

106, 120; Koehn, 1984:72-73). In David Abernethy's
(1983:15) assessment, "the salary structure of African
governments has helped to transform these governments into
powerful interest groups whose overriding interest is
themselves."[22] With specific reference to Algeria, John
Nellis (1980:421) found that "it is quite clear that the
... administrative system exists mainly for the benefit of
those who have succeeded in enfolding themselves within it,
rather than for the benefit of the public it supposedly
serves." David Gould (1980:xiii) concluded that "the
ruling class in Zaire has effectively 'privatized' the
public bureaucracy and converted it into an instrument for
self-enrichment."[23] Along with low levels of productivity,
Edward Schumacher (1975:86) referred to the "absence of
public service orientations among civil service personnel
at virtually every level" in Senegal. In short, public
administrators have assembled a record of failed will and
dismal performance. The failings of the public service
account in large measure for the perpetual crises from
which the state in Africa struggles unsuccessfully to
escape.

The recent literature on African administration is
revealing in terms of the underlying obstacles to improved
public sector performance. One encounters three particularly recurring themes. From a diverse set of country
studies, there are frequent references to (1) the constraints posed by prevailing orientations, (2) the failure
of technocratic leadership, and (3) the primacy of resource
inadequacy over structural maladjustment.

Management Orientations

The principal problem with contemporary public management in Africa is not a scarcity of skilled and competent
public managers. Indeed, Ibbo Mandaza (1988:64) maintains
that several countries compare "favorably with many a
developed country with regard to the number and quality of
senior administrators in the public sector." Although the
specific reasons vary, most scholars hold the prevailing
orientations adhered to by public administrators responsible for today's dismal state of affairs. Managers are
the target for the brunt of such criticism because of their
influence as role models to mid- and low-level personnel.

The most serious attitudinal constraints affecting
public policy outcomes and the public sector's performance
are centered around the loyalty dimension. The lack of

"public regard" documented in the Nigerian case throughout
this book is too often a central feature of public administration elsewhere in Africa. Disregard for performance
criteria, especially in evaluation and reward systems,
constitutes another major problem (see Chapter 1; Vengroff,
1983:26).

The prevalent authoritarian, non-participatory
approach to public management in Africa also is frequently
faulted by scholars. Donald Rothchild (1987:145) refers to
the resistance to decentralization exhibited by national
government bureaucrats who "retain a corporate interest in
continued central direction and control." Louise White
(1987:163) maintains that administrative refusal to share
control over the planning process denies community groups
the opportunity for meaningful involvement in local development policy (see also Esman and Uphoff, 1984:36;
Uwazurike, 1987:232). Richard Vengroff (1983:3, 26;
1988b:7) directs his criticism at management's overreliance
in Zaire and elsewhere on formal hierarchical command
within public organizations (also see Balogun and
Oshionebo, 1985:304). Lack of internal flexibility and the
widespread unwillingness of public administrators to
delegate authority stifle the initiative and resourcefulness needed to bring about change and to serve the poor.

Technocratic Leadership

The Nigerian experience also is indicative of the substitution of technical in place of political values which
is occurring across the continent within the administocracy.[24] The ascendancy of technocratic leadership in
Tanzania and Senegal, two socialist countries with highly
developed political parties, is particularly noteworthy.
In Tanzania, individuals with technical skills and bureaucratic experience have become more prominent in positions
of national leadership while those with political backgrounds are now less numerous (Samoff, 1986:12). In
Senegal, there has been a growing tendency for technical
feasibility, efficiency, and profitability criteria to
determine governmental policy decisions (Schumacher,
1975:xviii-xix).[25] The rise to power of Abdou Diouf
resulted in dominance of the technocratic element of the
ruling class and encouraged the breakdown of long-standing
decision-making processes based on "norms ensuring extensive consultation, consensus building, conciliatory
bargaining, and behind-the-scenes deliberations involving

most interests likely to be affected by particular policy decisions" (Schumacher, 1975:228).

These examples reveal that the transformation of state and society in Africa will not result from the rejection of patronage and the neglect of key political ties. In neither Tanzania nor Senegal have moves in a techno-bureaucratic direction produced the predicted attainment of basic economic goals (Schumacher, 1975:xix). The common outcomes under technocratic leadership are continued deterioration of the economy, increased political insta-bility, greater insensitivity to the needs of the poor, resistance to decentralization, and further external penetration and control (also see Nelson, 1984:993). Furthermore, citizen and NGO participation in policy making and planning are largely "precluded by the ideology of 'government as expertise'" upon which the technocratic element within the public bureaucracy bases its claim to exercise authority (Smith, 1985:196).

Resource Shortages and Structural Maladjustment

The debate over the relative importance of internal versus external factors as explanations for the poverty of African economies is far from settled. In terms directly related to public administration, those who view domestic considerations as the central source of Africa's problems currently emphasize structural maladjustment. Specifi-cally, the public sector needs to be reduced in size and pulled back from extensive involvement in the economy (see Chapter 1). From the perspective of analysts who attribute underdevelopment in Africa primarily to external influences, the principal constraint at the present time is insufficient resources. National resources continue to be exploited by multinational firms or world commodity markets and/or extracted in the form of debt repayments, while the promised infusion of foreign funding has not been forth-coming (Hodges, 1988:22-24; Brooke, 1989:4).

Where firm commitment to public service exists, devel-opment administrators in Africa typically are frustrated by lack of sufficient resources to accomplish their objec-tives. In Vengroff's assessment (1988a:6-7), the most cru-cial problem affecting all dimensions of development performance is "lack of resources." In Tanzania, local bureaucrats failed to promote adoption of a socialist stra-tegy for self-reliant agricultural development largely because they did not possess material incentives which

could be offered to the peasantry (see Hyden, 1980:106).
Schumacher (1975:220-221) suggests that "resource constraints" have consistently ranked among the key factors
limiting policy outcomes in Senegal throughout the
post-colonial period.[26] When inadequate resources prevent
delivery of promised inputs or result in program abandonment, the long-term consequence is the destruction of
incentives for popular local participation in government-
sponsored development projects (see Esman and Uphoff,
1984:37).

Under conditions of inadequate or non-existent
resources, dedicated public servants are unfairly criticized for poor public sector performance. It is more
honest and fruitful to recognize that administrative
failure in such cases stems from underlying resource constraints. Today, the major source of capital resource
exhaustion in Africa is payment to foreign creditors on the
massive external debt (see Chapter 4). Without rescheduling, the countries of Sub-Saharan Africa would have been
forced to devote 47 per cent of all export earnings to servicing the combined $218 billion external debt in 1987.
The net financial transfer from Africa to the International
Monetary Fund alone amounts to almost $1 billion per year.
Moreover, Africa's estimated 1987 debt service obligation
of $28 billion is projected to reach $45 billion per annum
by 1995 (Hodges, 1988:23-24).

In light of their heavy debt-financing burden, poor
countries now find it necessary to rely on foreign donors
for the bulk of their expenditures on development projects
(see Hyden, 1986:3). However, overseas development assistance to Africa from the United States and the European
Economic Community has stagnated or declined over recent
years (Hodges, 1988:24).[27] On a continent-wide basis,
public health and education programs have experienced the
most negative consequences due to the low priority placed
upon human resource development under the prevailing
structural-adjustment policy (Hodges, 1988:22; Africa
Recovery 2, No. 4, December 1988, p. 25).[28]

ALTERNATIVE FUTURES

Although the critical analyses are compelling, little
agreement exists regarding the most promising avenue for
arresting the reported slide in administrative behavior.
One approach focuses on change in the political economy.
For some, this involves severing bonds of external

dependency. Certain influential theorists argue that
socialism is necessary for making such a break (see
Beckman, 1979:5, 10).[29] Breaking bonds of dependency is a
step that is likely, following Gould's (1980:7-8) formulation, to result in a dramatically reduced incidence of
bureaucratic corruption. Moreover, critical analysis indicates that enhanced national self reliance is a prere-
quisite for introducing the most promising changes in
public administration and public policy in Africa.

Donor Designs

The public sector reform tactic, which has been
favored by donor agencies pushing structural adjustment,
emphasizes changes in the size, functions, and efficiency
of the national bureaucracy. The limits of this approach
are evident in the discussion of structural reforms
presented in Chapter 1. To summarize, advocates have
underestimated societal constraints on bureaucratic behavior as well as the extent of management overload,
neglected the needs of the poor, and ignored the vital role
of grass-roots institutions and local participation in
development decision making and policy implementation.
Most important, from the perspective of the patient, public
sector reform along with the other economic policy ingredients contained in the bitter structural-adjustment medi-
cine have not succeeded in reversing Africa's deteriorating
economic situation (Hodges, 1988:22). Neither increased
productivity nor more equitable distribution can be found
among the policy outcomes of structural adjustment. The
familiar lesson concerns the failure of imported economic
models.

The standard donor agency "reform" designs which dominate government- and Bank-sponsored foreign assistance
projects have proven to be detrimental to recipient
countries because of their high-cost, large-scale, complex,
inappropriate, and information-intensive nature (Hyden,
1983:176-179; Beckman, 1982:47). The external supply of
funds to cover recurrent costs reinforces dependency, while
the absence of such assistance virtually ensures that
existing large-scale and technically complex projects will
rapidly become inoperational (Hyden, 1983:166-167).
Corruption is only one of several serious environmental
constraints on the execution of structural-adjustment
policy (see, for instance, Gould, 1988; Rondinelli,
1981:600).

Moreover, privatization moves often result in further deterioration in service provision for the poor and allow deeper penetration of the host economy by foreign capital (Nigeria, Political Bureau, 1987:63). The roots of the problem with public administration in Africa involve lack of accountability to the local populace. It is naive to rely upon private firms when the need is to secure greater institutional responsiveness and the reduction of poverty and dependency.

The demand by donors for public service staff cutbacks also must be treated cautiously. The complexity of this issue is revealed when one attempts to identify specific positions for retrenchment. Although managers are the most expensive staff on the public payroll, they tend to be the state's most qualified personnel. Consequently, they already are overloaded with policy-decision and supervisory responsibilities. Where it exists, overstaffing tends to be limited to lower levels of the bureaucracy. Furthermore, accommodating informal privatization is likely to require increased governmental efforts to coordinate and regulate a vast and diverse array of public/private service providers (Stren, 1988b:243).

When administrative retrenchment is required, Phillip Morgan (personal communication, Chicago, 29 October 1989) recommends that staff personnel should be targetted for reductions and that service-providing line operations should be left in place. On the other hand, the crucial importance of policy analysts in government staff positions becomes apparent if one accepts Lawrence Graham's (1988) argument that the key to improved managerial performance lies in overcoming constraints imposed by the existing power structure, local political conditions, process barriers, and the politics of donor organizations. A similar justification for preserving high-level staff positions can be found in assertions that Africa requires "a superior diplomatic corps intimate with economic issues to gain heightened respect and improved terms from international financial institutions" (Shaw, 1988:36) and a core of professional policy analysts capable of formulating "plausible and ... defensible alternative packages of their own ..." (Helleiner, 1986:10).

Thus, privatization of many government functions remains impractical, indefensible, and/or undesirable in much of Africa. In such situations, radical changes in the nature of public administration merit serious attention. One key to success in such ventures is the ability to break away from the troubling tendency to imitate Western models

(Hodder-Williams, 1984:172). Neither parastatals nor
multinational corporations constitute appropriate organizational forms in Africa. One too rarely encounters mention
that the widely acknowledged failure of parastatal organizations reveals the bankruptcy of the business-management
model for the generation and distribution of national
resources.

Self-Reliant Approaches

The most promising institutional changes are designed
to maximize the advantages associated with small size,
local knowledge and concerns, and community involvement
discussed elsewhere in this volume (especially in Chapters
3 and 6; also see Bryant and White, 1982:218; Oyovbaire,
1985:273; Wilks, 1985:271; Montgomery, 1987:358; Hellinger,
et al., 1983:44-45). The primary unit of organization and
service provision in such a system would be the rural
village and the urban neighborhood (see Hyden, 1983:95-96;
Yahaya, 1982:62-64). This book has attempted to demonstrate the advantages in terms of social justice as well as
productive capacity that result from discarding technically
complex and pre-packaged project blueprints (also see
Hyden, 1983:177-79) and adopting simple, locally based, and
labor-intensive approaches in the conduct of public administration and in the content of public policy. Innovative,
community-determined methods are useful in areas as diverse
as responding to the need for international debt reduction,
selecting appropriate agricultural practices, structuring
participation in land-allocation processes and development
planning, and organizing local government activities.

Access and Process. In times of scarcity, such as
those which prevail in most African countries at the start
of the 1990s, access to bureaucratic policy making assumes
crucial importance as a factor determining the distribution
of resources (Stren, 1982:89). Economic crisis makes it
even more urgent that the process changes identified in
this book be adopted. In particular, the administrative
regulations which govern resource-allocation must involve
less complexity and fewer entry requirements based upon
income, employment, and educational level so that the poor
are no longer excluded from participation (also see Stren,
1982:90).

NGO Contributions. In an effective system of development administration, moreover, local government officials
would work closely with and through non-governmental

organizations (NGOs). NGOs are neither state organs nor profit-making institutions. Their officers usually have close ties to the poor, are well-informed about local conditions, are cost-conscious and disciplined, and interact personally in small-scale organizations. Non-governmental associations tend to employ labor-intensive approaches, to be adaptable and suited to learning from past mistakes, to manifest strong commitment to goals, to be positioned to mobilize vital local resources, and to be perceived as politically independent and dedicated to the welfare of local residents (Hyden, 1983:120-121, 130-131; L. White, 1987:64, 160-161, 164; Esman and Uphoff, 1984:39, 274-275; Curry, 1987:285-286; Hellinger, et al., 1983:21-23, 26; also see Osuji, 1982:180-182; Ibodje, 1980:286, 290).[30] Most NGOs provide the further advantage of a permanent institutional presence in the local area and, therefore, can assume long-term maintenance functions. Peasants and urban dwellers have demonstrated their willingness to participate in informal local associations (Hyden, 1980:215). The challenge is to link such participation with the policy-making process (Koehn and Waldron, 1978:47-86, 91-93; Hellinger, et al., 1983:40-41) and to join NGO activities with efforts to address basic human needs (Curry, 1987:288).[31]

OD and Training. At the same time, there are benefits to be secured in terms of administrative performance from management attention to organizational development and training. Chris Ukaegbu (1985:510) persuasively argues, based upon his systematic study of the motivating strength of extrinsic and intrinsic factors among professionals employed in the Nigerian public service, that the most effective reorganization proposals would aim to reduce the chain of command, define tasks clearly, and create conditions conducive to professional autonomy and individual initiative. From research on public management in Zaire and elsewhere in francophone Africa, we learn of the need to tie the organizational reward structure tightly to individual and collective performance and professional commitment (Vengroff, 1983:70; 1988:5, 9; Gould, 1988:40).

Based upon extensive consultancy experience in francophone Africa, James McCullough (1988), a professor of marketing, concludes that appropriate training in management skills is a more important determinant of effective performance than is shifting ownership from the public to the private sector. To be successful, management training must emphasize flexible and innovative responses to development challenges (Vengroff, 1983:70) and diverse local conditions. Public administrators cannot, as Goran Hyden (1980:227) recommends, "cut themselves off from their social environment." To the contrary, administrative training programs must prepare participants to utilize local culture and political relations for public ends. This requires a reliable base of knowledge built upon careful local research (see Schaeffer, 1985:246) and effective training in ethics.

Accountability and Orientations

Structural reorientation must be accompanied by much tighter internal auditing procedures (Abdullah, 1980:246) and by tough and strictly enforced penalties for illegal, unethical, and unproductive behavior.[32] In order to institutionalize accountability, however, changes in bureaucratic orientations are of primary importance. Without attitudinal change, structural reforms remain hollow. To begin with, public servants must possess a firm sense of purpose and strong reasons for caring about their work if they are to refrain from and be willing to expose corrupt behavior. They must come to understand that "the bureaucracy cannot have it both ways -- it cannot accept the theory and fact of power and still avoid the moral consequences of the exercise of that power" (Dvorin and Simmons, 1972:37, 35).

Ronald Cohen (1980:78, 80-82) reminds us that the bureaucratic attitudes of privileged status, self-enrichment, and state illegitimacy which are entrenched in government offices are absent in voluntary organizations (also see Ibodje, 1980:283; Hyden, 1980:26-27).[33] Indeed, the largest proportion (50%) of the surveyed members of community organizations in Irepodun LGA (Kwara State, Nigeria) ranked "highly placed government officials" as most influential in terms of initiating self-help development projects in their home communities (Adedayo, 1985:29).[34] This shows us that the initial challenge is to nurture similar attitudes of respect and enthusiasm for public organizations (also see Schumacher, 1975:230). Clearly, public administration training in Africa cannot be value-free. It must emphasize the ethical principles of loyalty to the constitution rather than to the ruler or to one's patron, and devotion to public service rather than to personal acquisition.

The paramount goal of bureaucratic reorientation must be serving the poor in a development capacity and not the

inculcation of capitalistic motives in an administrative
ruling class. In this effort, public administrators are
likely to find that they have a great deal to learn from
their NGO counterparts, who generally live among and
empathize with the poor, seek small-scale approaches that
rely upon local resources when attacking development problems and obstacles, are highly motivated and altruistic,
understand the importance of personal relations (with
clients and co-workers) that are based upon trust, expect
modest remuneration for their service, and value maintenance operations over the acquisition of new capital and
equipment (see Hyden, 1983:120-121, 146; Adedayo, 1985:29).

It is still possible for higher public administrators
in Africa to carve out a legitimate and popular role for
themselves in the shaping of public policy. Concomitantly,
they can strengthen the power of their countries to resist
external penetration and restore the public servant's
diminished stature and morale. They can achieve these
results by forcefully demonstrating that they are willing
to suspend attention to their own interests and to the ex-
tractive demands of foreign firms. Instead, they must concentrate on collectively delivering services and developing
resources that address the primary needs of the most disadvantaged rural and urban citizens (see Otobo, 1986:126;
Petras, 1969:283-284; Esman and Uphoff, 1984:279). A long
list of commentators on Nigerian public administration
have noted that development leadership through example,
sacrifice, and primary commitment to the needs of the
poorer classes of society must become the rule rather than
the exception among the administrative and professional
elite (see Okoli, 1980:12; Anise, 1980:23; Ikoiwak,
1979:93, 247-248; Harris, 1978:305-307; Osoba, 1979:64-67,
70, 74; Joseph, 1977:11; Collins, Turner, and Williams,
1976:191-192; Nigeria, Political Bureau, 1987:111).
According to Chinua Achebe (1983:25), there is no alternative to dedicated public service:

> We have no option really; if we do not move, we
> shall be moved. The masses whose name we take in
> vain are not amused; they do not enjoy their
> punishment and poverty.

Attitudinal change within the public sector offers the most
promising avenue for leading public administration out of
trouble. Once committed public servants are in place,
however, the focus of attention must shift to expanding the

resources available to local administrators responsible for
encouraging increased production and meeting basic needs.

Empowerment, Community Participation, and Public Service

The crucial outstanding issue concerning the role of
public administrators in public policy making in Africa,
then, is not whether they will continue to be actively and
centrally involved in initiating, formulating, and implementing capacities, but how they can be transformed from an
essentially "parasitic class" that is primarily oriented
toward preserving and enhancing multinational corporate
interests and pursuing self-aggrandizement into a committed, effective force for social change and independent,
mass-based economic development. Although leadership by
example and devotion to ethical behavior are crucial in the
long run,[35] the historical record offers no indication that
further education and retraining, or approaches that rely
on bureaucratic initiative and voluntary compliance, will
provide sufficient impetus for reorientation of the public
services. Increasing democratization of structures at all
levels of government and tighter external control over
career bureaucrats offer more promising avenues for effec-
tuating such a transformation (see Riggs, 1963:129, 144;
Osoba, 1979:71; Shaw and Fasehun, 1980:570; Anise, 1980:22,
35; Nigeria, Political Bureau, 1987:111, 158, 216-217;
Esman and Uphoff, 1984:39, 259, 279-280).

Two problems concerning external control of bureaucracy must first be overcome, however. As long as elected
representatives and military governors pursue narrow, self-
interested goals of their own,[36] it is unlikely that the
exercise of firmer political direction over the activities
of career administrators will yield radically different
results from those experienced in the past. In any event,
"today's constituencies are too broad and public issues too
complex to allow moral leadership to be exercised solely by
elected executives" (Dvorin and Simmons, 1972:39). Given
that elected officials and political appointees typically
rely heavily upon public administrators, it is essential
that the latter also provide initiatives for ethical
action.

Secondly, the overwhelming influence which administrative officers exercise over public policy making results,
in part, from the intellectual and technical dependency of
political actors.[37] Therefore, moving the bureaucracy in a
more locally responsive direction awaits the arrival of

informed, reliable, forceful, and effective representation
of the rural and urban poor. An indispensable first step
in this direction is political mobilization and education
of the peasantry and poor city dwellers.[38] Political education at the local level includes giving people a clear
idea of the new standards of behavior and performance
expected of local government councillors and staff. For
example, Bala Takaya (1980:67) maintains that "because
Local Government Staff used to be seen as representatives
of the person of the ruler ...," most local residents in
the northern states of Nigeria "still consider the employees of local government as rulers rather than servants of
public will." Professor Adamolekun (1983:213) emphasizes
the importance of removing restrictions on citizen access
to governmental information. A further step is the
training of local leaders. Existing training programs
typically concentrate on administrative officials and
ignore community leaders. Leadership training should be
designed to enhance the ability of elected officers to
resist efforts by career public administrators to circum-
vent and/or undermine their authority and must be responsive to expressed local needs (see Esman and Uphoff,
1984:36-38, 229, 250, 279). For Hellinger, et al.
(1983:22-23, 30, 35-36), beneficiary training must include
consciousness raising designed to inspire "awareness of the
need to organize and participate" among the poor along with
programs that impart problem-solving, organizational,
management, planning, and technical skills.

D. L. Sheth (1987:163-165) points out that "the consumers of development cannot be kept out of the process of
formulation of norms of alternative development -- however
inconvenient this might prove ...."[39] Empowerment of the
rural and urban masses for effective participation in
policy making would generate a persistent and powerful
countervailing force against bureaucratic privilege, non-
accountable behavior, and external exploitation (also see
Yahaya, 1982; Sani, 1980:119; L. White, 1987:165; Ibodje,
1980:286).[40] In the absence of this kind of check at the
base of the political system, public servants will continue
to act primarily as self-serving agents of underdevelopment.

NOTES

1. The former super perm sec, P. C. Asiodu (1979:74),
presents a strident defense of the Nigerian public service.

He asserts, for instance, that civil servants have been "misunderstood and abused, underpaid, undefended against libel and slander, unpraised for much dedicated and useful work ... [and] used as scapegoats by the rulers and their critics and would-be dispossessors alike" (also see Nigeria, Political Bureau, 1987:104-105 where both sides of the issue are reported).

2. Shehu Othman (1984:452) defines contractocracy as "government of contractors, for contractors and by contractors" and notes that Governor Balarabe Musa first popularized the term in Nigeria. Yusuf Bangura (1986a:47) cites and then critiques other proponents of the contractocracy school.

3. Richard Sklar (1981:14-15) emphasizes the "salutary" nature of these changes in federal arrangements for the future of constitutional government in Nigeria. Other analysts, including a majority of the members of the panel set up in 1975 to study the matter, maintain that division of the former regions into states enhanced political stability in Nigeria (see Elaigwu, 1980:4-6; Phillips, 1980:11-12; Luckham, 1971:221; Welch and Smith, 1974:134). A. D. Yahaya (1978:220-221) argues, however, that more fundamental cleavages in Nigerian society are not addressed by this action and that movements to create new states and realign boundaries are artificially inspired by elites seeking to "consolidate their influence in their own localities" and to grasp "an opportunity to share booty."

4. Thomas Biersteker's (1987:160-198) discussion of the Second Indigenization Decree lends some support to this view.

5. The public sector's share of Nigeria's total projected investment of 82 billion naira for the Fourth National Development Plan period (1981-85) came to 70.5 billion naira (Adedeji, 1981:805).

6. Biersteker (1983:202) notes that domestic capitalists are "engaged in a dual alliance with the state, on one hand, and with foreign capital, on the other." He also argues (1987:273) that the largest indigenous firms "have better contacts with the state and are better able to obtain lucrative contracts than their multinational counterparts" (also see Beckman, 1982b:44).

7. One example is the devaluation policy pursued by the Babangida regime (see Chapter 4).

8. With regard to the Second Indigenization Decree, "no real change took place in the effective control (and operations) of the vast majority of the enterprises affected ...." The principal exception is that "the state

has assumed managerial control of most of the domestic
financial sector, especially the banks" (Biersteker,
1987:245, 269, 277-279, 298-299).

9. An accounting of bureaucracy-designated expenditures must include specialized services such as modern
health-care facilities. In the Kano Metropolitan Area, for
instance, two restricted-access hospitals cater to high-
level civil servants and officers in the armed forces,
respectively (Stock, 1985:472). Robert Stock (1985:481)
graphically describes the hierarchy of access to medical
treatment in Kano State in this passage:

> Kano civil servants have the former European
> hospital to cater to their needs. Private clin-
> ics look after those able to afford their ser-
> vices. The urban masses join the crowd at the
> public hospital and hope for the best. They are
> still better off than the rural poor who usually
> have little or no alternative to traditional
> healers except ill-equipped dispensaries, often
> several kilometers from home. The public health
> care offered at these facilities is too often
> iatrogenic -- literally dangerous to health.

10. The 1976 local government reform, moreover, is
viewed by Graf (1986:115-116) as an attempt to rationalize
elite control over local-level largess.

11. For a similar conclusion regarding the ruling
class of state employees in Senegal, see D. O'Brien
(1979:219-220); on Zaire, see Gould (1980:xiii, 52, 122).

12. One recent study indicated that the University of
Lagos subsidized the housing costs of its lecturers at the
tune of 70 per cent of the employee's annual salary.
UNILAG spent ₦3,660,159 on the costs of renting off-campus
accommodations alone (Thisweek, 4 May 1987, p. 12).

13. Pita Agbese (1988a:278) notes that "it is common
to find a military officer earning no more than 7,000 naira
in annual salary and wages living in a house whose annual
rent is at least 40,000 naira."

14. One of the least publicized methods involves the
practice of not collecting debts to parastatals owed by
influential government functionaries and later writing such
debts off as non-recoverable (see Ihimodu, 1986:237).

15. Communication from the Secretary, BSSHLS, State
Ministry of Finance and Economic Development, Treasury
Division, MOF/TD/HOU/C.27/2/213, 1 July 1980; also see

Barnes (1982:8) on housing loans for government employees in Lagos.

16. For public allegations regarding illegal land allocations and transactions involving government officials in Kano, see New Nigerian, 15, 26 February 1980; 21, 25 March 1980.

17. Richard Joseph (1982:6a) has labelled this type of behavior "prebendal."

18. Samoff (1981:305) argues, with reference to Tanzania, that the bureaucratic ruling class allies itself at times with international capital and on other occasions with the aspiring indigenous bourgeoisie.

19. This is not always the case, however. Biersteker (1987:277-279, 291) shows how local businessmen, in alliance with foreign capital, enhanced their position in Nigeria by decisively influencing the indigenization program at the policy-implementation stage.

20. This means that Hyden's "economy of affection" is not the only plausible obstacle to capitalist development in Africa.

21. Agbese (1988c:3, 29) maintains that Nigeria's current economic crisis makes it imperative that the ruling class "discipline itself (and the rest of society) in a desperate attempt to save its privileged position ...."

22. According to Rothchild and Olorunsola (1983:8), expenditures on civil servant salaries consume about 85 per cent of government revenues in the Central African Republic.

23. Richard Hodder-Williams (1984:171) traces the tendency for civil servants to use their position to extract personal benefits to "the cumulative inequality that was one of the central legacies of the imperial intermission ...." Claude Meillassoux (1970:107) published an early report, with reference to Mali, describing the tendency of the post-independence bureaucracy to resort to exploitation and repression -- characteristics of a ruling class. In Mali, public managers continued to act as "an instrument of western interests" under conditions of dependency (Meillassoux, 1970:108).

24. For a general discussion of the behavior of the "technobureaucracy," see Batley (1978:67-70).

25. In general, French-influenced post-colonial educational systems produce "administrators and technicians who are both out of touch and sympathy with ... contemporary socioeconomic reality" (Le Brun, 1979:200-201).

26. Unequal access to available capital resources exacerbates this problem. In 1982, for instance, public

enterprises provided 1 per cent of all bank deposits,
received 49 per cent of all bank credit, and accounted for
82 per cent of the Senegalese banking sector's total outstanding debt (Nellis, 1986:24).

27. In terms of foreign aid, Ghana currently serves
as the exceptional case. Led by a determined World Bank
effort to make this West African country a structural-
adjustment showcase, the level of aid per capita reached
$40 in 1988 -- roughly double the average figure for the
rest of the continent. Even in Ghana, nevertheless, structural adjustment has not resulted in improved living conditions for the poor masses (Brooke, 1989:4).

28. This outcome is ironic because public education
and health schemes usually rank among the top demands
advanced by local residents when citizen participation in
policy formulation is allowed (see, for instance, Vengroff,
1974:308; Samoff, 1986).

29. A different formulation of the political economy
approach insists on transformation from the pre-capitalist
"economy of affection" to one built on capitalist economic
and bourgeois political and bureaucratic relations (Hyden,
1983:11, 29, 103, 107, 197, 212). Hyden (1980:232-233,
22-23) contends that "development in a direction towards
greater national self-reliance is improbable without a successful subordination of the peasantry to the demands of
the ruling classes, ... by forcing the peasants into more
effective relations of dependence."

30. Sheth (1987:166-167) argues that one must carefully distinguish between apolitical local associations and
politically oriented grassroots movements and organizations. The non-political organizations "function as voluntary agencies in charge of implementing 'development
programmes' which their governments have often got from the
international development or credit agencies." Their
efforts "remain by and large confined to the dispensation
of goods, rather than to creating assets and entitlements
for the betterment of the poor." The "non-party political
formations," in contrast, "generally invest their energies
in issues affecting the poor ...." Unfortunately, "their
concrete political action on these issues is often located
in conventional politics ... [and] they often end up being
no more than spade-workers for the political parties."

Milton Esman and Norman Uphoff (1984:18, 60-67) develop a different typology of local organizations. On the
basis of goals, authority, functions, membership, and
resources, they distinguish local government from (1) local

development associations, (2) cooperatives, (3) interest
associations, and (4) local political organizations.

31. However, as Esman and Uphoff (1984:209, 36, 277)
warn, "government linkage to the point of domination has
negative consequences for LO [local organization] performance. Capture by more privileged elements of the local
community will usually be equally unfortunate" (also see
Migdal, 1974:201; Koehn and Waldron, 1978:92-93).

32. Professor Adamolekun (1982:34) argues for the
establishment of a permanent investigative body in the
chief executive's office which would be responsible for
uncovering and prosecuting bureaucratic corruption.

33. The legitimacy attached to local, non-governmental associations in comparison to the lack of legitimacy
afforded state institutions reveals the negative "public
service" legacy of colonialism.

34. Community associations in Kwara State have compiled an enviable record of involvement in development
projects (Kwara State, 1980:25-29). Umar Benna (1975:132)
suggests that this can be explained in part by the degree
of local autonomy historically exercised at the village
level in Ilorin Province.

35. The behavior of senior administrators is particularly important because it establishes the boundaries of
acceptable conduct within the public service (see Diamond,
1984:2, 11).

36. In Nigeria, for instance, the newly elected
Second Republic legislators devoted priority attention to
fixing their own salary levels. They also occupied the
complex of high-rise flats which had been earmarked for
high-level federal civil servants and gave themselves car
loans (Otobo, 1986:117; Forrest, 1986:20). Similarly, the
Nigerian Association of Local Governments pushed for
massive improvements in the salaries and benefits enjoyed
by LG councillors (Graf, 1986:118). On the illegal enrichment of state governors during the Second Republic, see
Diamond (1984:7, 47) and Bangura (1986a:45); for the military rule period, consult Dudley (1982:317-319).

37. I am indebted to Christopher Ukaegbu for this
point.

38. Recognition of this need can be found in Nigeria,
Political Bureau (1987:111, 206-207; also see Esman and
Uphoff, 1984:209, 280) and in the goals of the new Directorate for Social Mobilisation (Shekwo, 1988). One now
finds a social mobilization officer (SMO) and assistant SMO
in each LG area in Nigeria and there are plans to organize
mobilization committees at the ward and community develop-

ment area levels (Shekwo, 1988:12). These steps are important given that structural decentralization alone is prone to "reinforce the position of those already dominant in local society ..." (Smith, 1985:199).

39. This understanding can be contrasted with Hyden's (1980:227) position that participation by outside groups in policy and plan formulation dilutes government goals and results in conflicts which waste human energy.

40. This step is important because even weak state institutions determine the beneficiaries of allocation processes (Herbst, 1987:412-414). At the same time, citizen participation offers the prospect of expanding the resources available for national distribution.

Bibliography

Abalu, George O. I., and D'Silva, Brian. 1979. "Integrated
    Rural Development and the Environment: Some Lessons
    from an Integrated Rural Development Project in
    Northern Nigeria." Paper presented at the Inter-
    national Conference on the Environment, Arlon,
    Belgium, September.
Abdullah, Z. A. [1980]. "Accounting System for Effective
    Management of Local Government Finance," in Kumo,
    Suleiman and Aliyu, A. Y. (eds.), Local Government
    Reform in Nigeria. Zaria: Institute of Administra-
    tion, A.B.U. pp. 242-259.
Abernethy, David B. 1983. "Bureaucratic Growth and Economic
    Decline in Sub-Saharan Africa: A Drama in Twelve
    Acts." Paper presented at the 26th Annual Meeting
    of the African Studies Association. Boston, December.
Abu, Bala Dan. 1988. "The Rain of Confetti." Newswatch
    7 (17):19-21.
Abubakar, Ahmad. [1979]. "Administrative Machinery and
    Planning Process: A Case Study of Bauchi Local
    Government," in Bauchi State, Ministry of Finance and
    Economic Development, Outline of Decisions on Planning
    Seminar for Local Government Officials. Bauchi:
    Ministry.
Achebe, Chinua. 1983. The Trouble with Nigeria. London:
    Heinemann Educational Books.
Achimu, V. 1977. "The Draft Constitution: The Division of
    Legislative Powers," in Kumo, S. and Aliyu, A. Y.
    (eds.), Issues in the Nigerian Draft Constitution.
    Zaria: Institute of Administration, Ahmadu Bello
    University. pp. 163-173.

304

Adamolekun, 'Ladipo. 1974. "Staff Development in Nigerian
    Public Service: Trends and Prospects." The Bureaucrat
    [Bendel State] 2 (2):411-419.
_____. 1978a. "High Level Ministerial Organization
    in Nigeria and the Ivory Coast," in Murray, D. J.
    (ed.), Studies in Nigerian Administration. Second
    Edition. London: Hutchinson & Co., Ltd. pp. 11-42.
_____. 1978b. "Postscript: Notes on Developments in
    Nigerian Administration since 1970," in Murray, D. J.
    (ed.), Studies in Nigerian Administration. Second
    edition. London: Hutchinson & Co., Ltd. pp. 310-
    327.
_____. 1979a. "The Idea of Local Government as a
    Third Level of Government" in Adamolekun, 'Ladipo, and
    Rowland, L. (eds.), The New Local Government System
    in Nigeria; Problems and Prospects for Implementation.
    Ibadan: Heinemann Educational Books. pp. 3-13.
_____. 1979b. "Introduction: A Tentative Profile of
    the Higher Civil Servants." Quarterly Journal of
    Administration (April-July):191-194.
_____. 1979c. "The Political Context," in
    Adamolekun, 'Ladipo, and Gboyega, Alex (eds.), Leading
    Issues in Nigerian Public Services. Ile-Ife:
    University of Ife Press. pp. 209-229.
_____. [1982]. "Reforming the Nigerian Public
    Service: The Udoji Commission and Its Aftermath."
    Unpublished paper in the author's possession.
_____. 1983. Public Administration; A Nigerian and
    Comparative Perspective. London: Longman.
_____. 1985a. "Conclusion: Retrospect and
    Prospect," in Adamolekun, 'Ladipo (ed.), Nigerian
    Public Administration 1960-1980: Perspectives and
    Prospects. Lagos: Heinemann Educational Books, Ltd.
    pp. 327-336.
_____. 1985b. "Nigeria's Ombudsman System: A
    National Network of Public Complaints Commissions,"
    in Adamolekun, 'Ladipo (ed.), Nigerian Public Admini-
    stration 1960-1980: Perspectives and Prospects.
    Lagos: Heinemann Educational Books, Ltd. pp. 307-
    326.
Adamolekun 'Ladipo, and Rowland, L. 1979. "Epilogue," in
    Adamolekun, 'Ladipo, and Rowland, L. (eds.), The New
    Local Government System in Nigeria; Problems and
    Prospects for Implementation. Ibadan: Heinemann
    Educational Books. pp. 293-301.
Adebayo, Akanmu G. 1988. "Postwar Economy and Foreign
    Policy: Gowon and the Oil Boom, 1970-1975," in

Falola, Toyin and Ihonvbere, Julius O. (eds.), *Nigeria
and the International Capitalist System*. GSIS
Monograph Series in World Affairs. Boulder: Lynne
Rienner. pp. 57-80.

Adebayo, Augustus. 1979. "Policy-Making in Nigerian Public
Administration, 1960-1975." *Journal of Administration
Overseas* 18 (January):4-14.

_____. 1981. *Principles and Practice of Public
Administration in Nigeria*. Chichester: John Wiley
& Sons.

Adebo, Simeon. 1979. "Personal Profile." *Quarterly
Journal of Administration* (April-July):195-199.

Adedayo, Adebisi. 1985. "The Implications of Community
Leadership for Rural Development Planning in Nigeria."
Community Development Journal 20(1):24-31.

Adedeji, Adebayo. 1968a. "The Evolution, Organization, and
Structure of the Nigerian Civil Services," in Adedeji,
A. (ed.), *Nigerian Administration and its Political
Setting*. London: Hutchinson Educational. pp. 2-10.

_____. 1968b. "The Future of Nigerian Admini-
stration," in Adedeji, A. (ed.), *Nigerian Adinistra-
tion and Its Political Setting*. London: Hutchinson
Educational. pp. 140-55.

_____. 1981. "Restoring Civil Service Morale." *West
Africa*, No. 3324 (13 April):803-805.

_____. 1988a. "The Evolution of the Public Service
in Africa." *Courier* 109 (May-June):60-63.

_____. 1988b. "Structural Adjustment with a Human
Face." *African Farmer* (1):52-55.

Adegboye, Rufus. 1977. "Problems of Land Tenure and Land
and Water Use Rights at the Kainji Dam." *Land Tenure
Center Newsletter* (Madison), No. 57 (July-Sept.):
41-42.

Adeniji, Kunle. 1984. "Local Planning Authorities and the
Crisis of Urban Growth Management in Nigeria: The
Oyo State Experience." *Planning & Administration* 11,
No. 1 (Spring):24-29.

Adeniyi, Eniola O. 1980. "National Development Planning
and Plan Administration in Nigeria." *Journal of
Administration Overseas* (July):160-174.

Adeogun, Adesola. 1980. "Contractors in Politics: A
Comparative Study of the Social Background of
Councillors in the Local Government Areas of Nigeria
with Reference to 1976 Local Government Elections."
Nigerian Journal of Public Affairs 9 (May):111-133.

_____. [1982]. "The Constraint of Personnel on Local
Government Functioning: Some Suggested Solutions,"

in Aliyu, Abubakar Yaya (ed.), The Role of Local Government in Social, Political and Economic Development in Nigeria 1976-79. Zaria: Institute of Administration, A.B.U. pp. 176-192.

Adewumi, J. B. 1977. "Local Government and Development Planning and Plan Implementation." Planning and Administration 4 (Spring):7-18.

_____. 1979. "Local Government Service Board and Local Governments: Cooperation or Conflict?" in Adamolekun, 'Ladipo, and Rowland, L. (eds.), The New Local Government System in Nigeria; Problems and Prospects for Implementation. Ibadan: Heinemann Educational Books. pp. 105-113.

Agbese, Pita O. 1988a. "Defense Expenditures and Private Capital Accumulation in Nigeria." Journal of Asian and African Studies 23 (3-4):270-286.

_____. 1988b. "Demise of Nigeria's Forthcoming 3rd Republic." Paper presented at the Political Science Lecture Series, Grinnell College, 11 November.

_____. 1988c. "Politics Without 'Old' Politicians: Transition to Civil Rule and the Ban on Politicans in Nigeria." Paper presented at the 31st Annual Meeting of the African Studies Association, 28 October.

Aina, Sola. 1982. "Bureaucratic Corruption in Nigeria: The Continuing Search for Causes and Cures." International Review of Administrative Sciences 48 (1):70-76.

Ajala, D.A. 1978a. "Administrative Machinery for Development at the Local Level." Paper presented at the National Seminar on Planning for Local Government Officials, Kaduna, October.

_____. 1978b. "Planning at the Local Level: Case Studies and Exercises in Project Selection and Preparation." Paper presented at the National Seminar on Planning for Local Government Officials, Kaduna, October.

Ake, Claude. 1985a. "Indigenization: Problems of Transformation in a Neocolonial Economy" in Ake, Claude (ed.), Political Economy of Nigeria. London: Longman. pp. 173-200.

_____. 1985b. "The Nigerian State: Antinomies of a Periphery Formation," in Ake, Claude (ed.), Political Economy of Nigeria. London: Longman. pp. 9-32.

Akeredolu-Ake, E. O. 1985. "Values and Underdevelopment in Nigeria" in Adamolekun, 'Ladipo (ed.), Nigerian Public

Administration 1960-1980: Perspectives and Prospects.
Lagos: Heinemann Educational Books.

Akinola, Amos A. 1987. "Prospects and Problems for Foreign
Investment in Nigerian Agriculture." Agricultural
Administration and Extension (24):223-232.

Akinrinade, Soji. 1988. "Politics of Reforms." Newswatch
7 (17):14-16.

Akinsanya, A. A. 1976. "The Military Regime, Top
Bureaucrats and What Next? The Nigerian Case."
Geneve-Afrique 40 (1):57-77.

Akinyele, T.A. 1979. "On Being a Higher Civil Servant."
Quarterly Journal of Administration (April-July):
231-242.

Akpan, Ntieyong U. 1982. Public Administration in Nigeria.
Lagos: Longman Nigeria.

Aliyu, Abubakar Yaya. 1977. "The Executive Presidential
System: A Political Imperative for Future Nigeria,"
in Kumo, S., and Aliyu, A.Y. (eds.), Issues in the
Nigerian Draft Constitution. Zaria: Department of
Research and Consultancy, Institute of Administration,
Ahmadu Bello University. pp. 261-269.

_____. 1979. "The Role of the Chief Executive,
Bureaucracy, and Assembly in Public Policy Formulation
and Execution under the New Nigerian Constitution."
Paper presented at the Seminar for Commissioners and
Top Civil Servants in Niger State held in Minna, 19-23
November.

_____. 1980a. "Local Government and the Administra-
tion of Social Services in Nigeria - The Impact of the
Local Government Reform." Paper presented at the
National Conference on Local Government and Social
Services Administration in Nigeria held at the
University of Ife, 18-20 February.

_____. 1980b. "The Nature and Composition of the
Legislature: Some Selected States." Paper presented
at the National Conference on the Return to Civilian
Rule held at the Institute of Administration, Ahmadu
Bello University, Zaria, 26-30 May.

Aliyu, Abubakar, and Joye, E. Michael. 1980. "The New
Nigerian Constitution: Some Problems." Paper pre-
sented at the National Conference on the Return to
Civilian Rule held at the Institute of Administration,
A.B.U., Zaria, 26-30 May.

Aliyu, Abubakar, and Koehn, Peter H. 1980. "Proposal for a
Unified System of Administration for the New Federal

Capital Territory of Nigeria." Nigerian Journal of Public Affairs 9 (May):1-21.

Aliyu, A. Y., and Koehn, P. H. 1982. Local Autonomy and Inter-Governmental Relations in Nigeria: The Case of the Northern States in the Immediate Post Local Government Reform Period (1976-79). Zaria: Institute of Administration, Ahmadu Bello University.

Aliyu, A. Y.; Koehn, P. H.; and Hay, R. A. 1980. "The Involvement of Local Governments in Social, Political and Economic Development in the Northern States: 1976-79." Unpublished paper presented at the National Seminar on the Role of Local Government in Social, Political, and Economic Development held at the Institute of Administration, Ahmadu Bello University, 28-30 April.

_____. 1982. "The Involvement of Local Government in Social, Political and Economic Development in the Northern States: 1976-1979," in Aliyu, A. Y. (ed.), The Role of Local Government in Social, Political and Economic Development in Nigeria 1976-79. Zaria: Institute of Administration, A. B. U. pp. 9-38.

Aliyu, A. Y.; Koehn, P. H.; et al. 1979. First Report on the Establishment of a Unified System of Administration for the Federal Capital Territory (Abuja). Zaria: Department of Research, Management, and Consultancy, Ahmadu Bello University.

Aluko, T. M. 1968. "Administration in Our Public Services: A Professional Officer Speaks Up on Bureaucracy," in Adedeji, A. (ed.), Nigerian Administration and Its Political Setting. London: Hutchinson Educational. pp. 68-78.

Ambroggi, Robert P. 1980. "Water." Scientific American 243 (September):101-116.

Anao, A. R. 1985. "Performance and the Structure of Control in State-owned Companies: A Case Study of Bendel Construction Company Limited," in Adamolekun, 'Lapido (ed.), Nigerian Public Administration 1960-1980: Perspectives and Prospects. Lagos: Heinemann Educational Books, Ltd. pp. 269-285.

Andrae, Gunilla, and Beckman, Bjorn. 1985. The Wheat Trap; Bread and Under-Development in Nigeria. London: Zed Books Ltd.

_____. 1986. "The Nigerian Wheat Trap," in Lawrence, Peter (ed.), World Recession and the Food Crisis in Africa. London: James Currey. pp. 213-30.

Anise, Ladun. 1980. "Desubsidization: An Alternative Approach to Governmental Cost Containment and Income

Redistribution Policy in Nigeria." _African Studies Review_ 33 (September):17-37.

Anosike, B. J. O. 1977. "Education and Economic Development in Nigeria: The Need for a New Paradigm." _African Studies Review_ 20 (September):27-51.

Aquaisua, N. A. [1980]. "Development and Management of Human Resources at the Local Level," in Kumo, Suleiman and Aliyu, A.Y. (eds.), _Local Government Reform in Nigeria_. Zaria: Institute of Administration, A.B.U. pp. 93-101.

Aronson, Dan R. 1978. Capitalism and Culture in Ibadan Urban Development," _Urban Anthropology_ 7(3):253-270.

Asabia, S. O. 1968. "The Role of the Administrator in the Nigerian Public Services," in Adedeji, A. (ed.), _Nigerian Administration and Its Political Setting_. London: Hutchinson Educational. pp. 112-117.

Asiama, Seth O. 1984. "The Land Factor in Housing for Low Income Urban Settlers," _Third World Planning Review_ 6, No. 1 (February):170-184.

Asiodu, Phillip. 1970. "The Future of the Federal and State Civil Services in the Context of the Twelve States Structure," in Tukur, M. (ed.), _Administrative and Political Development; Prospects for Nigeria_. Zaria: Institute of Administration, Ahmadu Bello University. pp. 124-146.

Asiodu, P. Chiedo. 1979. "The Civil Service: An Insider's View," in Oyediran, Oyeleye (ed.), _Nigerian Government and Politics Under Military Rule, 1966-79_. New York: St. Martin's Press. pp. 73-95.

Awa, Njoku E. 1980. "Food Production Problems of Small Farmers in Low-Technology Nations: Some Evidence from Nigeria." Cornell International Agriculture Mimeograph 79. Ithaca: Cornell University, June.

Ayeni, Victor. 1987. "Nigeria's Bureaucratized Ombudsman System: An Insight into the Problem of Burueacratization in a Developing Country," _Public Administration and Development_ 7 (4):309-324.

Ayida, A. A. 1968. "The Contribution of Politicians and Administrators to Nigeria's National Economic Planning," in Adedeji, A. (ed.), _Nigerian Administration and Its Political Setting_. London: Hutchinson Educational. pp. 45-65.

_____. 1979. "The Federal Civil Service and Nation Building." _Quarterly Journal of Administration_ (April-July):217-229.

Ayua, Ignatius A. 1980. "The Operation of the New Nigerian Constitution; Its Strengths and Weaknesses." Paper

presented at the National Conference on Return to
Civilian Rule held at the Institute of Administration,
Ahmadu Bello University, Zaria, Nigeria from 26-30
May.

Bach, Daniel C. 1980. "The Role of the Permanent Secretary
in Nigeria's New Executive Presidential System."
Paper presented at the Imo State Top Management
Seminar held at Oguta, 21-24 January.

Balogun, M. J. 1983. Public Administration in Nigeria: A
Developmental Approach. London: Macmillan Nigeria.

_____. 1988. "The Nature and Effectiveness of
Training for Decentralised Administrative Systems in
Africa." International Journal of Public Sector
Management 1 (1):51-64.

Balogun, M. J., and Oshionebo, B. O. 1985. "Personnel
Management in the Public Sector: A Survey of
Developments, 1960-1980," in Adamolekun, 'Ladipo
(ed.), Nigerian Public Administration 1960-1980:
Perspectives and Prospects. Lagos: Heinemann
Educational Books, Ltd. pp. 287-305.

Bangura, Yusuf. 1986a. "The Nigerian Economic Crisis," in
Lawrence, Peter (ed.), World Recession and the Food
Crisis in Africa. London: James Currey. pp. 40-58.

_____. 1986b. "Structural Adjustment and the
Political Question." Review of African Political
Economy 37 (December):24-37.

Barnes, Sandra T. 1975. "Political Participation in Lagos,
Nigeria." Paper presented at the SSRC Regional
African Urban Seminar, New York City, December.

_____. 1979. "Migration and Land Acquisition: The
New Landowners of Lagos." African Urban Studies
4 (Spring):59-70.

_____. 1982. "Public and Private Housing in Urban
West Africa: The Social Implications," in Morrison,
Minion K. C. and Gutkind, Peter C. W. (eds.),
Housing the Urban Poor in Africa. Syracuse: Maxwell
School, Syracuse University. pp. 5-30.

Bates, Robert H. 1987. "The Politics of Agricultural
Pricing in Sub-Saharan Africa," in Ergas, Zaki (ed.),
The African State in Transition. New York: St.
Martin's Press. pp. 237-261.

Batley, R. A. 1978. "Technocracy and Exclusion."
International Journal of Urban and Regional Research
2, No. 1 (March):53-77.

Bauchi Local Government. [1980]. "Development Plan
1981/85." Unpublished internal document.

Bauchi State Government. 1979. "Bauchi State Progress Report," in Adamolekun, 'Ladipo, and Rowland, L. (eds.), The New Local Government System in Nigeria; Problems and Prospects for Implementation. Ibadan: Heinemann Educational Books. pp. 133-144.

Bauchi State, Ministry for Local Government and Community Development (MLGCD). 1977. "Local Government Estimates 1978/79." Directive from the permanent secretary dated 22 August. Ref. No. MLG/LD/EST/1/121, p. 123.

Bauchi State, Ministry for Local Government. 1980. "Financial Position of the Local Governments for 1978-79 and 1979-1980." Unpublished mimeo in the author's possession.

Beckett, Paul, and O'Connell, James. 1977. Education and Power in Nigeria. New York: Africana Publishing Company.

Beckman, Bjorn. 1979. "Causes of Underdevelopment." Paper delivered at the Bauchi College of Arts and Science, Bauchi, 2 November.

_____. 1982a. "Public Investment and Agrarian Transformation in Northern Nigeria." Paper presented at the Workshop on State and Agriculture in Nigeria held at the University of California in Berkeley, 7-9 May; revised and expanded version.

_____. 1982b. "Whose State? State and Capitalist Development in Nigeria." Review of African Political Economy, No. 23 (Jan.-April):37-51.

_____. 1984. "Public Investment and Agrarian Transformation in Nigeria," in Watts, Michael (ed.), The State, Oil, and Agriculture in Nigeria. Berkeley: Institute of International Studies. pp. 110-137.

Benna, Umar G. 1975. "Effectiveness of Planning Strategies in Northern Nigeria." Unpublished Ph.D. dissertation, University of North Carolina.

Bennett, Valerie P., and Kirk-Greene, A. H. M. 1978. "Back to the Barracks: A Decade of Marking Time," in Panter-Brick, Keith (ed.), Soldiers and Oil; The Political Transformation of Nigeria. London: Frank Cass. pp. 13-26.

Bernal, Victoria. 1988. "Coercion and Incentives in African Agriculture: Insights from the Sudanese Experience." African Studies Review 31 (2):89-108.

Berry, Sara S. 1983. "Work, Migration, and Class in Western Nigeria: A Reinterpretation," in Cooper, Frederick (ed.), Struggle for the City: Migrant

Labor, Capital, and the State in Urban Africa.
Beverly Hills: Sage Publications. pp. 247-273.

Bienen, Henry, and Fitton, Martin. 1978. "Soldiers,
Politicians, and Civil Servants," in Panter-Brick,
Keith (ed.), Soldiers and Oil; The Political
Transformation of Nigeria. London: Frank Cass.
pp. 27-57.

Biersteker, Thomas J. 1980. "Indigenization and the
Nigerian Bourgeoisie: Dependent Development in an
African Context." Paper presented at a conference
held in Dakar, Senegal.

_____. 1983. "Indigenization in Nigeria: Re-
nationalization or Denationalization?" in Zartman,
I. William (ed.), The Political Economy of Nigeria.
New York: Praeger. pp. 185-206.

_____. 1987. Multinationals, the State, and Control
of the Nigerian Economy. Princeton: Princeton
University Press.

Bisrat Aklilu. 1980. "The Diffusion of Fertilizer in
Ethiopia: Pattern, Determinants, and Implications."
Journal of Developing Areas 14 (April):387-399.

Blackburn, Peter. 1986. "The Year of the IMF?" Africa
Report 31 (6):18-20.

Boidin, J.-C. 1988. "The Structural Adjustment Process."
Courier 111 (September-October):52-53.

Brau, Eduard H. 1986. "External Debt Management in the
African Context," in Helleiner, Gerald K. (ed.),
Africa and the International Monetary Fund.
Washington, D. C.: I.M.F. pp. 160-180.

Bridger, G. A., and Winpenny, J. T. 1983. Planning
Development Projects; A Practical Guide to the Choice
and Appraisal of Public Sector Investments. London:
Her Majesty's Stationery Office.

Brooke, James. 1987. "Nigeria's Flying Elephant; National
Airline Faces Criticism." New York Times, 3 August
1987.

_____. 1988. "Ailing Nigeria Opens Its Economy."
New York Times, 15 August 1988. p. D4.

_____. 1989. "In Western Eyes, Ghana is Regarded as
African Model." New York Times, 3 January. pp. 1, 4.

Browne, Robert S. 1988. "Evaluating the World Bank's Major
Reports: A Review Essay." Issue 16 (2):5-10.

Bryant, Coralie, and White, Louise G. 1982. Managing
Development in the Third World. Boulder: Westview
Press.

Caiden, Naomi, and Wildavsky, Aaron. 1980. Planning and Budgeting in Poor Countries. New Brunswick: Transaction Books.

Callaghy, Thomas M. 1987. "The State as Lame Leviathon: The Patrimonial Administrative State in Africa," in Ergas, Zaki (ed.), The African State in Transition. New York: St. Martin's Press. pp. 87-116.

_____. 1988. "Debt and Structural Adjustment in Africa: Realities and Possibilities." Issue 16 (2):11-18.

Callaway, Barbara J. 1987. "Women and Political Participation in Kano City." Comparative Politics 19 (July): 379-393.

Campbell, Alan K. 1979. "Testimony on Civil Service Reform and Reorganization," in Thompson, F.J. (ed.), Classics of Public Personnel Policy. Oak Park, Ill.: Moore Publishing Co., Inc. pp. 77-102.

Campbell, Ian. 1978. "Army Reorganization and Military Withdrawal," in Panter-Brick, Keith (ed.), Soldiers and Oil; The Political Transformation of Nigeria. London: Frank Cass. pp. 58-100.

Chambers, Robert. 1978. "Project Selection for Poverty-focused Rural Development: Simple is Optimal." World Development 6 (February):210-215.

Chick, John D. 1969. "Some Problems of Administrative Training: The Northern Nigerian Experience." Journal of Administration Overseas 8, No. 2 (April):97-110.

Christelow, Allan. 1987. "Three Islamic Voices in Contemporary Nigeria," in Roff, William R. (ed.), Islam and the Political Economy of Meaning; Comparative Studies of Muslim Discourse. Berkeley: University of California Press. pp. 226-253.

Ciroma, Adamu Liman. 1979. "Personal Profile." Quarterly Journal of Administration (April-July):209-215.

_____. 1980. "The Civil Service Today (1-3)." New Nigerian, 9, 10, 12 May.

Clausen, A. W. 1985. Poverty in the Developing Countries - 1985. Hunger Project Papers No. 3. San Francisco: Hunger Project, March.

Cline, William R. 1987. Mobilizing Bank Lending to Debtor Countries. Washington, D. C.: Institute for International Economics.

Clough, Paul, and Williams, Gavin. 1983. "Marketing With and Without Marketing Boards: Cocoa, Cotton and Grain Marketing in Nigeria." Paper presented at the Seminar

on Marketing Boards in Tropical Africa, Leiden,
September.

_____. 1984. "Decoding Berg: The World Bank in
Northern Nigeria," in Watts, Michael (ed.), The State,
Oil, and Agriculture in Nigeria. Berkeley: Institute
of International Studies. pp. 168-201.

Cohen, John M. 1984. "Participatory Planning and Kenya's
National Food Policy Paper." Food Research Institute
Studies 19 (2):187-213.

Cohen, John M., and Koehn, Peter H. 1977. "Rural and Urban
Land Reform in Ethiopia." African Law Studies 14
(1):3-62.

_____. 1980. Ethiopian Provincial and Municipal
Government; Imperial Patterns and Postrevolutionary
Changes. Monograph No. 9. East Lansing: Michigan
State University, African Studies Center.

Cohen, John M., and Uphoff, Norman T. 1977. Rural Develop-
ment Participation: Concepts and Measures for Project
Design, Implementation, and Evaluation. Ithaca:
Center for International Studies, Cornell University.

Cohen, Ronald. 1979. "Corruption in Nigeria: A Structural
Approach," in Ekpo, M.U. (ed.), Bureaucratic
Corruption in Sub-Saharan Africa. Washington, D.C.:
University Press of America. pp. 291-305.

_____. 1980. "The Blessed Job in Nigeria," in
Britan, Gerald M., and Cohen, Ronald (eds.), Hierarchy
& Society; Anthropological Perspectives on Bureau-
cracy. Philadelphia: Institute for the Study of
Human Issues. pp. 73-88.

Cole, Taylor. 1960. "Bureaucracy in Transition:
Independent Nigeria." Public Administration 38
(Winter):321-337.

Collins, Paul. 1977. "Public Policy and the Development of
Indigenous Capitalism: The Nigerian Experience."
Journal of Commonwealth and Comparative Politics
15 (July):127-150.

_____. 1980a. "Current Issues of Administrative
Reform in the Nigerian Public Services: The Case of
the Udoji Review Commission," in Collins, P. (ed.),
Administration for Development in Nigeria. Lagos:
African Education Press. pp. 310-28.

_____. 1980b. "Introduction," in Collins, Paul
(ed.), Administration for Development in Nigeria.
Lagos: African Education Press. pp. 1-14.

Collins, Paul; Turner, Terisa; and Williams, Gavin. 1976.
"Capitalism and the Coup," in Williams, Gavin (ed.),

Nigeria; Economy and Society. London: Rex Collings.
pp. 185-92.

Commission of the European Communities [C.E.C]. 1986.
"Financing and Debt." Courier 97 (May-June):91-96.

Conyers, Diana, and Hills, Peter. 1984. An Introduction to
Development Planning in the Third World. New York:
John Wiley & Sons.

Cooper, Frederick. 1983. "Urban Space, Industrial Time,
and Wage Labor in Africa," in Cooper, Frederick (ed.),
Migrant Labor, Capital, and the State in Urban
Africa. Beverly Hills: Sage Publications. pp. 7-50.

Crocker, Chester A. 1981a. "Strengthening U.S.-African
Relations." Current Policy No. 289. Washington,
D.C.: U.S. Department of State, Bureau of Public
Affairs, June 20.

_____. 1981b. "U.S. Interests in Africa." Current
Policy No. 330. Washington, D.C.: U.S. Department
of State, Bureau of Public Affairs, October 5.

_____. 1983. "Our Development Dialogue with
Africa." Current Policy No. 462. Washington, D.C.:
U.S. Department of State, Bureau of Public Affairs,
March 3.

Curry, Robert L., Jr. 1987. "Basic Human Needs and the
African State," in Ergas, Laki (ed.), The African
State in Transition. New York: St. Martin's Press.
pp. 263-293.

Dahlberg, Kenneth A. 1979. Beyond the Green Revolution;
The Ecology and Politics of Global Agricultural
Development. New York: Plenum Press.

Dar al-handasah Consultants. 1978. Preparation of a Feasi-
bility Study for Bauchi State Sites and Services and
Slum Upgrading Project. Interim Report, Vols. I, III.
Beirut: Shair and Partners.

Dash, Leon. 1980. "Restive Unions Pose a Threat to
Nigeria's Plans." Washington Post, 27 December.

_____. 1981. "Sahelian Herders Rebuilding after
Drought." Washington Post, 1 September.

Daudu, P. C. A. 1977. "Nigerian Draft Constitution:
Analysis of Powers in Relation to the Financial
Provisions," in Kumo, S., and Aliyu, A.Y. (eds.),
Issues in the Nigerian Draft Constitution. Zaria:
Institute of Administration, Ahmadu Bello University.
pp. 131-141.

Davies, Fela. 1981. "Interview with the Minister of
Agriculture on the Green Revolution in Nigeria."
Africa Agriculture (1):67-71.

Denhardt, Robert B. 1984. Theories of Public Organization. Monterey: Brooks/Cole Publishing Company.

Dent, M. J. 1978. "Corrective Government: Military Rule in Perspective," in Panter-Brick, Keith (ed.), Soldiers and Oil; The Political Transformation of Nigeria. London: Frank Cass. pp. 101-137.

de Onis, Juan. 1981. "Rice Shortage in Nigeria Brings Charges of Corruption." New York Times, 18 January. p. 9.

Diamond, Larry. 1981. "Society and Democracy in the Second Nigerian Republic: Party, Class, Ethnicity and Region." Paper presented at the Annual Meeting of the African Studies Association, Bloomington, 22 October.

_____. 1983. "Social Change and Political Conflict in Nigeria's Second Republic," in Zartman, I. William (ed.), The Political Economy of Nigeria. New York: Praeger. pp. 25-84.

_____. 1984. "The Political Economy of Corruption in Nigeria." Paper presented at the 17th Annual Meeting of the African Studies Association, Los Angeles, October.

D'Silva, Brian, and Raza, M. Rafique. 1980. "Integrated Rural Development in Nigeria: The Funtua Project." Food Policy (November):282-297.

Dudley, Billy J. 1968. Parties and Politics in Northern Nigeria. London: Frank Cass.

_____. 1982. An Introduction to Nigerian Government and Politics. London: Macmillan.

Dvorin, Eugene P., and Simmons, Robert H. 1972. From Amoral to Humane Bureaucracy. San Francisco: Canfield Press.

Eckholm, Erik. 1979. The Dispossessed of the Earth: Land Reform and Sustainable Development. Worldwatch Paper 30. Washington, D.C.: Worldwatch Institute.

Edoh, Tony. 1988. "The 1987 Local Government Election: The Northern Zonal Overview." Report to the National Electoral Commission, Lagos.

Ega, L. Alegwu. 1979. "Security of Tenure in a Transitory Farming System: The Case of Zaria Villages in Nigeria." Agricultural Administration 10(4): 287-298.

_____. 1984. "Land Acquisition and Land Transfer in Zaria Villages in Nigeria," Agricultural Administration 15 (2):87-100.

_____. 1987. "Local Policy and Land Administration in Nigeria: Some Conceptual and Practical Issues

Affecting Implementation." Urban Law and Policy 8, No. 5 (December):423-434.

Egwurube, Joseph O. 1989. "The Functioning of Local Governments During the Political Transition Period in Nigeria: 1987-1989." Unpublished paper in the author's possession.

Eicher, Carl K., and Johnson, Glenn L. 1970. "Policy for Nigerian Agricultural Development in the 1970's," in Eicher, Carl K., and Liedholm, Carl (eds.), Growth and Development of the Nigerian Economy. East Lansing, Mich.: Michigan State University Press. pp. 376-392.

Ekwueme, Alex I. 1980. "Keynote Address." Presented at the National Conference on the Return to Civilian Rule held at the Institute of Administration, Zaria, 26-30 May.

Elaigwu, Jonah Isawa. 1977. "Federal-State Relations in Nigeria's New Federalism: A Review of the Draft Constitution," in Kumo, S., and Aliyu, A.Y. (eds.), Issues in the Nigerian Draft Constitution. Zaria: Institute of Administration, Ahmadu Bello University. pp. 143-161.

_____. 1980. "Federal-State Relations in Nigeria's Presidential Federalism: Problems and Prospects." Paper presented at the National Conference on Return to Civilian Rule held at the Institute of Administration, Ahmadu Bello University, Zaria, 26-30 May.

_____. 1982. "Local Government and Political Development; The Challenges of Participation and Control in Grassroot Government in Nigeria," in Aliyu, Abubakar Yaya (ed.), The Role of Local Government in Social, Political and Economic Development in Nigeria 1976-79. Zaria: Institute of Administration, A.B.U. pp. 126-150.

Esman, Milton J., and Uphoff, Norman T. 1984. Local Organizations; Intermediaries in Rural Development. Ithaca: Cornell University Press.

Evans, Peter. 1979. Dependent Development; The Alliance of Multi-national, State, and Local Capital in Brazil. Princeton: Princeton University Press.

Falola, Toyin, and Ihonvbere, Julius O. 1985. The Rise & Fall of Nigeria's Second Republic: 1979-84. London: Zed Books Ltd.

_____. 1988. "Shagari: Oil and Foreiqn Policy in the Second Republic, 1979-1983," in Falola, Toyin and

318

Ihonvbere, Julius O. (eds.), Nigeria and the International Capitalist System. Boulder: Lynn Rienner
Pubishers. pp. 103-119.

Famoriyo, Segun. 1979a. "Institutional Perspectives for
Food Production Policies in Nigeria in the 1980's."
Paper presented at the 12th Annual Conference of the
Nutrition Society of Nigeria, Benin City, 1 December.

Famoriyo, Segun. 1979b. "Private and Public Ownership of
Land in Nigeria." Nigerian Journal of Political
Science 1, No. 1 (June):7-19.

Fantu Cheru. 1987a. "Adjustment Problems and the Politics
of Economic Surveillance: Tanzania and the IMF."
Paper presented at the 1987 Annual Meeting of the
African Studies Association, Denver, 19-23 November.

_____. 1987b. "Debt and Famine in Africa: The Year
of Living Dangerously." Africa and the World 1, No. 1
(October):1-9.

Feit, Edward. 1968. "Military Coups and Political Development; Some Lessons from Ghana and Nigeria." World
Politics 20 (January):179-193.

Fischer, Frank, and Sirianni, Carmen. 1984. "Organization
Theory and Bureaucracy: A Critical Introduction," in
Fischer, Frank, and Sirianni, Carmen (eds.), Critical
Studies in Organization & Bureaucracy. Philadelphia:
Temple University Press. pp. 3-20.

Fletcher, Frederick J. 1978. "The Executive Class in
Nigeria: Introduction, Problems, Prospects," in
Murray, D. J. (ed.), Studies in Nigerian Administra-
tion. Second edition. London: Hutchinson & Co.,
Ltd. pp. 140-176.

Forrest, Thomas G. 1977. "The Economic Context of
Operation Feed the Nation." Savanna 6 (June):75-80.

_____. 1981. "Agricultural Policies in Nigeria
1900-78," in Heyer, Judith; Roberts, Pepe; and
Williams, Gavin (eds.), Rural Development in Tropical
Africa. New York: St. Martins. pp. 222-258.

_____. 1986. "The Political Economy of Civil Rule
and the Economic Crisis in Nigeria (1979-84)."
Review of African Political Economy (35):4-26.

_____. 1987. "State Capital, Capitalist Development,
and Class Formation in Nigeria," in Lubeck, Paul M.
(ed.), The African Bourgeoisie; Capitalist Development
in Nigeria, Kenya, and the Ivory Coast. Boulder:
Lynne Rienner. pp. 307-342.

Forrest, Thomas G., and Odama, J. S. 1978/1979. "Nigerian
Budget Policy in the '70's." Nigerian Journal of
Public Affairs (8):115-130.

Francis, Paul. 1984. "'For the Use and Common Benefit of
All Nigerians': Consequences of the 1978 Land
Nationalization." Africa 54(3):5-27.

Franke, Richard W., and Chasin, Barbara H. 1980. Seeds of
Famine: Ecological Destruction and the Development
Dilemma in the West African Sahel. Montclair, N.J.:
Allanheld/Universe.

Freund, William. 1979. "Oil Boom and Crisis in Contemporary Nigeria." Review of African Political Economy
13 (May-Aug.):91-100.

Frisch, Dieter. 1988. "Adjustment, Development and
Equity." Courier 111 (September-October):67-72.

Frishman, Alan I. 1977. "The Spatial Growth and Residential Location Pattern of Kano, Nigeria." Unpublished
Ph.D. dissertation, Northwestern University.

_____. 1980. "The Changing Revenue Base and Budget
Crisis of the Kano Metropolitan Local Government."
African Urban Studies 8 (Fall):11-20.

_____. 1986. "Transportation Decisions in Kano,
Nigeria." African Urban Quarterly 1 (January):54-64.

_____. Forthcoming. "The Rise of Squatting in Kano,
Nigeria," in Obudho, R.A. (ed.), The Squatter
Settlement in Africa: Towards a Planning Strategy.
New York: Praeger.

Fromont, Michel. 1988. "The Poor: Bearing the Burden of
Adjustment." Courier 111 (September-October):94-95.

Gana, Jerry A. 1980. "Diffusion of Agricultural Innovations and Rural Development: A Case Study of Funtua
and Malumfashi Areas." Paper presented at the
Seminar on Change in Rural Hausaland, Bagauda Lake,
February.

Garba, Joseph N. 1979. "The Military Regime and the
Nigerian Society." New Nigerian, 28 September.
pp. I-III.

Gauhar, Altaf. 1984. "Julius K. Nyerere: Interview."
Third World Quarterly 6, No. 4 (October):815-838.

Gboyega, E. Alex. 1979a. "The Making of the Nigerian
Constitution," in Oyediran, Oyeleye (ed.), Nigerian
Government and Politics Under Military Rule, 1966-79.
New York: St. Martin's Press. pp. 235-250.

_____. 1979b. "Prospects for the New 'Cabinet
System' for Local Governments," in Adamolekun,
'Ladipo, and Rowland, L. (eds.), The New Local
Government System in Nigeria; Problems and Prospects
for Implementation. Ibadan: Heinemann Educational
Books. pp. 35-41.

320

_____. 1979c. "Summary of Discussions," in Adamolekun, L., and Gboyega, A. (eds.), Leading Issues in Nigerian Public Services; Proceedings of a National Symposium. Ile-Ife: University of Ife Press. pp. 230-235.

George, Susan. 1987. Food Strategies for Tomorrow. Hunger Project Papers 6. San Francisco: Hunger Project, December.

Gilbert, Alan, and Healey, Patsy. 1985. The Political Economy of Land; Urban Development in an Oil Economy. Brookfield, Vermont: Gower Publishing Company.

Gill, Gerard J. 1975. "Improving Traditional Ethiopian Farming Methods." Rural Africana 28 (Fall):107-118.

Gobir, Yusuf. 1970. "Discussant's Comments," in Tukur, M. (ed.), Administrative and Political Development: Prospects for Nigeria. Zaria: Institute of Administration, Ahmadu Bello University. pp. 159-162.

Goonesekere, R. K. W. [1980]. Land Tenure Law and Land Use Decree. Zaria: Institute of Administration, Ahmadu Bello University.

Gould, David J. 1976. "The Role of Dependent Public Administration in Underdevelopment and Its Impact on Decision Making and Administrative Reform in Local Zairian Administration." Paper presented at the Annual Meeting of the African Studies Association held in Boston.

_____. 1980. Bureaucratic Corruption and Underdevelopment in the Third World: The Case of Zaire. New York: Pergamon Press.

_____. 1988. "Combatting Administrative Corruption in Africa: Incidence, Causes and Remedial Strategies." Paper presented at the 31st Annual Meeting of the African Studies Association, Chicago, 2 October.

Goulet, Dennis. 1978. The Cruel Choice; A New Concept in the Theory of Development. New York: Atheneum.

Graf, William. 1986. "Nigerian 'Grassroots' Politics: Local Government, Traditional Rule and Class Domination." Journal of Commonwealth & Comparative Politics 24, No. 2 (July):99-130.

Graham, Lawrence S. 1988. "Strategic Management Considerations in the Improvement of Organizational Performance." Paper presented at the 31st Annual Meeting of the African Studies Association, Chicago, 29 October.

Gran, Guy. 1983. Development by People. New York: Praeger Publishers.

Gruhn, Isebill V. 1983. "The Recolonization of Africa: International Organizations on the March." Africa Today 30(4):37-48.

Gusau, Ibrahim. 1981. "The Green Revolution Policy of the Federal Government." Africa Agriculture (1):72-79.

Hamma, Sule Y. 1975. "Government Policy and Development Opportunities in Metropolitan Kano." Unpublished M.P.A. thesis, Institute of Administration, Ahmadu Bello University.

Hardy, Chandra S. 1986. "Africa's Debt: Structural Adjustment with Stability," in Berg, Robert G. and Whitaker, Jennifer S. (eds.), Strategies for African Development. Berkeley: University of California Press. pp. 453-475.

Harmon, Michael M. 1987. "What People Can Do: Managing Despite the Organization," in Hummel, Ralph P., The Bureaucratic Experience. Third edition. New York: St. Martin's Press. pp. 53-56.

Harris, Lawrence. 1986. "Conceptions of the IMF's Role in Africa," in Lawrence, Peter (ed.), World Recession and the Food Crisis in Africa. London: James Currey. pp. 83-95.

Harris, Richard L. 1978. "The Role of Higher Public Servants in Nigeria: As Perceived by the Western-educated Elite," in Murray, D. J. (ed.), Studies in Nigerian Administration. Second edition. London: Hutchinson & Co., Ltd. pp. 283-309.

Hassan, Musa. 1979. "Local Government at Bauchi, 1979." Unpublished report. Zaria: Department of Local Government Studies, Ahmadu Bello University.

Hawkins, Tony. 1987. "How Debt Chokes Development." Africa Recovery 2 (May-August):27-29.

Hay, Richard; Koehn, Eftychia; and Koehn, Peter. 1980. "Attitudes of Local Government Administrators in the Northern States Toward Rural and Community Development." Paper presented at the National Seminar on the Role of Local Government in Social, Political and Economic Development held at the Institute of Administration, Ahmadu Bello University, 28-30 April.

Heady, Ferrel. 1966. Public Administration: A Comparative Perspective. Englewood Cliffs, N.J.: Prentice-Hall, Inc.

Hecht, Robert. 1981. "U.S. Looks for Increased Involvement in Nigeria's Agriculture." African Economic Digest, 27 February.

Heinecke, Patrick. 1979. "A Dependent Civil Service in an 'Independent' Country." Paper presented at the

Faculty of Administration Seminar, Ahmadu Bello University, 13 November.

Helleiner, Gerald K. 1983. "The IMF and Africa in the 1980s," Essays in International Finance 152 (Princeton University, Department of Economics, July): 1-23.

_____. 1986. "Introduction," in Helleiner, Gerald K. (ed.), Africa and the International Monetary Fund. Washington, D. C.: I.M.F. pp. 1-11.

Hellinger, Douglas; O'Regan, Fred; Hellinger, Stephen; Lewis, Blane; and Bendavid-Val, Avrom. 1983. "Building Local Capacity for Sustainable Development." Washington, D. C.: Development Group for Alternative Policies, December.

Herbst, Jeffrey I. 1987. "Policy Formulation and Implementation in Zimbabwe: Understanding State Autonomy and the Focus of Decision-Making." Unpublished Ph.D. dissertation, Yale University.

Hicks, Ursula K. 1965. Development Finance; Planning and Control. Oxford: Clarendon Press.

Hodder-Williams, Richard. 1984. An Introduction to The Politics of Tropical Africa. London: George Allen & Unwin.

Hodges, Tony. 1988. "African, International Efforts Fail to Halt Economic Slide." Africa Recovery 2, No. 4 (Dec.):22-24.

Honadle, George. 1982. "Development Administration in the Eighties: New Agendas or Old Perspectives?" Public Admnistration Review 42, No. 2 (March/April):174-179.

Hossain, Mahabub. 1988. "Credit for Alleviation of Rural Poverty: The Grameen Bank in Bangladesh." IFPRI Abstract (February):1-4.

Hummel, Ralph P. 1987. The Bureaucratic Experience. Third edition. New York: St. Martin's Press.

Huntington, Richard; Ackroyd, James; and Deng, Luka. 1981. "The Challenge for Rainfed Agriculture in Western and Southern Sudan: Lessons from Abyei." Africa Today 28 (2):43-53.

Hurley, Judith. 1986. "U.S. Farmers and the Third World: The Crisis Points toward Unity." Hunger Notes 11, No. 6 (January):10-13.

Hutchful, Eboe. 1981. "The Political Economy of International Debt Renegotiation: Ghana 1966-1974." Paper presented at the School of Social Sciences Seminar, University of Port Harcourt.

Hyden, Goran. 1980. Beyond Ujamaa in Tanzania; Under-
development and an Uncaptured Peasantry. Berkeley:
University of California Press.
_____. 1983. No Shortcuts to Progress; African
Development Management in Perspective. Berkeley:
University of California Press.
_____. 1986. "Beyond Hunger in Africa--Breaking the
Spell of Mono-culture." Paper presented at the
Annual Meeting of the African Studies Association.
Ibodje, S. W. E. [1980]. "Towards a More Realistic
Approach to Local Government Reforms in Nigeria," in
Kumo, Suleiman and Aliyu, A. Y. (eds.), Local Govern-
ment Reform in Nigeria. Zaria: Institute of Admini-
stration, A.B.U. pp. 282-295.
Ibraheem, Umar Toungo. 1981. "Political Participation and
Development in Gongola State of Nigeria." Unpublished
Ph.D. dissertation, Northwestern University,
Department of Political Science.
Ibrahim, Mohammed, M.; Mosau, M. S.; and Gadu, S. A. 1979/
1980. "Research on Administrative Aspects of Bauchi
Local Government." Unpublished ADLG paper. Zaira:
Department of Local Government Studies, Ahmadu Bello
University.
Ibrahim, Umar Ibn. [1980]. "Local Government and Rural
Development," in Kumo, Suleiman, and Aliyu, A.Y.
(eds.), Local Government Reform in Nigeria. Zaria:
Institute of Administration, Ahmadu Bello University.
pp. 161-168.
Idris, Tukur. [1980]. "Discussion Paper," in Kumo,
Suleiman, and Aliyu, A.Y. (eds.), Local Government
Reform in Nigeria. Zaria: Institute of Adminis-
tration, Ahmadu Bello University. pp. 89-92.
Ihimodu, Ifeyori I. 1986. "Managing Public Commercial
Enterprises in Nigeria: The Case of Kwara State
Commercial Parastatals," Public Administration and
Development 6 (July/Sept.):223-238.
Ikoiwak, Ebong A. 1979. "Bureaucracy in Development: The
Case of the Nigerian Federal Civil Service."
Unpublished Ph.D. dissertation, Atlanta University.
_____. 1981. "Role Constraints in the Operations
of the Nigerian Federal Civil Service 1960-1979:
Some Lessons in Retrospect." Nigerian Journal of
Public Affairs 10 (May):95-113.
International Bank for Reconstruction and Development.
1974. Nigeria: Options for Long-Term Development.

Report of a mission sent to Nigeria by the World Bank. Baltimore: Johns Hopkins University Press.

International Bank for Reconstruction and Development. 1981. Accelerated Development in Sub-Saharan Africa. Washington, D.C.: I.B.R.D.

Izah, Paul P. 1987. "Review of Works by Ibrahim L. Bashir." African Affairs 86 (April):285-286.

Joseph, Richard A. 1977. "National Objectives and Public Accountability: An Analysis of the Draft Constitution," in Kumo, S., and Aliyu, A.Y. (eds.), Issues in the Nigerian Draft Constitution. Zaria: Institute of Administration, Ahmadu Bello University. pp. 1-22.

_____. 1978. "Affluence and Underdevelopment: The Nigerian Experience." Journal of Modern African Studies 16 (June):221-239.

_____. 1982. "Ethnicity and Prebendal Politics in Nigeria: A Theoretical Outline." Paper presented at the Annual Meeting of the American Political Science Association, Denver.

Kaduna State Government. 1979. "Kaduna State Progress Report," in Adamolekun, 'Ladipo, and Rowland, L. (eds.), The New Local Government System in Nigeria; Problems and Prospects for Implementation. Ibadan: Heinemann Educational Books. pp. 191-199.

Kaduna State Government. [1981]. White Paper on the Report of the Land Investigation Commission. Volumes I, V, XII. Kaduna: Government Printer.

Kaduna State. Ministry of Economic Development. 1977. "Population Census of Kaduna State 1963 and Projections for 1976-80 by Local Government Councils and Districts." Kaduna: Ministry, Statistics Division.

Kaduna State. Ministry of Internal Affairs and Information. 1981. The Kaduna Political Stalemate; Factors Responsible for the Conflict between the Executive and the Legislature in Kaduna State. Kaduna: Government Printer, May.

Kaduna State. Office of the Governor. 1980. "Building the Foundation of a New Social Order: The First Progress Report of the Government of Kaduna State, 1st October 1979-1st October 1980." Kaduna: Office of the Governor, 2 October.

Kalu, Kalu I. 1987. "Adjustment, Recovery, and Regional Integration." Courier 106 (November-December):24-27.

Kang, Darshan S. 1982. "Environmental Problems of the Green Revolution with a Focus on Punjab, India," in Barrett, Richard N. (ed.), International Dimensions

of the Environmental Crisis. Boulder: Westview.
pp. 191-215.

Kaunda, Kenneth. 1988. "'We Want to Develop Our Own
Resources.'" Courier 111 (September-October):76-77.

Kehinde, Bayo. 1968. "The Politics and Administration of
Public Corporations in Nigeria," in Adedeji, A. (ed.),
Nigerian Administration and Its Political Setting.
London: Hutchinson Educational. pp. 92-99.

Kent, Tom, and McAllister, Ian. 1985. Management for
Development; Planning and Practice from African and
Canadian Experience. Washington, D. C.: University
Press of America.

Kerkvliet, Benedict J. 1979. "Land Reform: Emancipation
or Counterinsurgency?" in Rosenberg, David A. (ed.),
Marcos and Martial Law in the Philippines. Ithaca:
Cornell University Press. pp. 113-144.

Kigera, Musa Muhammed III. [1980]. "Traditional Rulers and
Local Government," in Kumo, Suleiman and Aliyu, A. Y.
(eds.), Local Government Reform in Nigeria. Zaria:
Institute of Administration, A.B.U. pp. 312-317.

Kilborn, Peter T. 1989. "Shift is Seen in U.S. Policy to
Ease Repayment Costs for Third World." New York
Times, 9 March. pp. 1,28.

Killick, Tony, and Martin, Matthew. 1989. "African Debt:
The Search for Solutions." U.N. African Recovery
Programme Briefing Paper, No. 1 (June):1-8.

King, Roger. 1981. "Cooperative Policy and Village
Development in Northern Nigeria," in Heyer, Judith;
Roberts, Pepe; and Williams, Gavin (eds.), Rural
Development in Tropical Africa. New York: St.
Martins. pp. 259-280.

Kingsley, J. Donald. 1963. "Bureaucracy and Political
Development, with Particular Reference to Nigeria,"
in La Palombara, J. (ed.), Bureaucracy and Political
Development. Princeton, N.J.: Princeton University
Press. pp. 301-317.

Kirk-Greene, Anthony, and Rimmer, Douglas. 1981. Nigeria
Since 1970; A Political and Economic Outline. New
York: Africana Publishing Company.

Koehn, Peter. 1979. "Ethiopia: Famine, Food Production,
and Changes in the Legal Order." African Studies
Review 22, No. 1 (April):51-71.

_____. 1980. "The Involvement of Local Governments
in Nigeria's National Development Planning for the
1980's; A Comparative Analysis of Bauchi and Kaduna
Capital Project Proposals." Paper presented at the
Department of Research, Management, and Consultancy

326

Seminar, Institute of Administration, Ahmadu Bello
University, Zaria, 5 June.

_____. 1982a. "African Approaches to Environmental
Stress: A Focus on Ethiopia and Nigeria," in Barrett,
Richard (ed.), International Dimensions of the
Environmental Crisis. Boulder: Westview Press.
pp. 253-298.

_____. 1982b. "The Evolution of Public Bureaucracy
in Nigeria," in Tummala, Krishna K. (ed.), Administrative Systems Abroad. Washington, D.C.: University
Press of America. pp. 188-228.

_____. 1982c. "Government Land Allocation in Kano
and Bauchi States of Nigeria: An Analysis of Process
and Beneficiaries in Pre and Post Land Use Decree
Periods." Paper presented at the 25th Annual Meeting
of the African Studies Association held in Washington,
D.C.

_____. 1983a. "The Role of Public Administrators in
Public Policy Making: Practice and Prospects in
Nigeria." Public Administration and Development 3
(January-March):1-26.

_____. 1983b. "State Land Allocation and Class
Formation in Nigeria." Journal of Modern African
Studies 21 (3):461-481.

_____. 1984. "'Development' Administration and Land
Allocation in Nigeria." Rural Africana 18 (Winter):
59-75.

_____. 1986. "Agricultural Policy and Environmental
Destruction in Ethiopia and Nigeria." Rural Africana
25-26 (Spring-Fall):25-53.

_____. 1987. "Political Access and Capital Ac-
cumulation: An Analysis of State Land Allocation
Processes and Beneficiaries in Nigeria." Afrique
et Developpement 12 (1):163-186.

_____. 1988a. "Political Development in Nigeria --
A Review Essay," Africa Today 35 (1):49-56.

_____. 1988b. "State-Local Relations in Nigeria."
Paper presented at the International Political Science
Association Regional Conference in New Delhi, 15-20
March.

_____. Forthcoming. "Development Administration in
Nigeria: Inclinations and Results." In Farazmand,
Ali (ed.), Handbook of Comparative and Development
Public Administration. New York: Marcel Dekker, Inc.

Koehn, Peter, and Waldron, Sidney R. 1978. Afocha: A Link
Between Community and Administration in Harar,

Ethiopia. Foreign and Comparative Studies, African Series No. 31. Syracuse: Syracuse University.

Koester, Ulrich; Valdes, Alberto; and Schafer, Hartwig. 1988. "Commentary: Demand-side Constraints and Structural Adjustment in Sub-Saharan Africa." IFPRI Report 10, No. 4 (October):1, 4.

Kolo, J. T. 1985. "Development in Niger State," in Adamolekun, 'Ladipo (ed.), Nigerian Public Administration 1960-1980: Perspectives and Prospects. Lagos: Heinemann Educational Books, Ltd. pp. 143-156.

Korten, David C. 1980. "Community Organization and Rural Development: A Learning Process Approach," Public Administration Review 40 (5):480-511.

Koto, Yahya I. I. 1980. "My Attachment with Bauchi State Ministry for Local Government, Inspectorate Division." Unpublished A. M. T. C. field attachment report. Zaria: Institute of Administration, Ahmadu Bello University, May.

Kwara State. 1980. "State Background Paper." Paper presented at the National Conference on the Role of Local Government in Social, Political and Economic Development held at the Institute of Administration, Ahmadu Bello University, Zaria, 28-30 April.

Laishley, Roy. 1988. "Bank, IMF Meetings Bring Debt Relief Closer." Africa Recovery 2 (4):1, 16.

Lancaster, Carol. 1987. "Foreign Exchange and the Economic Crisis in Africa," in Ergas, Zaki (ed.), The African State in Transition. New York: St. Martin's Press. pp. 217-235.

_____. 1988. "Policy Reform in Africa: How Effective?" Issue 16 (2):30-35.

Lappé, Frances M.; Collins, Joseph; and Kinley, David. 1980. Aid as Obstacle. San Francisco: Institute for Food and Development Policy.

Lar, S.D. 1980. "The Relationship Between the Executive and the Legislature." Paper presented at the National Conference on the Return to Civilian Rule held at the Institute of Administration, A.B.U., Zaria, 26-30 May.

Le Brun, Oliver. 1979. "Education and Class Conflict," in O'Brien, Rita C. (ed.), The Political Economy of Underdevelopment; Dependence in Senegal. Beverly Hills: Sage Publications. pp. 175-208.

Lehman, Howard P. 1988. "Coalition Politics and Adjustment Strategies in Kenya and Zimbabwe." Paper presented

328

at the 31st Annual Meeting of the African Studies
Association, Chicago, October.

Le Houerou, H.N., and Lundholm, B. 1976. "Complementary
Activities for the Improvement of the Economy and
the Environment in Marginal Drylands," in Can Desert
Encroachment Be Stopped? A Study with Emphasis on
Africa. Ecological Bulletins No. 24. Stockholm:
Swedish Natural Science Research Council.

Lele, Uma. 1975. The Design of Rural Development; Lessons
from Africa. Baltimore: Johns Hopkins University
Press.

Leonard, David K. 1973. "Why Do Kenya's Agricultural
Extension Services Favor the Rich Farmers?" Paper
presented at the Sixteenth Annual Meeting of the
African Studies Association, Syracuse, New York.

Le Prestre, Philippe G. 1981. "The Role of the World Bank
Group in the Formation of International Environmental
Policy." Paper presented at the Annual Meeting of
the International Studies Association, Philadelphia,
18-21 March.

Lewis, A. 1985. "Developments in Lagos State," in
Adamolekun, 'Ladipo (ed.), Nigerian Public Administra-
tion 1960-1980: Perspectives and Prospects. Lagos:
Heinemann Educational Books, Ltd. pp. 129-141.

Lofchie, M. F. 1967. "Representative Government, Bureaucracy, and Political Development: The African Case."
Journal of Developing Areas 2 (October):37-56.

Loxley, John. [1985]. "Alternative Approaches to Stabliza-
tion in Africa," in Helleiner, Gerald K. (ed.),
Africa and the International Monetary Fund.
Washington D. C.: I.M.F.

_____. 1986. "IMF and World Bank Conditionality and
Sub-Saharan Africa," in Lawrence, Peter (ed.), World
Recession and the Food Crisis in Africa. London:
James Currey. pp. 96-103.

Lubeck, Paul M. 1979. "Labour in Kano Since the Petroleum
Boom." Review of African Political Economy, No. 13
(May-August):37-46.

_____. 1987. "Structural Determinants of Urban
Islamic Protest in Northern Nigeria: A Note on
Method, Mediation and Materialist Explanation," in
Roff, William R. (ed.), Islam and the Political
Economy of Meaning; Comparative Studies of Muslim
Discourse. Berkeley: University of California
Press. pp. 79-107.

Luckham, Robin. 1971. The Nigerian Military; A Socio-
    logical Analysis of Authority & Revolt, 1960-67.
    Cambridge: Cambridge University Press.
Mabogunje, Akin L. 1972. "Regional Planning and the
    Development Process: Prospects in the 1970-74 Plan,"
    in Barbour, K. M. (ed.), Planning for Nigeria; A
    Geographical Approach. Ibadan: Ibadan University
    Press.
Maine, Abbe. 1988. "Zambia Blazes Own Economic Trail, But
    Does Recovery Lie Ahead?" Christian Science Monitor,
    23 May.
Makgetla, Neva S. 1986. "Theoretical and Practical Impli-
    cations of I.M.F. Conditionality in Zambia."
    Journal of Modern African Studies 24, No. 3
    (September):395-422.
_____. 1988. "New Faces of Foreign Investment: The
    Zambian Experience." Paper presented at the 31st
    Annual Meeting of the African Studies Association,
    Chicago, November.
Maloney, Joseph F. 1968. "The Responsibilities of the
    Nigerian Senior Civil Servant in Policy Formulation,"
    in Adedeji, A. (ed.), Nigerian Administration and Its
    Political Setting. London: Hutchinson Educational.
    pp. 118-126.
Mandaza, Ibbo. 1988. "Public Administration in Africa: A
    Brief Overview." Courier 109 (May-June):64-65.
Marenin, Otwin. 1987. "Implementing Deployment Policies in
    the National Youth Service Corps of Nigeria: Goals
    and Constraints." Paper presented at the 1987 Annual
    Meeting of the African Studies Association, Denver,
    November.
_____. 1988. "The Nigerian State as Process and
    Manager." Comparative Politics 20, No. 2 (January):
    215-232.
Markakis, John. 1987. National and Class Conflict in the
    Horn of Africa. Cambridge: Cambridge University
    Press.
Markovitz, Irving L. 1977. Power and Class in Africa.
    Englewood Cliffs, N.J.: Prentice-Hall, Inc.
Matlon, Peter J. 1983. Income Distribution Among Farmers
    in Northern Nigeria: Empirical Results and Policy
    Implications. African Rural Economy Paper No. 18.
    East Lansing, Mich.: Department of Agricultural
    Economics, Michigan State University, 1979, revised as
    "The Structure of Production and Rural Incomes in
    Northern Nigeria: Results of Three Village Case

Studies," in Bienen, H., and Diejemoah, V.P. (eds.),
*The Political Economy of Income Distribution in
Nigeria*. New York: Holmes and Meier. pp. 261-310.

Mawakani, Samba. 1986. "Fund Conditionality and the Socio-
economic Situation in Africa," in Helleiner, Gerald
K. (ed.), *Africa and the International Monetary Fund*.
Washington, D. C.: I.M.F.

McCullough, James M. 1988. "Can Management Training
Develop the Private Sector?" Paper prepared for the
31st Annual Meeting of the African Studies Association, October.

McHenry, Dean E. 1984. "The Government Versus Management
Controversy: Ascribing Responsibility for Failure
of a Public Corporation in Nigeria." *Public Administration and Development* 4 (3):259-274.

Meillassoux, Claude. 1970. "A Class Analysis of the
Bureaucratic Process in Mali." *Journal of Development
Studies* 6, No. 2 (January):97-110.

Mellor, John W. 1985. "Agricultural Change and Rural
Poverty." *Food Policy Statement*, No. 3 (October):1-4.

Migdal, Joel S. 1974. *Peasants, Politics, and Revolution;
Pressures Toward Political and Social Change in the
Third World*. Princeton: Princeton University Press.

Mittelman, James H., and Will, Donald. 1987. "The International Monetary Fund, State Autonomy and Human
Rights." *Africa Today* 34 (1&2):49-68.

Mkandawire, Thandika. 1988. "Political Independence and
the Capitalist Transformation of African Agriculture."
Paper presented at the Conference on the Crisis of
African Agriculture, Dakar, 19-23 December.

Mohammed, Disina. 1980. "Address Delivered to the Bauchi
State House of Assembly on 7th January." Bauchi:
Ministry for Local Government and Cultural Affairs.
Ref. No. MLG/ADM/S/60/75, p. 7.

Mohammed, Haruna Nda. 1983. "Decentralized Staffing System
Versus Centralized (Unified) Local Government Staffing
System in Nigeria - Benue State." Unpublished term
paper in the author's possession, May.

Mohammed, Salihu (Chairman) (n.d.). "Report of the
Committee of Inquiry to Investigate Allegations of
Land Racketing within the [Kano] Metropolitan Area."
Kano: unpublished report.

Montgomery, John D. 1979. "Decisions, Nondecisions, and
Other Phenomena: Implementation Analysis for
Development Administrators" in Honadle, George, and
Klauss, Rudi (eds.), *International Development
Administration; Implementation Analysis for Develop-*

ment Projects. New York: Praeger Publishers.
pp. 55-72.
_____. 1986. "Life at the Apex: The Functions of
Permanent Secretaries in Nine Southern African
Countries." Public Administration and Development
6 (3):211-221.
_____. 1987. "How African Managers Serve Development
Goals." Comparative Politics 19, No. 3 (April):
347-360.
Morelos, P. C. 1979. "Local Government Budgeting in
Nigeria: Some Suggestions for Improvement" in
Adamolekun, L., and Rowland, L. (eds.), The New Local
Government System in Nigeria; Problems and Prospects
for Implementation. Ibadan: Heinemann Educational
Books. pp. 67-84.
Muhammed, Muhammed A. [1979]. "Opening Address," in Bauchi
State, Ministry of Finance and Economic Development,
Outline of Decisions on Planning Seminar for Local
Government Officials, 13th-15th November, 1978.
Bauchi: Ministry.
Murray, David J. 1968a. "The Impact of Politics on
Administration," in Adedeji, A. (ed.), Nigerian
Administration and Its Political Setting. London:
Hutchinson Educational. pp. 11-23.
_____. 1968b. "Interest Groups and Administration
in Nigeria," in Adedeji, A. (ed.), Nigerian Admini-
stration and Its Political Setting. London:
Hutchinson Educational. pp. 34-44.
_____. 1978. "Nigerian Field Administration: A
Comparative Analysis," in Murray, David J. (ed.),
Studies in Nigerian Administration. Second edition.
London: Hutchinson & Co., Ltd. pp. 90-139.
_____. 1983. "The World Bank's Perspective on
How to Improve Administration." Public Administration
and Development 3 (Oct.-Dec.):291-297.
_____. Forthcoming. "Appointing Permanent Secre-
taries under the Constitution." Quarterly Journal
of Administration.
Musa, Abdulkadir Balarabe. 1980. Factors Responsible for
the Conflict Between the Executive and Legislature in
Kaduna State: Memorandum Submitted to the Peace
Mission of the Plateau State House of Assembly by the
Kaduna State Governor. Kaduna: Government Printer,
10 November.
Musa, S. A. 1985. "Developments at the Federal Level," in
Adamolekun, 'Ladipo (ed.), Nigerian Public Administra-

tion 1960-1980: Perspectives and Prospects. Lagos:
Heinemann Educational Books, Ltd. pp. 111-119.

Ndegwa, Philip. 1986. "The Economic Crisis in Africa," in
Helleiner, Gerald K. (ed.), Africa and the International Monetary Fund. Washington, D. C.: I.M.F.
pp. 45-51.

Nellis, John R. 1980. "Maladministration: Cause or Result
of Underdevelopment? The Algerian Example," Canadian
Journal of African Studies 13 (3):407-422.

_____. 1986. Public Enterprises in Sub-Saharan
Africa. Washington, D. C.: World Bank.

Nelson, Joan M. 1984. "The Political Economy of Stabilization: Commitment, Capacity, and Public Response."
World Development 12 (10):983-1006.

Nigeria, Central Planning Office. [1979]. "Selection and
Preparation of Development Projects at the Local
Government Level," in Bauchi State, Ministry of
Finance and Economic Development, Outline of Decisions
on Planning Seminar for Local Government Officials.
Bauchi: Ministry.

Nigeria, Federal Department of Information. 1982.
"Nigeria." New York Times, 11 October 1982. pp. D5-
D8.

Nigeria, Federal Government. 1987. Government's Views and
Comments on the Findings and Recommendations of the
Political Bureau. Lagos: Federal Government Printer.

Nigeria, Federal Ministry of National Planning (FMNP).
1979. Guidelines for the Fourth National Plan
1981-85. Lagos: FMNP.

Nigeria, Federal Republic. 1974. Public Service Review
Commission: Main Report. Lagos: Government
Printers.

_____. 1988. Implementation Guidelines on the Civil
Service Reforms. Lagos: Federal Government Printer.

Nigeria, Funtua Agricultural Development Project. 1977.
The Funtua Agricultural Development Project. Funtua:
FADP.

Nigeria, National Committee on Arid Zone Afforestation.
1978. Report of the National Committee on Arid Zone
Afforestation to the Federal Military Government of
Nigeria. Lagos: NCAZA.

Nigeria, National Conference on Local Government. 1979.
"Conclusions and Recommendations of the Conference
held September 19-23, 1977" in Adamolekun, L., and
Rowland, L. (eds.). The New Local Government System
in Nigeria; Problems and Prospects for Implementation.
Ibadan: Heinemann Educational Books. pp. 275-291.

Nigeria, Political Bureau. 1987. Report of the Political
    Bureau. Lagos: Federal Government Printer.
Norman, David W.; Pryor, David H.; and Gibbs, Christopher
    J. N. 1979. Technical Change and the Small Farmer in
    Hausaland, Northern Nigeria. African Rural Economy
    Paper No. 21. East Lansing, Mich.: Department of
    Agricultural Economics, Michigan State University.
Nsingo, Kapembe. 1988. "Problems and Prospects of Economic
    Structural Adjustment in Zambia." Courier 111
    (September-October):78-84.
Nuru, Saka. [1980]. "Local Governments and Their Role in
    the Development of Livestock Industry," in Kumo,
    Suleiman and Aliyu, A. Y. (eds.), Local Government
    Reform in Nigeria. Zaria: Institute of
    Administration, A.B.U. pp. 184-191.
Nwabueze, Ben O. 1985. Nigeria's Presidential Constitu-
    tion: The Second Experiment in Constitutional
    Democracy. New York: Longman.
Nwankwo, G. Onyekwere. 1984. "Management Problems of the
    Proliferation of Local Government in Nigeria." Public
    Administration and Development 4 (1):63-76.
Nwanwene, Omorogbe. 1978. "The Nigerian Public Service
    Commissions," in Murray, David J. (ed.), Studies in
    Nigerian Administration. Second edition. London:
    Hutchinson & Co., Ltd. pp. 177-208.
Nwoke, Chibuzo. 1986. "Towards Authentic Economic
    Nationalism in Nigeria." Africa Today 33 (4):51-69.
Nwosu, Humphrey N. 1977a. "Nigeria's Third National
    Development Plan, 1975-80: Major Problems to Imple-
    mentation." Africa Today 24 (Oct.-Dec.):23-38.
_____. 1977b. Political Authority and the Nigerian
    Civil Service. Enugu: Fourth Dimension Publishers.
Nyaburerwa, Bernard. 1988. "Absorption Potential - The
    Problems of ACP Authorities." Courier 109 (May-
    June):68-69.
Nyirabu, C. M. 1986. "A View from Africa: 2" in
    Helleiner, Gerald K. (ed.), Africa and the Inter-
    national Monetary Fund. Washington, D. C.: I.M.F.
    pp. 32-42.
O'Brien, Donal B. Cruise. 1979. "Rural Class and Peasantry
    in Senegal, 1960-1976: The Politics of a Monocrop
    Economy," in O'Brien, Rita Cruise (ed.), The Political
    Economy of Underdevelopment; Dependence in Senegal.
    Beverly Hills: Sage Publications. pp. 209-227.
O'Brien, Rita Cruise. 1979. "Foreign Ascendance in the
    Economy and the State: The French and Lebanese," in
    O'Brien, Rita Cruise (ed.), The Political Economy of

334

    Underdevelopment; Dependence in Senegal. Beverly
    Hills: Sage Publications. pp. 100-125.
Oculi, Okello. 1979. "Dependent Food Policy in Nigeria,
    1975-1979." *Review of African Political Economy* 15/16
    (May-Dec.):63-74.
Odama, J. S. 1977. "The Draft Constitution and Mixed
    Economy," in Kumo, S., and Aliyu, A.Y. (eds.), *Issues
    in the Nigerian Draft Constitution*. Zaria: Institute
    of Administration, Ahmadu Bello University. pp. 95-
    101.
Ofoegbu, Ray. 1980. "The Chief Executive in a Presidential
    System." Paper presented at the Imo State Top
    Management Seminar held at Oguta, 21-24 January.
_____. 1985. Developments in Imo State," in
    Adamolekun, 'Ladipo (ed.), *Nigerian Public Administra-
    tion 1960-1980: Perspectives and Prospects*. Lagos:
    Heinemann Educational Books, Ltd. pp. 121-127.
Ogbonna, M. N. 1984. "Manpower Planning Problems in
    Nigeria Since Independence." *Nigerian Journal of
    Economic and Social Studies* 8 (July):231-246.
Ojo, A. T. 1977. "Financing Urban Housing in Nigeria," in
    *Urbanization and Nigerian Economic Development;
    Proceedings of the 1977 Annual Conference of the
    Nigerian Economic Society*. Ibadan: Department of
    Economics, University of Ibadan.
Ojo, Folayan. 1980. "Youth Employment and the Impact of
    the National Youth Service Corps on Labor Mobility in
    Nigeria." *African Studies Review* 23 (September):
    51-62.
Ojo, Olatunde J. B. 1985. "Self-Reliance as a Development
    Strategy," in Ake, Claude (ed.), *Political Economy
    of Nigeria*. London: Longman. pp. 141-172.
_____. 1988. "The 1987 Local Government Elections:
    Perspectives from Port Harcourt, Owerri, Brass and
    Ikono Local Government Areas." Report to the National
    Electoral Commission, Lagos.
Ojo, Olatunde, and Koehn, Peter. 1986. "Nigeria's Foreign
    Exchange Controls: An Alternative to IMF Conditions
    and Dependency?" *Africa Today* 33 (No. 4):7-32.
Okafor, Francis C. 1984. "Dimensions of Community Develop-
    ment Projects in Bendel State, Nigeria." *Public
    Administration and Development* 4 (3):249-258.
Okoli, Fidelis. 1978/1979. "Relationship Between the
    Secretary to the Government and the Head of the Civil
    Service under the 1979 Nigerian Constitution."
    Nigerian Journal of Public Affairs (8):7-23.

_____. 1980. "The Dilemma of Premature Bureau-
cratization in the New States of Africa: The Case of
Nigeria." African Studies Review 23 (September):1-16.

Okongwu, Chu. 1987. "Review and Appraisal of SAP (1)."
West Africa (5 October):1961-1963.

Okpala, Don C. I. 1977. "Strengthening Urban Local
Governments for Development Planning: The Critical
Need for Urban Chief Executives," in Urbanization
and Nigerian Economic Development; Proceedings of the
1977 Annual Conference of the Nigerian Economic
Society. Ibadan: Department of Economics, University
of Ibadan. pp. 151-171.

_____. 1979. "Accessibility Distribution Aspects of
Public Urban Land Management: A Nigerian Case."
African Urban Studies 5 (Fall):25-44.

Okpala, Ifebueme. 1979. "The Land Use Decree of 1978: If
the Past Should Be Prologue. . .!" Journal of
Administration Overseas 18 (January):15-21.

Okunoren, Z. O. 1968. "The Administration of Public
Corporations and the Political Factors," in Adedeji,
A. (ed.), Nigerian Administration and Its Political
Setting. London: Hutchinson Educational. pp.
100-109.

Olayemi, Olusegun A. 1979. "Sub-standard Housing and Slum
Clearance in Developing Countries: A Case Study of
Nigeria." Habitat International 4 (3):345-354.

Olinger, John P. 1979. "The World Bank and Nigeria."
Review of African Political Economy 13 (May-August):
101-107.

Olomolehin, O. G., and Ndanusa, Y. A. [1980]. "A Proposal
for Basic Health Service Programme in Local Government
Areas," in Kumo, Suleiman, and Aliyu, A.Y. (eds.),
Local Government Reform in Nigeria. Zaria: Institute
of Administration, A.B.U. pp. 192-198.

Olukoshi, Adebayo, and Abdulraheem, Tajudeen. 1985.
"Nigeria, Crisis Management Under the Buhari Administration." Review of African Political Economy
(34):95-101.

Olurode, Olayiwola. 1986. "Grass Roots Politics, Political
Factions, and Conflict in Nigeria: The Case of Iwo,
1976-1983." Rural Africana 25-26 (Spring/Fall):113-
124.

Onajide, M. O. 1979. "The Civil Service - Some Comments,"
in Adamolekun, L., and Gboyega, A. (eds.), Leading
Issues in Nigerian Public Services. Ile-Ife:
University of Ife Press. pp. 26-35.

Onibokun, Adepoju. 1976. "The Challenge of Housing Low
    Income People in Nigeria." Housing (November):21-25.
_____. 1989. "World Bank Assisted Site and Services
    Projects in Africa: A Critical Assessment." Seminar
    presented at the University of Montana, 11 May.
Onitiri, H. M. A. [1979]. "Planning from Below: An
    Essential Component of a New Development Strategy," in
    Bauchi State, Ministry of Finance and Economic
    Development, Outline of Decisions on Planning Seminar
    for Local Government Officials, 13th-15th November,
    1978. Bauchi: Ministry.
Onokerhoraye, Andrew G. 1984. Social Services in Nigeria;
    An Introduction. London: Kegan Paul International.
Onyeledo, Godwin-Collins. 1980. "Appointive Public Service
    in Presidential Democracy." New Nigerian, 14 March.
    pp. 5+.
Orewa, George O. (Committee Chairman). 1978. Report of the
    Committee on Local Government Training. Lagos:
    Secretary to the Federal Military Government,
    Cabinet Office.
_____. 1979. "Government Grants to Local Government
    in Comparative Perspective," in Adamolekun, L., and
    Rowland, L. (eds.), The New Local Government System in
    Nigeria; Problems and Prospects for Implementation.
    Ibadan: Heinemann Educational Books. pp. 53-66.
Orewa, George O., and Adewumi, J. B. 1983. Local Govern-
    Government in Nigeria; The Changing Scene. Benin
    City: Ethiope Publishing Corporation.
Oronsaye, Andrew O. 1984. "Pathologies of Nigerian
    Bureaucracy." Nigerian Journal of Public Affairs 11
    (May/Oct.):31-47.
Osoba, Segun. 1979. "The Deepening Crisis of the Nigerian
    National Bourgeoisie." Review of African Political
    Economy 13 (May-August):63-77.
Ostheimer, John M. 1973. Nigerian Politics. New York:
    Harper & Row.
Ostheimer, John M., and Buckley, Gary J. 1982. "Nigeria,"
    in Kolodziej, Edward A. and Harkavy, Robert E. (eds.),
    Security Policies of Developing Countries. Lexington:
    D. C. Heath and Company.
Osuji, Emmanuel E. 1982. "Local Government Administration
    Participation and Development in Nigeria: A Case
    Study," Quarterly Journal of Administration 16,
    Nos. 3 & 4 (April-July):169-186.
Othman, Shehu. 1984. "Classes, Crises and Coup: The
    Demise of Shagari's Regime." African Affairs 83
    (October):441-461.

Otobo, Dafe. 1986. "Bureaucratic Elites and Public-Sector Wage Bargaining in Nigeria." Journal of Modern African Studies 24, No. 1 (March):101-126.

Oyediran, O. 1979. "Local Government and Administration," in Adamolekun, 'Ladipo, and Gboyega, Alex (eds.), Leading Issues in Nigerian Public Service. Ile-Ife: University of Ife Press. pp. 44-58.

Oyeyipo, Ezekiel A. O. 1979. "Factors Affecting Decision Making in Public Organizations." Paper presented at the Seminar for Commissioners and Top Civil Servants in Niger State held in Minna, 19-23 November.

Oyinloye, Olatunji. 1970. "New Perspective for the Nigerian Civil Services," in Tukur, Mahmud (ed.), Administrative and Political Development; Prospects for Nigeria. Zaria: Institute for Administration, Ahmadu Bello University. pp. 147-158.

Oyovbaire, Sam Egite. 1980. "Politicization of the Higher Civil Service in the Nigerian System of Government." Quarterly Journal of Administration 14, No. 3 (April): 267-283.

_____. 1985. Federalism in Nigeria; A Study in the Development of the Nigerian State. New York: St. Martin's Press.

_____. 1987. "Prospects for Democracy in Nigeria." Paper presented at the 30th Annual Meeting of the African Studies Association, Denver, November.

Palmer-Jones, R. W. 1980. "Why Irrigate in the North of Nigeria?" Paper prepared for the Seminar on Change in Rural Hausaland, Bagauda Lake, February.

_____. 1984. "Irrigation and Agricultural Development in Nigeria: Social and Historical Origins of an Irrigation Policy," in Watts, Michael (ed.), The State, Oil, and Agriculture in Nigeria. Berkeley: Institute of International Studies. pp. 138-167.

Panter-Brick, Keith. 1978. "The Constitution Drafting Committee," in Panter-Brick, Keith (ed.), Soldiers and Oil; The Political Transformation of Nigeria. London: Frank Cass. pp. 291-350.

Parfitt, Trevor W., and Riley, Stephen P. 1986. "Africa in the Debt Trap: Which Way Out?" Journal of Modern African Studies 24(3):519-527.

Paul, Samuel. 1982. Managing Development Programs: The Lessons of Success. Boulder: Westview Press.

Payer, Cheryl. 1980. "The World Bank and the Small Farmer." Monthly Review 32 (Nov.):30-46.

_____. 1983. The World Bank. New York: Monthly Review Press.

338

_____. 1986. "The World Bank: A New Role in the Debt Crisis?" Third World Quarterly 8 (April):658-676.

Peil, Margaret. 1976. "African Squatter Settlements: A Comparative Study." Urban Studies 13, No. 2 (June): 155-166.

Peil, Margaret, and Sada, Pius O. 1984. African Urban Society. New York: John Wiley & Sons.

Perlez, Jane. 1989. "Zambia's on Its Uppers," New York Times, 23 January.

Permanent Secretary, Bauchi State Ministry of Works and Survey. 1979. "Implementation of Land Use Decree No. 6 of 1978 - Bauchi State." Memorandum to the Secretary of the FMG. 24 January. File No. MOW/ADM/LAND/S.17/70.

Peters, B. Guy. 1989. The Politics of Bureaucracy. 3rd edition. New York: Longman.

Petras, James. 1969. Politics and Social Forces in Chilean Development. Berkeley, California: University of California Press.

Phillips, A. O. 1985. "Two Decades of Government Budgetary Decision-Making," in Adamolekun, 'Ladipo (ed.), Nigerian Public Administration 1960-1980: Perspectives and Prospects. Lagos: Heinemann Educational Books, Ltd. pp. 247-268.

Phillips, Claude S. 1980. "Nigeria's New Political Institutions, 1975-9." Journal of Modern African Studies 18, No. 1 (March):1-22.

_____. 1981. "Bureaucracy and Regime Change in Nigeria." Paper presented at the 1981 Annual Meeting of the African Studies Association, Bloomington, 21-23 October.

_____. 1985. "Political vs. Administrative Development: What the Nigerian Experience Contributes." Paper presented at the 26th Annual Convention of the International Studies Association, Washington, D.C., March.

_____. 1989. "Political Versus Administrative Development: What the Nigerian Experience Contributes." Administration and Society 20, No. 4 (February):423-445.

Pirages, Dennis. 1978. The New Context for International Relations: Global Ecopolitics. North Scituate, Mass.: Duxbury Press.

Price, Robert M. 1975. Society and Bureaucracy in Modern Ghana. Berkeley, Calif.: University of California Press.

Richards, Paul. 1983. "Ecological Change and the Politics
of African Land Use." African Studies Review 26
(June):1-72.
_____. 1985. Indigenous Agricultural Revolution;
Ecology and Food Production in West Africa. London:
Hutchinson.
Riggs, Fred W. 1963. "Bureaucrats and Political Develop-
ment: A Paradoxical View," in La Palombara, J. (ed.),
Bureaucracy and Political Development. Princeton:
Princeton University Press. pp. 120-167.
Robertson, A. F. 1984. People and the State; An Anthro-
pology of Planned Development. Cambridge: Cambridge
University Press.
Roemer, Michael, and Stern, Joseph J. 1975. The Appraisal
of Development Projects; A Practical Guide to Project
Analysis with Case Studies and Solutions. New York:
Praeger.
Rondinelli, Dennis A. 1981. "Administrative Decentraliza-
tion and Economic Development: The Sudan's Experi-
ment with Devolution." Journal of Modern African
Studies 19 (4):595-624.
_____. 1982. "The Dilemma of Development Administra-
tion: Complexity and Uncertainty in Control-oriented
Bureaucracies." World Politics 35 (October):43-72.
Rothchild, Donald. 1987. "Hegemony and State Softness:
Some Variations in Elite Reponses," in Ergas, Zaki
(ed.), The African State in Transition. New York:
St. Martin's Press. pp. 117-148.
Rothchild, Donald, and Olorunsola, Victor A. 1983.
"Managing Competing State and Ethnic Claims," in
Rothchild, Donald and Olorunsola, Victor A. (eds.),
State Versus Ethnic Claims: African Policy Dilemmas.
Boulder: Westview Press. pp. 1-24.
Rowland, L. 1979. "The Challenge of Staffing the New Local
Governments," in Adamolekun, L., and Rowland, L.
(eds.), The New Local Government System in Nigeria;
Problems and Prospects for Implementation. Ibadan:
Heinemann Educational Books. pp. 85-104.
Rule, Sheila. 1987. "Zambia-I.M.F. Fallout Mirrors Third
World Woes," New York Times, 8 June.
Ruttan, Vernon W. 1974/1975. "Integrated Rural Development
Programs: A Skeptical Perspective." International
Development Review 17 (4):9-16.
_____. 1984. "Integrated Rural Development
Programmes: A Historical Perspective." World
Development 12 (4):393-401.

Saasa, Oliver S. 1985. "Public Policy-making in Developing Countries: The Utility of Contemporary Decision-making Models." Public Administration and Development 5 (4):309-321.

Sada, P.O. (n.d.). "Urbanization and Income Distribution in Nigeria," in Urbanization and Nigerian Economic Development; Proceedings of the 1977 Annual Conference of the Nigerian Economic Society. Ibadan: Department of Economics, University of Ibadan. pp. 57-80.

Saitoti, G. 1986. "A View from Africa: 1," in Helleiner, Gerald K. (ed.), Africa and the International Monetary Fund. Washington, D. C.: I.M.F. pp. 26-31.

Salau, Ademola. 1980. "Nigeria's Housing Policies and Programmes: A Preliminary Assessment." Planning and Administration 7, No. 1 (Spring): 49-56.

Samoff, Joel. 1981. "Crises and Socialism in Tanzania." Journal of Modern African Studies 19 (2):279-306.

_____. 1986. "Populist Initiatives and Local Government in Tanzania." Paper presented at the 29th Annual Meeting of the African Studies Association, Madison, November.

Sani, Habibu A. [1980]. "Management of Human and Material Resources at the Local Level: The Nigerian Experience," in Kumo, Suleiman, and Aliyu, A. Y. (eds.), Local Government Reform in Nigeria. Zaria: Institute of Administration, A.B.U. pp. 102-125.

Sano, Hans-Otto. 1983. The Political Economy of Food in Nigeria 1960-1982. Research Report No. 65. Uppsala: Scandinavian Institute of African Studies.

Sanusi, J. O. 1986. "African Economic Disequilibria and the International Monetary System," in Helleiner, Gerald K. (ed.), Africa and the International Monetary Fund. Washington, D. C.: I.M.F. pp. 52-67.

Schaeffer, Wendell G. 1985. "The Formation of Managers for Developing Countries: The Need for a Research Agenda," International Review of Administrative Sciences 51 (3):239-247.

Schatz, Sayre P. 1977. Nigerian Capitalism. Berkeley: University of California Press.

Schumacher, Edward J. 1975. Politics, Bureaucracy and Rural Development in Senegal. Berkeley: University of California Press.

Serageldin, Ismail. 1988. "The World Bank's Assistance to Africa; Adjustment for Growth and Equity." Courier 111 (September-October):54-61.

341

Seymour, Tony. 1975. "Squatter Settlement and Class
    Relations in Zambia." Review of African Political
    Economy 3 (May-October):71-77.
_____. 1979. Housing, Income and Occupational
    Activity in Selected Residential Areas of Kaduna.
    Research Report No. 5. Zaria: Center for Social
    and Economic Research, Ahmadu Bello University.
Shagari, Shehu. 1980. "Budget of Cautious Optimism." West
    Africa, No. 3307 (8 December):2469-2476.
Shaw, Timothy M. 1988. "Structural Readjustment: New
    Framework." Africa Recovery 2, No. 4 (Dec.):36.
Shaw, Timothy M., and Fasehun, Orobola. 1980. "Nigeria in
    the World System: Alternative Approaches, Explana-
    tions, and Projections." Journal of Modern African
    Studies 18 (December):551-573.
Shenton, Bob, and Watts, Mike. 1979. "Capitalism and
    Hunger in Northern Nigeria." Review of African
    Political Economy 15/16 (May-Dec.):53-62.
Sheth, D. L. 1987. "Alternative Development as Political
    Practice." Alternatives 12, No. 2 (April):155-171.
Singh, Ajit. 1986. "A Commentary on the IMF and World Bank
    Policy Programme," in Lawrence, Peter (ed.), World
    Recession and the Food Crisis in Africa. London:
    James Currey. pp. 104-113.
Sklar, Richard L. 1981. "Democracy for the Second
    Republic." Issue 11 (Spring/Summer):14-16.
Smith, Brian C. 1985. Decentralization: The Territorial
    Dimension of the State. London: George Allen &
    Unwin.
Spero, Joan E. 1981. The Politics of International
    Economic Relations. New York: St. Martin's Press.
Stamp, Patricia. 1986. "Local Government in Kenya:
    Ideology and Political Practice, 1895-1974." African
    Studies Review 29, No. 4 (December):17-42.
Stock, Robert. 1985. "Health Care for Some: A Nigerian
    Study of Who Gets What, Where, and Why?" Inter-
    national Journal of Health Services 15 (3):469-484.
Stren, Richard E. 1982. "Underdevelopment, Urban Squat-
    ting, and the State Bureaucracy: A Case Study of
    Tanzania." Canadian Journal of African Studies 16
    (1):67-91.
_____. 1985. "State Housing Policies and Class
    Relations in Kenya and Tanzania." Comparative Urban
    Research 10 (2):57-75.
_____. 1988a. "The African State and the Practice
    of Administrative Justice: Evidence from the Urban

342

Sector," in Simbi, Peter and Ngwa, Jacob (eds.),
Administrative Justice in Public Services: American
and African Perspectives. Stevens Point, Wisconsin:
Worzalla Publishing Company. pp. 103-131.
_____. 1988b. "Urban Services in Africa: Public
Management or Privatisation?" in Cook, Paul, and
Kirkpatrick, Colin (eds.), Privatisation in Less
Developed Countries. New York: St. Martin's Press.
pp. 217-247.

Stryker, Richard E. 1979. "The World Bank and Agricultural
Development: Food Production and Rural Poverty."
World Development 7 (March):325-336.

Tahir, Ibrahim. 1977. "A Critique and an Alternative to
the Unified State and Executive," in Kumo, S., and
Aliyu, A. Y. (eds.), Issues in the Nigerian Draft
Constitution. Zaria: Department of Research and
Consultancy, Institute of Administration, Ahmadu Bello
University. pp. 245-260.

Takaya, Bala J. [1980]. "Failures of Local Government
Reforms in Nigeria: In Search of Causal Factors," in
Kumo, Suleiman and Aliyu, A. Y. (eds.), Local Govern-
ment Reform in Nigeria. Zaria: Institute of
Administration, A.B.U. pp. 61-71.

Taylor, Robert W. 1987. "Urban Development Policies in
Nigeria: Planning, Housing and Local Policy."
Urban Law and Policy 8, No. 5 (December):435-442.

Temple, Frederick T., and Temple, Nelle W. 1980. "The
Politics of Public Housing in Nairobi," in Grindle,
Merilee S. (ed.), Politics and Policy Implementation
in the Third World. Princeton: Princeton University
Press. pp. 224-249.

Teriba, R. Owodunni. 1978. "Accountability and Public
Control of Public Corporations - the Experience of
Western Nigeria," in Murray, D. J. (ed.), Studies in
Nigerian Administration. Second edition. London:
Hutchinson & Co. Ltd. pp. 43-89.

Ter Kuile, C. H. H. 1983. "Farming in the Tropics."
Courier 82 (Nov.- Dec.):53-56.

Thiesenhusen, William C. 1978. "Reaching the Rural Poor
and the Poorest: A Goal Unmet." University of
Wisconsin Land Tenure Center Reprint, No. 136.

Toyo, Eskor. 1986. "Food and Hunger in a Petroleum
Neocolony: A Study of the Food Crisis in Nigeria," in
Lawrence, Peter (ed.), World Recession and the Food
Crisis in Africa. London: James Currey. pp. 231-
248.

Tukur, Mahmud, editor. 1970a. Administrative and Political Development; Prospects for Nigeria. Zaria: Institute of Administration, Ahmadu Bello University.

_____. 1970b. "Implications of the Development Administration Model for the Practioner [sic]." Nigerian Journal of Public Affairs 1 (October):18-24.

Turner, Terisa E. 1978. "Commercial Capitalism and the 1975 Coup," in Panter-Brick, Keith (ed.), Soldiers and Oil; The Political Transformation of Nigeria. London: Frank Cass. pp. 166-197.

_____. 1986. "Oil Workers and the Oil Bust in Nigeria." Africa Today 33 (4):33-50.

Uche, Chukwudum. 1977. "The Armed Forces, the Draft Constitution and the Next Republic," in Kumo, S. and Aliyu, A. Y. (eds.), Issues in the Draft Constitution. Zaria: Institute of Administration, Ahmadu Bello University. pp. 175-181.

Udo, Reuben K. 1977. Report of the Land Use Panel; Minority Report on Nationalisation of Land in Nigeria. Lagos: Federal Government Printer.

Udoji, Jerome. 1979. "Personal Profile." Quarterly Journal of Administration (April-July):201-208.

Ukaegbu, Chikwendu C. 1985. "Are Nigerian Scientists and Engineers Effectively Utilized? Issues on the Deployment of Scientific and Technological Labor for National Development." World Development 13 (No. 4):499-512.

United States, Council on Environmental Quality [C.E.Q.] and Department of State. 1980. The Global 2000 Report to the President of the U.S. Volume I. New York: Pergamon Press.

United States, Department of Commerce. 1985. Foreign Economic Trends and their Implications for the United States: Nigeria. Washington, D. C.: D.O.C., International Trade Administration, January.

United States, Department of State. 1979. "Basic Data on Sub-Saharan Africa." Special Report No. 61. Washington, D.C.: Department of State, Bureau of Public Affairs, December.

Usman, Yusufu Bala (ed.). 1982. Political Repression in Nigeria. Kano: Bala Mohammed Memorial Committee.

_____. 1986. Nigeria Against the I.M.F.; The Home Market Strategy. Kaduna: Vanguard.

Uwazurike, P. Chudi. 1987. "State, Knowledge and Development: On the Intellectual and Political Context of the Knowledge-Policy Interplay in Nigerian Development

Planning." Unpublished Ph.D. dissertation, Harvard University.

Vengroff, Richard. 1974. "Popular Participation and the Administration of Rural Development: The Case of Botswana." Human Organization 33, No. 3 (Fall): 303-309.

_____. 1982. "The Administration of Rural Development: The Role of Extension Agents in Upper Volta and Zaire." Paper presented at the Annual Meeting of the African Studies Association, Washington, D. C., November.

_____. 1983. Development Administration at the Local Level: The Case of Zaire. Foreign and Comparative Studies, African Series, No. 60. Syracuse: Syracuse University.

_____. 1988a. "Policy Reform and the Assessment of Management Training Needs in Africa: A Comparative Perspective." Paper presented at the 31st Annual Meeting of the African Studies Association, Chicago, 28 October.

_____. 1988b. "Training and the Public Sector in in Francophone Africa: Problems of Performance in Resource Poor Bureaucracies." Paper delivered at the NASPAA Workshop on Public Sector Reform, Management Training, and the Assessment of Training Needs in the Institutions of Developing Countries, Washington, D. C., 9-10 December.

Vengroff, Richard, and Johnson, Alan. "Decentralization and the Implementation of Rural Development in Senegal: The Role of Rural Councils." Public Administration and Development 7 (3):273-288.

Wallace, Tina. 1978-79. "Planning for Agricultural Development: A Consideration of Some of the Theoretical and Practical Issues Involved." Nigerian Journal of Public Affairs (8):54-78.

_____. 1979a. "Agriculture for What? Problems and Strategies of Nigeria's Food Policy in the Third Development Plan." Paper presented at the Department of Research, Management, and Consultancy Seminar, Ahmadu Bello University, Zaria.

_____. 1979b. Rural Development Through Irrigation: Studies in a Town on the Kano River Project. C.S.E.R. Research Report No. 3. Zaria: Center for Social & Economic Research, Ahmadu Bello University.

_____. 1980a. "Agricultural Bonanza? Some Crucial Issues Raised by the World Bank Agricultural

Development Projects in Nigeria." Nigerian Journal of Public Affairs 9 (May):61-78.

_____. 1980b. "Agricultural Projects and Land." Paper presented at the Seminar on Change in Rural Hausaland, Bagauda Lake, 29-30 February.

_____. 1980c. "Report on Some of the Social and Economic Issues and Problems Raised on the Bakalori Irrigation Project." Unpublished paper. Zaria: Department of Research and Consultancy. Ahmadu Bello University, March.

_____. 1981a. "Agricultural Projects in Northern Nigeria." Review of African Political Economy (17):59-70.

_____. 1981b. "The Challenge of Food: Nigeria's Approach to Agriculture 1975-80." Canadian Journal of African Studies 15 (2):239-258.

Watts, Michael. 1983. Silent Violence; Food, Famine and Peasantry in Northern Nigeria. Berkeley: University of California Press.

Watts, Michael, and Lubeck, Paul. 1983. "The Popular Classes and the Oil Boom: A Political Economy of Rural and Urban Poverty," in Zartman, I. William (ed.), The Political Economy of Nigeria. New York: Praeger. pp. 105-144.

Welch, Claude E., Jr., and Smith, Arthur K. 1974. Military Role and Rule: Perspectives on Civil-Military Relations. North Scituate, Mass.: Duxbury Press.

White, Louise G. 1987. Creating Opportunities for Change: Approaches to Managing Development Programs. Boulder: Lynne Rienner Publishers.

White, Rodney R. 1985. "The Impact of Policy Conflict on the Implementation of a Government-assisted Housing Project in Senegal." Canadian Journal of African Studies 19 (3):505-528.

_____. 1987. "Agricultural Policy Options in the Face of Uncertainty; The Case of Senegal." Paper presented at the Annual Meeting of the Canadian Association of African Studies, Edmonton, May.

Wilks, Stephen. 1985. "Review Article: Nigerian Administration -- in Search of a Vision?" Public Administration and Development 5 (3):265-276.

Williams, Gavin. 1976a. "Nigeria: A Political Economy," in Williams, Gavin (ed.), Nigeria; Economy and Society. London: Rex Collings. pp. 11-54.

_____. 1976b. "Taking the Part of Peasants: Rural Development in Nigeria and Tanzania," in Gutkind,

Peter C. W., and Wallerstein, Immanuel (eds.), The
Political Economy of Contemporary Africa. Beverly
Hills, Calif.: Sage Publications. pp. 131-154.
_____. 1980. State and Society in Nigeria. Indanre:
Afrografika.
_____. 1981a. "The World Bank and the Peasant
Problem," in Heyer, Judith; Roberts, Pepe; and
Williams, Gavin (eds.), Rural Development in Tropical
Africa. New York: St. Martins Press. pp. 16-51.
_____. 1981b. "Inequalities in Rural Nigeria."
Development Studies Occasional Paper No. 16. Norwich:
University of East Anglia, November.
_____. 1985. "Marketing Without and With Marketing
Boards: The Origins of State Marketing Boards in
Nigeria." Review of African Political Economy 34
(December):4-15.
_____. 1986. "Rural Development: Partners and
Adversaries." Rural Africana 25-26 (Spring):11-23.
Williams, Gavin, and Turner, Terisa. 1978. "Nigeria," in
Dunn, John (ed.), West African States: Failure and
Promise. Cambridge: Cambridge University Press.
pp. 132-172.
Wilson, Ernest J. 1988. "Privatization in Africa:
Domestic Origins, Current Status and Future
Scenarios," Issue 16 (2):24-29.
Yahaya, Ali D. 1978. "The Creation of States," in
Panter-Brick, Keith (ed.), Soldiers and Oil; The
Political Transformation of Nigeria. London: Frank
Cass. pp. 201-223.
_____. 1979a. "Local Government as an Agent of Rural
Development: An Evaluation." Nigerian Journal of
Political Science 1 (June):20-31.
_____. 1979b. "The Struggle for Power in Nigeria,
1966-79," in Oyediran, Oyeleye (ed.), Nigerian
Government and Politics under Military Rule, 1966-79.
New York: St. Martin's Press. pp. 259-272.
_____. [1980]. "Unified Local Authority Service: An
Attempted Solution to Local Authority Staff Problems,"
in Kumo, Suleiman, and Aliyu, A. Y. (eds.), Local
Government Reform in Nigeria. Zaria: Institute of
Administration, A.B.U. pp. 126-140.
_____. [1982]. "The Idea of Local Government in
Nigeria: The Need for a Redefinition," in Aliyu,
Abubakar Yaya (ed.), The Role of Local Government
in Social, Political and Economic Development in
Nigeria 1976-79. Zaria: Department of Local

Government Studies, Institute of Administration,
A.B.U. pp. 55-66.

Yeye, Evelyn. 1980. "Local Government Development Planning
in Kaduna State: An Experiment in Planning from the
Grass Roots." Unpublished A.M.T.C. field attachment
report, Institute of Administration, A.B.U., Zaria.

You, Nicholas, and Mazurelle, Jean. 1987. "The Impoverish-
ment of Large Francophone West African Cities is not
an Inevitability," Planning and Administration 14
(Autumn):14-26.

Young, Andrew. 1981. "Gone Fishing in Argungue."
Washington Post, 9 March.

Zahradeen, Usman. 1980. "Public Accountability Under
Civilian Administration in Nigeria." New Nigerian,
7 August. p. 5.

Index

Abidjan, 26
Absentee farmers, 90, 93-95,
    97, 103
Abuja, 67, 123
Accelerated Development Area
    Programme (ADA), 96-97
Accountability, 292-294
  absence of, 46, 53n, 288
Accra, 173
Achebe, Chinua, 1, 293
Adamolekun Commission, 52n
Adebayo, Augustus, 36
Adebo Commission, 30, 52n
Agricultural production, 7,
    85-88, 91-98, 101-104,
    105n, 107n, 124-125,
    130, 135, 154, 169,
    174n, 196-202, 223n,
    277-278
Ahmadu Bello University,
    xvii, 6, 49n, 224n,
    228n
Airport Road New Layout, 167
Akwa Ibom State, 39
Algeria, 284
Aliyu, A. Y., xviii, 6
Anambra State, 267n
Armed forces, 46, 51n, 57,
    65-66, 74, 154, 157,
    159, 161-162, 166, 170,
    174n, 276, 297n
  salaries of, 51n

Armed forces (cont'd),
    size of, 17
Armed Forces Ruling Council
    (AFRC), 25, 63, 79n,
    221, 261n, 264n, 267n
Attorney General, 79n
Ayida, Allison, 68, 79n
Azare, 148, 177n

Babangida, Ibrahim, 3, 23,
    25, 27, 39, 44, 50n,
    67, 74-75, 101, 112,
    114, 117-127, 132-135,
    138n-140n, 242, 260,
    296n
Badagry Local Government,
    261n, 262n
Bakalori. See Irrigated
    farming schemes
Bank of the North, Ltd.,
    157-158, 167-168, 177n
Banks. See Commercial banks
Basic needs, 191, 271, 282
  strategy, 135, 291-294
Bauchi, 148, 151-152, 177n,
    192, 227n-228n, 233,
    246
Bauchi Local Education
    Authority, 244
Bauchi Local Government,
    69-70, 182-223,
    223n-231n, 233-259,

349

Economic Stabilisation Act
    of 1982, 138n
Economy of affection, 298n,
    299n
Education, 113, 115, 128,
    132, 134, 141n, 154,
    194-203, 223n,
    227n-229n, 257, 259-
    260, 266n, 267n, 287,
    298n, 299n
Ekwueme, Alex, 35, 100
Elections
    1979, 47, 50n, 73, 105n
Electricity, 46, 138n, 151,
    194-196
Emirate council, 148
Enviromental protection,
    86-90, 93-98, 102-104,
    107
Equipment purchases, 92,
    100, 204-205, 230n, 277
Ethiopia, 54n, 107n, 143,
    177n, 264n, 267
Exchange Control Decree of
    1984, 113, 115,
    120-121
Executive office, 34
Expatriates, 18-20, 43, 49n,
    54n, 76-77, 82n, 89,
    91-92, 96-97, 99, 101,
    103, 105n, 113-114,
    132, 179, 278
Export marketing, 85, 98,
    101-102, 105n, 106n,
    111, 123-125, 133,
    135-136, 137n, 141n,
    286

Fadama, 90, 95
Farmers. See Absentee
    farmers; Progressive
    farmers
Feasibility studies,
    187-188, 221, 222,
    225n, 226n
Federal character, 36-38,
    40

Federal Executive Council
    (FEC), 23-24, 38, 50,
    52n, 61, 63-67, 79n
Federalism, 5, 271, 296n
Federal Mortgage Bank, 177n
First Republic, 13, 18-20,
    22-23, 61-62
Food security, 135, 139n
Foreign aid/assistance, 29,
    81n-82n, 92, 100,
    135-136, 137n, 141n,
    241, 299n
    EEC, 287
    U.S., 287
Foreign donors, 29, 43-44,
    77, 85-86, 91, 131,
    136, 286-289
Foreign exchange, 92,
    109-117, 119-122, 124,
    126-129, 131, 133-135,
    138n, 139n, 206
    control policy, 112-116,
    120-121, 124, 126-129,
    131, 133, 135, 140n
    crisis, 18
Foreign Exchange Market
    (FEM), 139n
Foreign investment, 45, 87,
    99, 121, 129-130,
    133-135, 139n, 141n,
    174, 273, 275-276, 288
Foreign service, 38, 50n
Fourth National Development
    Plan, 96, 98-99,
    181-188, 206, 212, 221,
    296n
France,
    public administration in,
    52n, 72
Fronting, 175n-176n
Funtua. See Integrated rural
    development projects

Garuba, Chris, 39
Ghana, 38, 44, 175n, 267,
    299n
Gombe, 96, 148, 177n

Urban poor, 6, 45, 55n,
75-76, 125-127, 129,
135, 142, 150-153, 155,
158-160, 168-174,
210-211, 229n, 230n,
271, 275, 283, 288,
290-295, 297n, 299n
Urban transportation, 54n,
55n, 196, 202, 229n
Villages, 108, 177n, 184,
214, 230n, 236, 238,
242, 290, 300n

Wages, 32-33, 52n, 121, 125-
128, 135, 140n, 158,
173, 261, 273, 278, 280
Water supply, 46, 55n, 106n,
151, 171, 173, 196,
203, 223n, 238, 265n
Whitehall model, 12, 34, 60,
81n

Women and development, 93,
103-104, 176n
World Bank, 3, 29, 43, 82n,
88, 91-93, 96-98,
102-103, 105n, 107-108,
110, 114, 123-124,
127-130, 132-134, 136,
137n-138n, 140n, 142,
177n, 288, 299n
Wudil, 167

Zaire, 107n, 123, 132n, 134,
137n, 140n, 267, 284,
285, 291, 297n
Zambia, 42, 54n, 122,
127-128, 132, 133,
137n, 140n, 143
Zaria, 6, 108
Zimbabwe, 44, 128, 133, 135